QCD AND HEAVY QUARKS

In Memoriam Nikolai Uraltsev

QCD AND HEAVY QUARKS
In Memoriam Nikolai Uraltsev

editors

Ikaros I Bigi
University of Notre Dame, USA

Paolo Gambino
University of Torino, Italy

Thomas Mannel
University of Siegen, Germany

World Scientific

NEW JERSEY • LONDON • SINGAPORE • BEIJING • SHANGHAI • HONG KONG • TAIPEI • CHENNAI

Published by

World Scientific Publishing Co. Pte. Ltd.

5 Toh Tuck Link, Singapore 596224

USA office: 27 Warren Street, Suite 401-402, Hackensack, NJ 07601

UK office: 57 Shelton Street, Covent Garden, London WC2H 9HE

Library of Congress Cataloging-in-Publication Data

QCD and heavy quarks : in memoriam Nikolai Uraltsev / edited by Ikaros I. Bigi (University of Notre Dame, USA), Paolo Gambino (University of Torino, Italy), Thomas Mannel (Universität Siegen, Germany).

    pages cm

  Includes bibliographical references.

  ISBN 978-9814602730

  1. Quantum chromodynamics. 2. Quarks. 3. Heavy particles (Nuclear physics). 4. CP violation (Nuclear physics). I. Uraltsev, Nikolai, 1957–2013, honouree. II. Bigi, I. I., editor. III. Gambino, Paolo, 1966–   editor. IV. Mannel, T. (Thomas), editor.

  QC793.3.Q35Q22 2015

  539.7'2167--dc23

                               2015012344

British Library Cataloguing-in-Publication Data

A catalogue record for this book is available from the British Library.

Printed in Singapore

Preface

This book is dedicated to our colleague and friend Prof. Dr. Nikolai "Kolya" Uraltsev who passed away early and unexpectedly on Feb. 25, 2013. The whole community of phenomenological particle physics consider this a great loss, since he was not only an outstanding scientist, but also a wonderful person.

The scientific community knew Kolya as a physicist with an enormous intuition; in particular he was famous for his understanding of QCD and its non-perturbative properties. For him a novel idea was wonderful, but only the first step; then, when necessary, he went into hard calculations, with all their subtleties. He had a very broad overview over phenomenological particle physics, ranging from electroweak and Higgs physics to flavor and CP violation, as well as QCD, on which he focussed over the last two decades of his life.

Kolya is one of the fathers of the heavy quark expansion. He approached this from the point of view of QCD by noticing that there is a small parameter, namely the ratio of the confinement scale over the heavy quark mass, which one could use for an expansion of the QCD matrix elements. He never appreciated (and in fact he never needed) the tool of effective theories, which is widely used in the description of heavy hadrons; he always extracted the relevant physics from QCD, usually without resorting to heavy quark effective theory.

Discussions with Kolya at conferences and seminars were legendary, and many colleagues from the scientific community may well remember this. Thanks to his deep insight into QCD he almost always had a point, and of course he insisted to get things right at the end. Many colleagues in particular from the West had a hard time to get acquainted with his style; he was never unfriendly nor insulting, but always straightforward. Once this was understood, working with Kolya was effective and very enjoyable.

To his friends he was known as a wonderful person, who was in many respects unconventional. Although he was a theorist, he was a very practical person: he could fix all kinds of machinery, including electronics and even cars, although his solutions were sometimes unorthodox. As an example, the German authorities eventually removed Kolya's 25-year old Toyota car from the streets; it still moved, but some of Kolya's 'fixes' were somewhat debatable. Nevertheless, he was a reliable friend and companion to many of us also outside physics.

Kolya was born in what was then called Leningrad in 1957 and got his PhD at the Leningrad Nuclear Physics Institute in 1983 under the supervision of Alexei Anselm. Shortly thereafter he obtained a permanent position at the same Institute, which he retained till his death. At first Kolya was very shy, but he slowly changed.

After 1986 he started visiting more and more frequently various western universities, spending long periods at Notre Dame, Minnesota, CERN, Milan. During his final years he was scientific staff in the theoretical particle physics group in Siegen, where he passed away on Feb. 25th, which was the Ash Wednesday of 2013. He left his wife Lilia and his son Gennady. Lilia was not only a wonderful wife, but also has standing in science by herself; in the summer she frequently went on archaeological excavations in the Kola Peninsula; Kolya followed her to help the expedition getting food by fishing. Gennady is closer to mathematics than physics, but he has an excellent future ahead of him there.

Kolya's departure was a great shock for us as his close friends, and we decided to gather in a volume dedicated to his memory contributions from some of his colleagues working in the same field. The book is on one side a memorial, on the other it is also a document representing Kolya's particular view on QCD and its anatomy, which in some respect has been unconventional. In this regard the book may also be useful for students and colleagues who did not know Kolya personally.

Ikaros Bigi, Notre Dame, USA
Paolo Gambino, Torino, Italy
Thomas Mannel, Siegen, Germany
January 2015

Left: Photograph supplied by Misha Shifman from his private collection.
Right: picture supplied by Gennady Uraltsev.

Cover picture supplied by Misha Shifman from his private collection.

Contents

Contents

New and Old About Renormalons

M. Shifman

William I. Fine Theoretical Physics Institute,
University of Minnesota, Minneapolis, MN 55455, USA
shifman@physics.umn.edu

I summarize what we know of renormalons from the 1970s and 80s: their uses and theoretical status. It is emphasized that renormalons in QCD are closely related to the Wilsonean operator product expansion (OPE) — a setup ideally suited for dealing with the factorially divergent series reflecting infrared dynamics. I discuss a breakthrough proposal due to Uraltsev *et al.* to use renormalons to evaluate nonperturbative (power) corrections in the processes without OPE. Some fresh ideas which were put forward recently are briefly discussed too, with emphasis on a possible relationship between resurgence via trans-series and OPE.

This article is devoted to the memory of my friend Kolya Uraltsev. I should emphasize that these are my personal recollections. Other people who closely knew Kolya may or may not agree with my opinions.

Foreword: Nikolai (Kolya) Uraltsev

Kolya, Alexei Anselm's student, was one of the most prominent theorists from the young generation of the Gribov Leningrad school. The heavy quark theory acquired the level of perfection it enjoys now to a large extent due to his works on quantum chromodynamics. In this area, there was no higher authority in the world than Kolya.

In science Kolya was a "slowpoke," in the sense that each new result or new assertion in his field — the theory of heavy quarks — had to be critically processed before being accepted (or not). Coming across something new Kolya pondered on all sides of this "something new" with incredible diligence. There was no way any ambiguity could be left after Kolya. He almost physically suffered from sloppy works and light-minded authors. Kolya gave no quarter to such speakers at seminars or conferences, no matter what regalia they may have possessed. With them he was aggressive and restless until he had exposed all loopholes in the arguments. This "inconvenient" style — science above everything else — that Kolya had learned from Gribov, he carried through his life, without changing it in the West, where it (the style) is almost extinct due to incompatibility with political correctness. Frankly speaking, physicists from the West slightly feared Kolya. None of the heavy quark theorists wanted to be ground by "millstones" in Kolya's mind.

It is ironic that in everyday life Kolya was not only shy, but rather super-shy. You can hardly find such shy people nowadays. For him it was a problem to talk to a stranger or to respond to the harsh words of an insolent fellow.

Every summer Kolya went on archaeological excavations at the Kola Peninsula. On one of these occasions he met his future wife. Lilia (that's her name) worked at the Institute of Archeology of the Russian Academy of Science and was the head and organizer of the expeditions.

When Kolya was thinking about physics, he did not notice anything around him. Once during a conference, after a session, we walked out of the conference hall to the street under heavy rain. Everybody opened umbrellas right away. Kolya did not react to a change in the environment from comfortable to dramatically uncomfortable, and continued the discussion as if nothing had happened... He kept a clean child's soul.

We — Kolya and I — published 17 joint works: the first in 1987 and the last in 1998. Especially productive was our collaboration during the academic year 1994/95 when Kolya spent the whole year with us at the University of Minnesota. Many ideas conceived during this year became parts of subsequent research on heavy quark theory. Here I would like to single out a particularly exciting insight: the use of renormalons as a tool for revealing power terms in the processes without the operator product expansion.

Kolya could repair with his own hands any damage to any vehicle, including those most modern and stuffed with electronics. It was his passionate hobby. In 1996, we spent six months together at CERN. For everyday commuting I bought a used Audi, which had problems all the time. In the Swiss garages they asked from me exorbitant prices for repairs. Kolya coped effortlessly.

In fact, Kolya could fix just about anything, not only cars. In this, like in physics, he was inquisitive; he loved the process of learning "how things work," be it a B-meson decay or a leak in a boat.

Striking thoroughness — that's how I would characterize Kolya's approach to every aspect of his life and work. During his 30-year career in theoretical physics Kolya closely and productively interacted with many colleagues on three continents, St. Petersburg, CERN, Technion, Milan, Orsay, FTPI, University of Notre Dame, and University of Siegen. I am sure that's how they will remember him — a deep thinker and a reliable friend.

Death is always untimely. When Kolya's heart stopped on February 13, 2013 he was only 54 years old, full of plans for the future both in science and life. Even now, six months after his tragic death, it is not easy for me to write this *in memoriam* article in a logically-ordered manner. Apparently, I will have to settle for less.

1. Introduction

One can say that Kolya burst onto heavy quark theory like a meteor. Our first (occasional) scientific encounter occurred in 1986.[1] Shortly after, our paths departed: he delved in the problem of CP violation for six long years,[2,3] while I returned to nonperturbative supersymmetry.[4] I was still heavily involved in this topic when Kolya appeared in our Institute[a] in 1992 full of enthusiasm with regards to a consistent theory of $1/m_Q$ expansion in heavy quarks based on the operator product expansion (OPE). Elements of this theory already existed.[5–7] However, they represented general guidelines rather than a theoretical construction worked out in detail. Applications were rather scarce. Kolya's enthusiasm was contagious, and shortly after both, Arkady Vainshtein and myself, got fully involved. One of the most elegant results established by Kolya and collaborators[8] (see also Ref. 9) was the absence of the $1/m_Q$ correction in the total inclusive decay widths of the heavy-flavor hadrons. This theorem (sometimes referred to as the CGG — BUV theorem) made its way into textbooks, let alone its practical importance for precision determination of V_{cb} and V_{ub} from data.

In twenty years that elapsed after 1992 Kolya managed to make definitive contributions to many topics from the heavy quark theory. It is fair to say that he left no stone unturned. His imprint is seen everywhere. Needless to say, I will be unable to cover all these topics. Instead, I will focus on one particular topic — determination of the power of $1/m_Q$ (or $1/Q$) terms from renormalons in the processes without OPE — in which Kolya was a trailblazer.[b]

2. Pioneers

Two papers, Refs. 15 and 16, which appeared on ArXiv on the same day, were the first to suggest the usage of renormalons for indication of the power of nonperturbative corrections (i.e. $1/Q$ or $1/m_Q$, or squares, cubes, etc. of the above parameters) in the processes *without* OPE.

The next relevant paper was Ref. 17, where the idea was first applied to hadronic event shapes.

Before explaining how this works, I will have to remind you what renormalon is. To this end I will have to start from the factorial divergences of the perturbative series.

3. Dyson Argument and Factorial Divergences

Sixty two years ago Freeman Dyson completed his famous paper entitled "Divergences of Perturbation Theory in Quantum Electrodynamics"[18] (reprinted in Ref. 19). He argued that the series in e^2 in QED could not be convergent due to the

---

[a]William I. Fine Theoretical Physics Institute, University of Minnesota.
[b]The reader unfamiliar with the range of questions associated with OPE, $1/m_Q$ corrections in heavy quark theory, renormalons and all that is advised to turn to reviews.[10–14]

M. Shifman

fact that analytic continuation to negative e^2 produced a theory with unstable vacuum. This became known as the Dyson argument. Shortly after, Thirring evaluated[20] the number of diagrams in $\lambda\phi^3$ field theory in high orders and came to the conclusion that the perturbative series in this theory is factorially divergent. In 1977 various field theories, including $\lambda\phi^4$, were thoroughly studied by Lipatov[21] who came to the same general conclusion: the perturbative series are asymptotic and characterized by the factorial divergence of the form

$$Z = \sum_k C_k \alpha^k k^{b-1} A^{-k} k! \,. \tag{1}$$

This is reviewed in some detail e.g. in Ref. 19. The notation in Eq. (1) is as follows: α is the expansion parameter,[c] k is the number of loops, C_k's are numerical coefficients of order one, and b and A are numbers.

Arkady Vainshtein was the first to point out[23] (see Ref. 24) that the factorial divergence in (1) is in one-to-one correspondence with the probability of the under-the-barrier penetration (vacuum instability in field theory language) for unphysical — negative — values of the expansion parameter. Ten years later this relation was rediscovered by Bender and Wu[25] in the quartic anharmonic oscillator or, which is the same, in $\lambda\phi^4$ theory.

The factorial divergence of the perturbative series discussed in Refs. 18, 19, 23–25 can be traced back to the factorially large number of multiloop Feynman diagrams (i.e. $k \gg 1$).

Renormalons which we will focus later have nothing to do with this mechanism. As was noted in Ref. 26, there exists a class of isolated graphs, in which each diagram grows factorially as we increase the number of loops. It is these graphs that are called renormalons. The theoretical feature responsible for the renormalon factorial divergence (1) is the logarithmic running of the effective coupling constant.

4. Borel Summability

Instead of the *asymptotic series* (1) let us introduce the Borel transform

$$B_Z(\alpha) = \sum_k C_k \, \alpha^k k^{b-1} A^{-k} \,. \tag{2}$$

In Eq. (2) the k-th term of expansion (1) is divided by $k!$, which implies, in turn, that the singularity of $B_Z(\alpha)$ closest to the origin in the α plain is at distance A from the origin. Thus, the sum (2) is convergent.

Mathematicians would say that the function defined by (2) is obtained from (1) by the inverse Laplace transformation.

---

[c]In QED it is customary to define $\alpha \equiv e^2/(4\pi)$. The asymptotic divergence of the coefficients in QED is somewhat more contrived[22] than in (1) due to the fact that the QED loops are due to fermions.

It is quite obvious that one can recover the original function Z performing the following integral transformation (the Laplace transformation):

$$Z(\alpha) = \int_0^\infty dt\, e^{-t} B_Z(\alpha t) \,, \qquad (3)$$

see e.g. Ref. 27, Sect. 37.3. The integral representation (3) is well-defined provided that $B_Z(\alpha)$ has no singularities on the real positive semi-axis in the complex α plane. This is the case if the asymptotic series (1) is sign-alternating, $C_k \sim (-1)^k$, (and then so is (2)). If $B_Z(\alpha)$ has singularities on the real positive semi-axis (as is the case if the coefficients C_k are all positive, or all negative), then the integral (3) becomes ambiguous. The ambiguity is of the order of $e^{-A/\alpha}$. One cannot resolve this ambiguity on the basis of purely mathematical arguments. More information is needed, which can be provided only by underlying physics.

In problems at weak coupling additional physical information can be obtained by quasiclassical methods. Indeed, at weak coupling deviations from perturbation theory are due to classical solutions with nonvanishing action, such as instantons or instanton–antiinstanton (IA) pairs. Say, in the quantal problem of the double-well potential, the contribution of the instanton–antiinstanton pair is ambiguous *per se*. However, one can combine (3) with the latter in such a way, that in the final answer these two ambiguities cancel, giving rise to a well-defined expression.[28,29] The next ambiguity occurs at the level of two instanton–antiinstanton pairs. It is canceled against the ambiguity in perturbation theory in the sector of a single instanton–antiinstanton pair plus a subleasing singularity[30] in (3). The process of cancellation of ambiguities is repeated *ad infinitum*. Continuing this procedure one arrives at the so-called trans-series combining perturbative and quasiclassical nonperturbative expansion at weak coupling. In a slightly simplified form the resurgence and trans-series can be expressed by the formula

$$Z(\alpha) = \sum_{k=0}^\infty \left\{ c_{0,k} + c_{1,k}\alpha + c_{2,k}\alpha^2 + c_{3,k}\alpha^3 + \cdots \right\} e^{-kA/\alpha} \,, \qquad (4)$$

where for each given k the coefficients $c_{n,k}$ are factorially divergent in n, and the sum in n in the braces (for each given k) is regularized in a well-prescribed manner. I will say a few words on the nature of the k series later.

In quantum mechanics the construction of the trans-series was explored in Refs. 28–33. Recently a progress along these lines was achieved in field theory too.[34,35]

To make sure that a field-theoretical model under consideration is weakly coupled, it was analyzed[34,35] in cylindrical geometry $R_1 \times S_1(r)$, with a compactified dimension of a very small size r. Then, in much the same way as in the above quantal problem, it proved to be possible to identify quasiclassical field configurations responsible for nonperturbative contributions,[36,37] to be combined with the Borel-resummed perturbative series.

It is quite plausible that in weakly coupled field theories a complete resurgence can be achieved along these lines, and at least some quantities are representable in the form of trans-series combining Borel-resummed perturbation theory with a (infinite) set of nonperturbative effects derivable from quasiclassical considerations. What remains to be seen is whether this program works in a more general setting of any weakly coupled field theory, for instance, in fully Higgsed Yang–Mills theory, and if yes, in which particular way. At the moment the idea of matching the factorial divergence to quasiclassical field configurations in fully Higgsed Yang–Mills theories is barely explored.[d]

If this idea survives in a more general formulation, the next intriguing question is obvious: whether or not a connection to strong coupling regime can be revealed. Note that at weak coupling continuous symmetries such as the chiral symmetry cannot be spontaneously broken. Therefore, a parallel between resurgence via trans-series in quantum mechanics on the one hand and OPE in QCD and similar theories on the other, which of course comes to one's mind, cannot be complete.

5. The First Source of Factorial Divergence

In quantum mechanics the coupling constant is fixed. In Yang–Mills field theory (e.g. QCD) the very notion of the smallness of the coupling constant is meaningless, since the coupling constant depends on scale; it runs and becomes strong at momenta of the order of dynamical scale Λ. At such momenta dynamics are by no means exhausted by perturbation theory and quasiclassical nonperturbative effects. In fact, in the infrared domain, at strong coupling, both cannot even be consistently defined. Below we will discuss what can be done under the circumstances.

For a short while, let us close our eyes at this feature pretending that somehow the blow off of α_s in the infrared (IR) domain is not essential. This neglect will be corrected shortly. In Yang–Mills theory one can identify at least two sources for the factorial divergence of the perturbative series. First, the number of various Feynman graphs with n loops grows as $n!$. This feature (similar to that one encounters in quantum mechanics) was known already to the explorers of QED from the times of the Dyson argument, see Sect. 3. As a result, even if each graph is of the order of unity in appropriate units, the contribution of the set of the n-loop graphs will be of the order of $n! \alpha^n$. At $n \sim 1/\alpha \gg 1$ multiloop graphs are typically represented by soft fields which can be viewed as quasiclassical field configurations, for instance, instantons. Instanton contributions to correlation functions are $\sim \exp\left(-\frac{2\pi}{\alpha}\right)$.

If this were the only source, the problem could be eliminated in an elegant way, which can be traced back to 't Hooft's observation[38] that in the limit

$$N \to \infty, \quad N\alpha \text{ fixed}, \tag{5}$$

[d]Some hypotheses are discussed in Sect. 12.

where N is the number of colors, only planar diagrams survive. The limit (5) is referred to as the 't Hooft limit. Three years after 't Hooft's original work a remarkable theorem was proved:[39] the number of planar diagrams with n loops $\nu(n)$ does *not* grow with n factorially, rather

$$\nu(n) \sim C^n, \quad n \gg 1, \tag{6}$$

where C is a numerical constant. In one-to-one correspondence with this fact is the vanishing of the instanton contribution at $N \to \infty$. Indeed, at weak coupling in the 't Hooft limit

$$\frac{2\pi}{\alpha} \sim \text{const } N,$$

and the instanton contribution is exponentially suppressed.

One can identify another source of the factorial divergence — unique diagrams of a special type present in Yang–Mills which produce $n!$ not because there are many of them, but because a single graph with n loops is factorially large. As was mentioned previously, such diagrams are called renormalons.[26,40] In the subsequent section we will consider them in more detail.

6. Renormalons

Both, ultraviolet (UV) and IR renormalons can be seen in the bubble diagram depicted in Fig. 1, where the dashed line represents an external (vector) fermion current, the solid lines show fermion propagation while the curvy lines stand for gluons. Consider the correlation functions of two vector currents of massless quarks

$$\Pi_{\mu\nu}(q) = i \int d^4x\, e^{-iqx} \langle T[j_\mu(x)j_\nu(0)] \rangle = (q_\mu q_\nu - q^2 g_{\mu\nu})\Pi(Q^2),$$

$$j_\mu = \bar{\psi}\gamma_\mu\psi, \tag{7}$$

where ψ is the quark field; we assume the number of flavors to be N_f, and denote

$$Q^2 = -q^2, \tag{8}$$

so that in the Euclidean domain Q^2 is positive. The number of colors $N_c = 3$. It is convenient to analyze the Adler function defined as

$$D(Q^2) = -4\pi^2 Q^2 \frac{d\Pi(Q^2)}{dQ^2} \tag{9}$$

and normalized to unity in the leading order. The bubble diagrams with the fermion loop insertions are gauge-invariant *per se*. Needless to say, in the given order there are many other graphs, but we will focus on those presented in Fig. 1, which will be sufficient for identification of renormalons.

The gauge coupling runs, and we must specify which particular coupling constant is used in the expansion. The Adler function, being expressed in terms of $\alpha_s(Q)$, is finite. It seems obvious that the external momentum Q^2 sets the scale of all virtual momenta in loops, and we should use $\alpha_s(Q)$. Is it indeed the case?

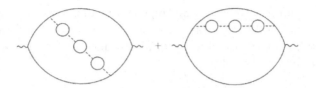

Fig. 1. Bubble diagrams for the Adler function consists of all diagrams with any number of fermion loops inserted into a single gluon line. Then we replace β_0^f, the fermion contribution to the first coefficient of the β function by the full β_0. This is a convenient computational device. The momentum flowing through the graph q is assumed to be large ($Q^2 \equiv -q^2 \gg \Lambda^2$). The momentum flowing through the bubble chain is k.

The answer to the above question is negative. In high orders of perturbation theory there appears an additional parameter n, the order of perturbation theory, which changes the naive estimate $k \sim Q$, see Fig. 1. To see that this is the case, let us have a closer look at Fig. 1 before integrating over k. The exact result for fixed k^2 was found by Neubert.[41] However, we will not need it since for our illustrative purposes it is sufficient to use a simplified interpolating expression[42] collecting all fermion bubble insertions[e] in the gluon propagator: $0, 1, 2$ and so on,

$$D = C \times Q^2 \int dk^2 \frac{k^2 \alpha_s(k^2)}{(k^2 + Q^2)^3} , \tag{10}$$

which coincides with the exact expression[41] in the limits $k^2 \ll Q^2$ and $k^2 \gg Q^2$, up to minor irrelevant details. The coefficient C in Eq. (10) is a numerical constant and $\alpha_s(k^2)$ is the running gauge coupling constant,

$$\alpha_s(k^2) = \frac{\alpha_s(Q^2)}{1 - \frac{\beta_0 \alpha_s(Q^2)}{4\pi} \ln(Q^2/k^2)} . \tag{11}$$

The definition of the coefficients in the β function is given in Appendix.

Now, let us examine the Adler function (10) paying special attention to the logarithmic dependence in (11), a crucial feature of QCD. We will first focus on the IR domain. Omitting the overall constant C, inessential for our purposes, we obtain

$$D(Q^2) = \frac{1}{Q^4} \alpha_s \sum_{n=0}^{\infty} \left(\frac{\beta_0 \alpha_s}{4\pi}\right)^n \int dk^2 \, k^2 \left(\ln \frac{Q^2}{k^2}\right)^n , \qquad \alpha_s \equiv \alpha_s(Q^2) \tag{12}$$

which can be rewritten as

$$D(Q^2) = \frac{\alpha_s}{2} \sum_{n=0}^{\infty} \left(\frac{\beta_0 \alpha_s}{8\pi}\right)^n \int dy \, y^n e^{-y} , \qquad y = 2 \ln \frac{Q^2}{k^2} . \tag{13}$$

[e]The fermion bubbles in Fig. 1 produce only the fermion contribution to $\alpha_s(k^2)$ usually denoted by β_0^f. However, then we can replace β_0^f by the full β_0. Note that adding the gluon and ghost *bubbles* is not sufficient (in particular, one would get a gauge noninvariant expression). The replacement $\beta_0^f \to \beta_0$ incorporates some additional contributions. Note that β_0^f and β_0 have opposite signs — a crucial feature as we will see below.

The y integration in Eq. (13) represents all diagrams of the type depicted in Fig. 1 after integration over the loop momentum k of the "large" fermion loop (and the angles of the gluon momentum).

The y integral from zero to infinity is $n!$. A characteristic value of k^2 saturating the integral is

$$y \sim n \quad \text{or} \quad k^2 \sim Q^2 \exp\left(-\frac{n}{2}\right). \tag{14}$$

Thus, if Q^2 is fixed and n is sufficiently large, the factorial divergence of the coefficients in (12) is indeed due to the infrared behavior in the integral (10). For what follows let us note that if at small $k^2 \sim \Lambda^2$ the diagram in Fig. 1 ceases to properly represent non-Abelian dynamics (which *is* the case in QCD due to strong coupling in the IR), then the integral must be cut off from below at $k^2 = \Lambda^2$, or at $y = n_*$ at large y. Here for each given Q^2,

$$n_* = 2\ln\frac{Q^2}{\Lambda^2}. \tag{15}$$

The summation of factorially divergent terms in the formula

$$D(Q^2) = \frac{\alpha_s}{2} \sum_{n=0}^{\infty} \left(\frac{\beta_0 \alpha_s}{8\pi}\right)^n n! \tag{16}$$

ceases to be valid at $n = n_*$. At $n > n_*$ the factorial growth is suppressed, see Fig. 2, and must be truncated,

$$D(Q^2) \rightarrow \frac{\alpha_s}{2} \sum_{n=0}^{n_*} \left(\frac{\beta_0 \alpha_s}{8\pi}\right)^n n!. \tag{17}$$

Note that n_* is also the critical value of the asymptotic series (16), i.e. the value at which the accuracy of approximation is the best. At $n = n_*$ the asymptotic series (16) achieves the highest accuracy. Truncation at $n = n_*$ ensures the deviation from the exact result to be $\exp\left(-\frac{8\pi}{\beta_0 \alpha_s}\right) \sim \Lambda^4/Q^4$, the same as the infrared sensitivity to the domain $k^2 \sim \Lambda^2$.

All terms in (16) have the same sign, which means that the asymptotic series *per se* is not Borel-summable.

Now let us briefly consider the large k^2 domain in (10). At large k^2

$$D(Q^2) = Q^2 \alpha_s \sum_{n=0}^{\infty} \left(\frac{\beta_0 \alpha_s}{4\pi}\right)^n (-1)^n \int dk^2 \frac{1}{(k^2)^2} \left(\ln\frac{k^2}{Q^2}\right)^n. \tag{18}$$

Introducing

$$\tilde{y} = \ln\frac{k^2}{Q^2} \tag{19}$$

we arrive at

$$D(Q^2) = \alpha_s \sum_{n=0}^{\infty} \left(\frac{\beta_0 \alpha_s}{4\pi}\right)^n (-1)^n \int d\tilde{y}\, \tilde{y}^n e^{-\tilde{y}} = \alpha_s \sum_{n=0}^{\infty} \left(\frac{\beta_0 \alpha_s}{4\pi}\right)^n (-1)^n n!. \tag{20}$$

$$y^n e^{-n}$$

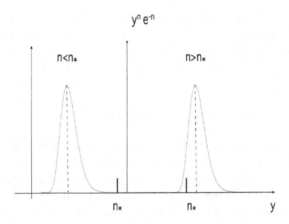

Fig. 2.   The plot of the integrand in Eq. (12) for two values of n, "small" and "large." A sharp peak at $y \sim n$ saturates the integral. In the left plot $n < n_* = 2\ln(Q^2/\Lambda^2)$ and the forbidden domain $k^2 \sim \Lambda^2$ does not contribute to the factorial factor. In the right plot $n > n_*$. The y integration has to be cut off at $y = n_*$, which tempers the factorial growth.

This series is sign-alternating and, hence, is Borel-summable. The characteristic value of $\tilde{y}$ saturating the integral is $\tilde{y} \sim n$ implying $k^2 \sim Q^2 e^n$. Thus, at large n we deal with large k^2 which explains why this contribution is referred to as the ultraviolet renormalon. It is well-defined *per se*. The best possible accuracy one can achieve with Eq. (20) compared to the exact result of the Borel transformation (3) is $\exp\left(-\frac{4\pi}{\beta_0 \alpha_s}\right) \sim \frac{\Lambda^2}{Q^2}$. I will not touch UV renormalons in what follows. Note that the singularities in the Borel-transform for the IR and UV renormalons have different separations from the origin.[f] Namely,

$$B_D(\alpha_s) \sim \begin{cases} \left(1 - \dfrac{\beta_0 \alpha_s}{8\pi}\right)^{-1}, & \text{IR}, \\[3mm] \left(1 + \dfrac{\beta_0 \alpha_s}{4\pi}\right)^{-1}, & \text{UV}. \end{cases} \qquad (21)$$

The positions of the singularities in the α_s plane are

$$(\alpha_s)_* = \begin{cases} \dfrac{8\pi}{\beta_0}, & \text{IR}, \\[3mm] -\dfrac{4\pi}{\beta_0}, & \text{UV}. \end{cases} \qquad (22)$$

They are depicted in Fig. 3 along with the singularities due to instantons.

---

[f] As was mentioned above, Eq. (10) is simplified. Working with the exact formula[41] we would have obtained for the UV renormalon contribution in B_D an expression that contains a single pole as well as a double pole at $(\alpha_s)_* = -\frac{4\pi}{\beta_0}$. In addition to the term in the second line in (21) we would get another term $\sim \left(1 + \frac{\beta_0 \alpha_s}{4\pi}\right)^{-2}$, see e.g. Ref. 11. No qualitative changes occur due to the presence of the double pole. I will not go into details, since the UV renormalons are mentioned only for completeness and will not be pursued further.

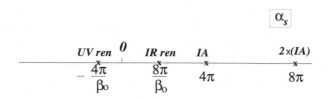

Fig. 3.   Singularities in the Borel plane.

7. Operator Product Expansion

I remember that after the first seminar on the SVZ sum rules[43] in 1978 Eugene Bogomol'nyi used to ask me each time we met: "Look, how can you speak of power corrections in the two-point functions at large Q^2 when even the perturbative expansion (i.e. the expansion in $1/\ln(Q^2/\Lambda^2)$) is not well defined? Isn't it an 'excessive force' abusive approach?"

At that time Wilson's wisdom only started to conquer the high-energy physics community. My understanding in 1978 missed some nuances too. In modern terminology what we mostly used in Ref. 43 is now known as the "practical" operator product expansion, its simplified version (see e.g. Ref. 14, Sect. 6, which is highly recommended to the reader unfamiliar with nuances of OPE in QCD). Although it is sufficient in solving problems arising in various applications and relatively easy to implement, in the present article in which I am after conceptual aspects, I have to be careful with formulations. Only then I will be able to answer the above Bogomoln'yi question in a positive way, namely:

> "Consistent use of Wilson's OPE makes everything well-defined at the conceptual level. Technical implementation may not always be straightforward, however."

The operator product expansion (OPE) in asymptotically free theories is a *bookkeeping device* separating short-distance (weak-coupling) contributions from those coming from large distances (strong coupling domain). To this end one introduces an auxiliary separation scale μ. OPE is applicable whenever one deals with problems that can be formulated in the Euclidean space-time and in which one can regulate typical Euclidean distances by a varying large external momentum Q (or m_Q in the heavy quark problems). Wilson's OPE is meaningful if one can choose $\mu \ll Q$ (or $\mu \ll m_Q$), but $\mu \gg \Lambda$.

Note that a practical version that draws a divide in OPE between perturbation theory and nonperturbative effects is a simplification which may or may not be approximately valid, depending on the theory under consideration. The correct divide is between short- and large-distance contributions. As a book-keeping device of this type it cannot fail,[13] provided no arithmetic mistake is made *en route*.

The idea of factorization of short and large distances, the central point of OPE, dates back to classical Wilson's work[44] (see also Ref. 45) where it was put forward in connection with theories of strong interaction with conformal invariance at short distances. Shortly after, Wilson formulated a very general procedure of the renormalization-group flow (e.g. Ref. 45) which became known as the Wilsonean renormalization group. Wilson's formulation makes no reference to perturbation theory, it applies both to strongly and weakly coupled theories. The focus of Wilson's work was on statistical physics, where the program is also known as the block-spin approach. Starting from the microscopic degrees of freedom at the shortest distances a, one "roughens" them, step by step, by constructing a sequence of effective (composite) degrees of freedom at distances $2a$, $4a$, $8a$, and so on. At each given step i one constructs an effective Hamiltonian, which fully accounts for dynamics at distances shorter than a_i in the coefficient functions.

Surprisingly, in high-energy physics of the 1970s the framework of OPE was narrowed down to a very limited setting. On the theoretical side, it was discussed almost exclusively in perturbation theory, as is seen, for instance, from Ref. 46. On the practical side, its applications were mostly narrowed down to deep inelastic scattering, where it was customary to work in the leading-twist approximation.

The general Wilson construction was adapted to QCD, for the systematic inclusion of power-suppressed effects, in Refs. 43 and 13. Vacuum expectation value of the gluon density operator and other vacuum condensates were introduced for the first time, which allowed one to analyze a large number of vacuum two- and three-point functions, with quite nontrivial results. A consistent Wilsonean approach requires an auxiliary normalization point μ which plays the role of a "regulating" parameter separating hard contributions included in the coefficient functions and soft contributions residing in local operators occurring in the expansion. The degree of locality is regulated by the same parameter μ.

Prevalent in the 1970s was a misconception that the OPE coefficients are determined exclusively by perturbation theory while the matrix elements of the operators involved are purely nonperturbative. Attempts to separate perturbation theory from "purely nonperturbative" condensates gave rise to inconsistencies (see e.g. Ref. 47; I will return to this paper later) which questioned the very possibility of using the OPE-based methods in QCD.

In the heavy quark theory, in which Kolya's contribution was instrumental, OPE acquired a new life constituting the basis of the heavy quark mass expansions (for a review see Ref. 10). In this range of questions one deals with expectation values of various operators over the heavy quark meson or baryon states, rather than vacuum expectation values. The overall ideology does not change, however.

The OPE formalism provides a natural framework for the discussion of IR renormalons and how they should be treated in theories with strong coupling regime.

8. An Illustrative Example

Despite the conceptual simplicity of OPE, it continues to be questioned in the literature, in particular, in connection with renormalons in strongly coupled theories. The statement which I would like to illustrate in this section is: if one introduces the boundary point μ (unavoidable in non-conformal field theories) and abandons the idea of separation along the line "perturbative vs. nonperturbative," all would-be inconsistencies disappear, and so does the problem of renormalons.

Following Ref. 13 I will consider here a relatively simple example of a two-dimensional model — the so-called $O(N)$ model — which has both, asymptotic freedom and renormalons, and at the same time is exactly solvable at large N. Classically excitations in this model are massless. A mass gap is generated at the quantum (nonperturbative) level. This example in the given context was suggested long ago in Ref. 48. In this paper OPE (in its "practical" version) was found to be perfectly consistent with the exact solution in the leading in $1/N$ approximation.

However, the subsequent exploration of composite operators[47] questioned the existence of consistently defined composite operators in OPE at the level of the first subleading correction (of the relative order of $1/N$). Now, I will demonstrate how inconsistencies are eliminated once μ is explicitly introduced.

The Lagrangian of the model has the form[49] (for a review see Ref. 50)

$$\mathcal{L} = \frac{N}{2\lambda}(\partial_\mu S^a)(\partial^\mu S^a), \quad \vec{S}^2 = 1, \tag{23}$$

where $\vec{S} = \{S^1, S^2, \ldots, S^N\}$ is an N-component real (iso)vector field, and λ is the 't Hooft coupling, which stays fixed in the limit $N \to \infty$. The $O(N)$ symmetry of this Lagrangian is evident. This model is asymptotically free,[50] in much the same way as Yang–Mills theory,

$$\lambda(p) = \frac{2\pi}{\ln \frac{p}{m}}, \quad \text{or} \quad m = p \exp\left[-\frac{2\pi}{\lambda(p)}\right], \tag{24}$$

where m is a dynamically generated mass gap. In perturbation theory the $O(N)$ symmetry is spontaneously broken implying $N-1$ Goldstone modes. The $O(N)$ symmetry is restored in the exact solution, in full accord with the Coleman theorem.[51] The vacuum condensates of the type $\langle (\partial_\mu S^a)(\partial^\mu S^a)\rangle \neq 0$ develop. To the leading order in N (for details see e.g. Ref. 50)

$$\left\langle \left[(\partial_\mu S^a)^2\right]^k \right\rangle = m^{2k}, \quad k = 1, 2, \ldots. \tag{25}$$

In this order the above matrix elements scale as N^0 and factorize. To order $O(N^0)$ each of them is μ independent because in this order the anomalous dimension of the operator $(\partial_\mu S^a)^2$ vanishes. Needless to say, there are nonfactorizable corrections scaling as $1/N$. For what follows it is convenient to introduce a special notation for the operator

$$(\partial_\mu S^a)^2 \equiv \alpha. \tag{26}$$

The operator basis in OPE to the order $O(N^0)$ consists of the composite operators of the type α^n. In the subleading orders operators with an entangled index structure appear, but we do not have to consider them here.

To discuss OPE let us consider the two-point function

$$P(q^2) = i \int d^2x \, e^{iqx} \langle T\{j_s(x)j_s(0)\}\rangle \,,$$

$$j_s = \sqrt{N}(\partial_\mu S^a)^2$$

(27)

at large (Euclidean) values of q^2 (i.e. q^2 negative and $Q^2 \equiv -q^2$ positive).

The general OPE formula for the two-point function (27) (at large Euclidean Q^2) has the form

$$P(Q^2) = c_0(Q^2,\mu^2)\, Q^2\mathbf{I} + c_1(Q^2,\mu^2)\alpha(\mu) + \frac{c_2(Q^2,\mu^2)}{Q^2}\,[\alpha(\mu)]^2 + \cdots\,, \quad (28)$$

where c_i are the coefficient functions.

With our normalization $P(Q^2) \sim N^0$ in the leading order in N. All coefficient functions and expectations values scale in the same way, as N^0, with subleading $1/N$ corrections. Moreover, to the leading order the OPE coefficients in (27) are μ independent, with no factorial divergences. As a result, at $N = \infty$ one can close one's eyes on subtleties and adhere to the simplified formula according to which the coefficient functions are determined exclusively by perturbation theory, and (large-distance) vacuum condensates exclusively by nonperturbative effects.[g] This simplified formula is self-consistent.[48]

An apparent inconsistency was noted at the level of $1/N$ corrections.[47] Among many additional computations at this level one has to define composite operators beyond factorization, the simplest of which is the operator α^2. Below I will show that introducing the normalization point μ — a necessary step not seen in Ref. 47 because of dimensional regularization in which the scale separation is not explicit — solves all would-be problems.

The vacuum expectation value of α^2 can be defined as follows:

$$\langle [\alpha(\mu)]^2 \rangle = m^4 + \langle [\alpha(\mu)]^2 \rangle_{\text{conn}}\,,$$

$$\langle [\alpha(\mu)]^2 \rangle_{\text{conn}} = \int_{\text{Eucl}\,p<\mu} \frac{d^2p}{(2\pi)^2} D(p^2)\,,$$

(29)

where the subscript conn means the connected (nonfactorizable) part and $D(p^2)$ is the propagator of the α field known from the exact solution of the model to the leading order in N,

$$D(p^2) = -\frac{4\pi}{N} \frac{\sqrt{p^2(p^2+4m^2)}}{\ln \frac{\sqrt{(p^2+4m^2)}+\sqrt{p^2}}{\sqrt{(p^2+4m^2)}-\sqrt{p^2}}}\,,$$

(30)

see Fig. 4.

---

[g]This exceptional situation specific to the $O(N)$ model has no parallel in QCD.

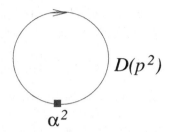

Fig. 4. The leading contribution to $\langle[\alpha(\mu)]^2\rangle_{\text{conn}}$. The propagator of α is presented in (30).

It is obvious that the connected part is suppressed by $1/N$ compared to the factorized part. For what follows it will be useful to rewrite the denominator in (30) at $p^2 \gg m^2$ as

$$\frac{1}{\ln\frac{\sqrt{(p^2+4m^2)}+\sqrt{p^2}}{\sqrt{(p^2+4m^2)}-\sqrt{p^2}}} \equiv \frac{\lambda(\mu)}{4\pi}\left[1 - \frac{\lambda(\mu)}{4\pi}\ln\frac{\mu^2}{p^2}\right]^{-1}\left[1 + O\left(\frac{m^2}{p^2}\right)\right]. \qquad (31)$$

This approximation certainly becomes meaningless at $p^2 \leq m^2$ since the expression $1 - \frac{\lambda(\mu)}{4\pi}\ln\frac{\mu^2}{p^2}$ vanishes at $p^2 = m^2$.

The integral in (29) is doable and can be expressed in terms of special functions.[13] We need to examine this integral in the limit $\mu \gg m$ because only in this limit the coefficient functions in OPE are predominantly perturbative. The result for $\langle[\alpha(\mu)]^2\rangle_{\text{conn}}$ includes terms $O(\mu^4)$, $O(\mu^2 m^2)$, $O(m^4)$ and $O(m^6/\mu^2)$. For simplicity will omit the latter and focus on the first three terms. The second and the third can be established from the exact result (30); as we will see shortly, they do not contain factorial divergences at all. The first term does, see below. We could find it from the exact result too. But it will be more instructive to calculate it from the approximate formula (31). Then the relation to renormalons will become more apparent. Equation (31), being integrated from m^2 to μ^2 (taking account of the remark after (31)), is perfectly sufficient to determine the μ^4 term in $\langle[\alpha(\mu)]^2\rangle_{\text{conn}}$. Indeed, substituting (31) in (30) and keeping m only in the argument of logarithms we obtain

$$\langle[\alpha(\mu)]^2\rangle_{\text{conn}} = -\frac{1}{N}\int_{m^2}^{\mu^2} p^2 dp^2 \frac{\lambda(\mu)}{4\pi}\sum_{k=0}^{\infty}\left(\frac{\lambda(\mu)}{4\pi}\ln\frac{\mu^2}{p^2}\right)^k$$

$$= -\frac{1}{N}\mu^4\sum_{k=0}^{\infty}\left(\frac{\lambda(\mu)}{8\pi}\right)^{k+1}\int_0^{k_*} dy\, y^k e^{-y}, \qquad (32)$$

where

$$k_* = 2\ln\frac{\mu^2}{m^2}. \qquad (33)$$

If $k \leq k_*$ the integral on the right-hand side of (32) can be extended to infinity since the saddle point lies at $y = k$, and then this integral produces $k!$,

$$\langle [\alpha(\mu)]^2 \rangle_{\text{conn}} = -\frac{1}{N} \mu^4 \sum_{k=0}^{k_*} \left(\frac{\lambda(\mu)}{8\pi} \right)^{k+1} k! . \tag{34}$$

If $k > k_*$ the integral is saturated at $y = k_*$, the factorial growth ceases to continue and the right-hand side of (32) reduces to $O(m^4)$.

The exact expression for $\langle [\alpha(\mu)]^2 \rangle_{\text{conn}}$ is well-defined and can be explicitly calculated. I omit a number of simple algebraic manipulations referring the reader to the original paper.[13] The result is[h]

$$\langle [\alpha(\mu)]^2 \rangle_{\text{conn}} = \frac{1}{N} \left\{ -\mu^4 \left[e^{-L} \operatorname{Ei}(L) \right] - \frac{4}{L} \mu^2 m^2 + 2m^4 \left(C + \ln L - \frac{1}{L} + \frac{5}{L^2} \right) \right\}, \tag{35}$$

where

$$L = 2 \ln \frac{\mu^2}{m^2} = \left[\frac{\lambda(\mu)}{8\pi} \right]^{-1}, \tag{36}$$

and $C \approx 0.5772\ldots$ is the Euler constant. The $O(\mu^4)$ terms in (34) and (35) perfectly match each other!

Needless to say, $P(Q^2)$ does not contain the auxiliary parameter μ; it depends only on physical parameters Q^2 and m^2. This means that on the right-hand side of (28) μ must cancel. As was mentioned, at order N^0 is does not appear at all. The first and the second term in (35) appearing at the level $O(1/N)$ must be canceled by the corresponding contributions coming from the first and the second term in (28). And they do, indeed! The coefficient c_0 has a correction $\frac{\mu^4}{Q^4} \frac{1}{N} [e^{-L} \operatorname{Ei}(L)]$ while c_1 has $\lambda(\mu) \frac{\mu^2}{Q^2}$.

The terms proportional to m^4 in Eq. (35) do not cancel. They still have a weak (logarithmic) dependence on μ through L^{-1} and $\ln L$. This is a manifestation of the anomalous dimension of the operator $[\alpha(\mu)]^2$ which shows up beyond the leading (factorization) order. It is canceled by the corresponding logarithmic terms in $c_2(Q^2, \mu^2)$.

The reader interested in additional details is referred to Ref. 52 for a later discussion of OPE in a particular correlation function in the $O(N)$ model at the subleasing level (i.e. $O(1/N)$ corrections).[i]

Concluding this section let me mention that the $O(N)$ sigma model is promising in one more aspect: In this model at $N > 3$ instantons disappear, while nothing dramatic happens to renormalons. Question: what replaces the instanton singularities in the Borel plane?

---

[h]I correct here a number of misprints in the expression for $\langle [\alpha(\mu)]^2 \rangle_{\text{conn}}$ given in Ref. 13.
[i]A remarkable feature making the model *different* from QCD is the fact of OPE convergence at the level of $O(N^0)$ and $O(N^{-1})$ terms. This is due the fact that at this level particle production thresholds do not extend to infinite energies in the $O(N)$ sigma model, unlike QCD.

9. OPE and Renormalons in QCD

After this brief digression intended to demonstrate peculiarities of perturbation theory in strongly coupled models with the known solution let us return to QCD where no exact solution is available. I will start from correlation functions of the type (7) at large Euclidean q^2 in which OPE can be consistently built through separation of large- and short-distance contributions. For simplicity, for our illustrative purposes, I will set the separation scale at $\mu = \Lambda$ rather than at $\mu \gg \Lambda$. This would be inappropriate in quantitative analyses; however, my task is to reveal qualitative aspects. For this purpose no harm will be done if I put $\mu = \Lambda$. With this convention all relevant expressions will dramatically simplify.

Let us have a closer look at Eqs. (10) and (11). The unlimited factorial divergence in (16) is a direct consequence of integration over k^2 in (12) all the way down to $k^2 = 0$. Not only this is nonsensical because of the pole in (11) at $k^2 = \Lambda^2$, this is *not* what we should do in calculating coefficient functions in OPE. The coefficients must include $k^2 > \Lambda^2$ by construction. The domain of small k^2 (below Λ^2) must be excluded from c_0 and referred to the vacuum matrix element of the gluon operator $G_{\mu\nu}^2$. Indeed, in the sum in Eq. (12) all terms with $n > n_*$ can be written as (see Fig. 2)

$$\Delta D(Q^2) = \frac{\alpha_s}{2} \sum_{n > n_*} \left(\frac{\beta_0 \alpha_s}{8\pi} \right)^n n_*^n e^{-n_*}$$

$$= \frac{\alpha_s}{2} \sum_{n > n_*} \frac{\Lambda^4}{Q^4}, \tag{37}$$

where I used the fact that $\frac{\beta_0 \alpha_s(Q^2)}{8\pi} = \frac{1}{2\ln(Q^2/\Lambda^2)} = 1/n_*$. Of course, we can*not* calculate the gluon condensate from the above expression for the tail of the series (12) representing the large distance contribution, for a number of reasons. In particular, the value of the coefficient in front of Λ^4/Q^4 remains uncertain in (37) because Eq. (11) is no longer valid at such momenta. We do not expect the gluon Green functions used in calculation in Fig. 1 and in Eq. (11) to retain any meaning in the nonperturbative domain of strong coupling dynamics. A qualitative feature — the power dependence $(\Lambda/Q)^4$ in (37) — is correct, however.

We note with satisfaction that the fourth power of the parameter Λ/Q which we find from this tail exactly matches the OPE contribution of the operator $\langle G_{\mu\nu}^2 \rangle$. In Sect. 8 where we analyzed an exactly solvable model we could convince ourselves that this is not a coincidence.

Summarizing this section I can say that consistent use of OPE cures the problem of the renormalon-related factorial divergence of the coefficients in the α_s series, absorbing the IR tail of the series in the vacuum expectation value of the gluon operator $G_{\mu\nu}^2$ and similar higher-order operators. Although the value of $\langle G_{\mu\nu}^2 \rangle$ cannot be calculated from renormalons, the very fact of its existence can be established.

10. Sources of Factorials and Master Formula

From quantum mechanics we learn that the factorial divergence can arise from "soft" fields, e.g. instantons (see Sect. 5). In QCD the instantons are ill-defined in the IR and, strictly speaking, nobody knows what to do with them.[j] There is a perfectly legitimate conceptual way out, however. If one considers QCD in the 't Hooft limit of large number of colors,[38] instantons decouple. At the same time, none of the essential features of QCD disappears. In addition to phenomenological arguments,[54] this statement is supported by an exact solution of a strongly coupled two-dimensional model with asymptotic freedom.[55]

*****

Now I will try to summarize the lessons we learned in a single (simplified) "master" formula. At large Euclidean momenta the correlation functions of the type (7) and similar can be represented as

$$
D(Q^2) = \sum_{n=0}^{n_*^0} c_{0,n} \left(\frac{1}{\ln Q^2/\Lambda^2} \right)^n
$$

$$
+ \sum_{n=0}^{n_*^1} c_{1,n} \left(\frac{1}{\ln Q^2/\Lambda^2} \right)^n \left(\frac{\Lambda}{Q} \right)^{d_1}
$$

$$
+ \sum_{n=0}^{n_*^2} c_{2,n} \left(\frac{1}{\ln Q^2/\Lambda^2} \right)^n \left(\frac{\Lambda}{Q} \right)^{d_2} + \cdots
$$

$$
+ \text{``exponential terms''} . \tag{38}
$$

Equation (38) is simplified in a number of ways. First, it is assumed that the currents in the left-hand side have no anomalous dimensions, and so do the operators appearing on the right-hand side. They are assumed to have only normal dimensions given by d_i for the i-th operator. Second, I ignore the second and all higher coefficients in the β function so that the running coupling is represented by a pure logarithm. All these assumptions are not realistic in QCD.[k] I stick to them to make the master formula concise. Inclusion of higher orders in the β function and anomalous dimensions both on the left- and right-hand sides will give rise to rather contrived additional terms and factors containing log log's, log log log's (log log / log)'s, etc. This is a purely technical, rather than conceptual, complication, however.

So far I discussed the convergence of the perturbative series (explaining that the regulating parameter μ in OPE allows one to make them meaningful). The expansion (38) runs not only in powers of $1/\ln Q^2$, but also in powers of Λ/Q. This

---

[j]This statement is an exaggeration. The inquisitive reader is referred to Ref. 53 for an alternative point of view on instantons in QCD vacuum.
[k]They could be made somewhat more realistic in $\mathcal{N} = 2$ super-Yang–Mills.

is a double expansion, and the power series in Λ/Q is also infinite in its turn. Does it have a finite radius of convergence?

Needless to say, this is an important question. The answer to it is *negative*.[1] As was argued in Refs. 56 and 14, power series are factorially divergent in high orders. This is a rather straightforward observation following from the analytic structure of $D(Q^2)$. In a nut shell, since the cut in $D(Q^2)$ runs all the way to infinity along the positive real semi-axis of q^2, the $1/Q^2$ expansion cannot be convergent. The last line in Eq. (38) symbolically represents a divergent tail of the power series.

The actual argument is somewhat more subtle than that, but the final conclusion — that high-order tail of the (divergent) power series gives rise to exponentially small corrections (exponentially small in Euclidean, oscillating in Minkowski) — still holds. The most instructive way to see it is provided by a toy model presented in Sect. 2.2 of Ref. 14 which refers to the 't Hooft limit. Then qualitatively one can saturate $\Pi(Q^2)$ by an infinite comb of equidistant infinitely narrow resonances. For simplicity one can assume that the couplings of these resonances to the current do not depend on the excitation number. Then[m]

$$\Pi(Q^2) = -\frac{N_c}{12\pi^2}\psi(z) + \text{const}\,, \tag{39}$$

where

$$z = \frac{Q^2 + m_\rho^2}{3m_\rho^2}\,, \tag{40}$$

and $\psi(z)$ is Euler's ψ function. In the Euclidean domain of positive Q^2

$$\ln z \sim -\frac{1}{2z} - \sum_{n=1}^{\infty} \frac{B_{2n}}{2n}\frac{1}{z^{2n}}\,, \tag{41}$$

B_{2n} stand for the Bernoulli numbers

$$B_{2n} = (-1)^n \frac{2(2n)!}{(2\pi)^{2n}} \zeta(2n)\,, \tag{42}$$

and ζ is the Riemann function. The tilde "$\sim$" in (41) means that the series in $1/Q^2$ is asymptotic: since $\zeta(2n) \sim 1$ the expansion coefficients in (41) are obviously factorially divergent. The tail of the $1/Q^2$ series after optimal truncation is exponentially small. Alternatively, one can apply the Borel procedure since the alternating signs in (42) indicate Borel summability,

$$\left(\frac{1}{Q^2}\right)^n \to \frac{1}{(n-1)!}\left(\frac{1}{M^2}\right)^n\,. \tag{43}$$

The position of the singularity in the $1/M^2$ plane is $2/(3m_\rho^2)$.

---

[1]See also footnote i.
[m]The factor 3 in the denominator of (40) is an approximate empiric number.

11. A Breakthrough Idea

Now I am finally ready to explain the idea first put forward in Refs. 15 and 16. As was elucidated above, in the processes with OPE renormalons play no special role as long as the operator basis in OPE is complete, no relevant operator is accidentally omitted. However, there exists a wide range of phenomena at high energies (or in heavy quark physics) which do not allow one to carry out OPE-based analyses. The most well-known example of this type is jet physics. Up to a certain time these processes were treated exclusively in the realm of perturbative QCD. An estimate of nonperturbative effects, even as approximate as it could be, was badly needed. A minimalistic and urgent task was to find the power of $1/E$ (or $1/m_Q$) which controls the degree of fall-off of the leading nonperturbative effect.

To this end it was suggested[15,16] to analyze the tails of the renormalon series. I hasten to add that renormalons by no means capture all nonperturbative effects. For instance, they are blind to any effects due to chiral symmetry breaking. Thus, they cannot guide us if chiral symmetry breaking plays a role. Hints associated with renormalons refer to gluons.

The first example of the "renormalon guidance" (that later proliferated to many other analyses) was the so-called heavy quark pole mass. The heavy quark mass is a key parameter in most aspects of heavy quark physics. The pole mass was routinely used in analyzing data. It is well-defined (infrared stable) and unambiguous to any finite order in perturbation theory. This infrared stability could give an impression that the pole mass is well-defined in general. This misinterpretation was quite common in the literature in the early 1990s.

The fact that the pole mass is *not* well-defined at the nonperturbative level was first noted and emphasized in Refs. 15 and 16. What is even more important, a rather powerful renormalon-based tool was suggested for evaluating the corresponding nonperturbative contribution. The problem arises because the pole mass is sensitive to large distance dynamics, although this fact is not obvious in perturbative calculations. Infrared contributions lead to an intrinsic uncertainty in the pole mass of order Λ, i.e. a Λ/m_Q power correction. Renormalons produce clear evidence for this non-perturbative correction to m_Q^{pole}. The signal comes from the factorial growth of the high order terms in the α_s expansion corresponding to a singularity residing at $2\pi/\beta_0$ in the Borel plane.

The renormalon contribution to the pole mass is shown in Fig. 5. The bubble chain generates the running of the strong coupling α_s. To leading order, it can be accounted for by inserting the running coupling constant $\alpha_s(k^2)$ in the integrand corresponding to the one-loop expression. In the non-relativistic regime, when the internal momentum $|k| \ll m_Q$, the expression is simple,

$$\delta m_Q \sim -\frac{4}{3} \int \frac{d^4k}{(2\pi)^4 i k_0} \frac{4\pi\alpha_s(-k^2)}{k^2} = \frac{4}{3} \int \frac{d^3\vec{k}}{4\pi^2} \frac{\alpha_s(\vec{k}^2)}{\vec{k}^2}, \qquad (44)$$

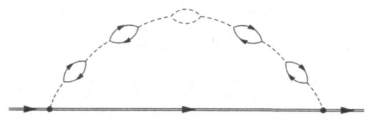

Fig. 5. Perturbative diagrams leading to the IR renormalon uncertainty in m_Q^{pole} of order Λ. The number of bubble insertions in the gluon propagator can be arbitrary. The horizontal line at the bottom is the heavy quark Green's function.

where $\alpha_s(\vec{k}^2)$ can be read off from Eq. (11) with the substitution $Q^2 \to m_Q^2$. Expressing the running $\alpha_s(k^2)$ in terms of $\alpha_s(m_Q^2)$ (note that $\vec{k}^2 < m_Q^2$), and expanding in $\alpha_s(m_Q^2)$ we arrive at

$$\frac{\delta m_Q^{(n+1)}}{m_Q} \sim \frac{4}{3}\frac{\alpha_s(m_Q^2)}{\pi}n!\left(\frac{\beta_0\alpha_s(m_Q^2)}{2\pi}\right)^n. \tag{45}$$

The right-hand side represents the renormalon series. This series is factorially divergent and is not Borel-summable. Moreover, in the case at hand there is no OPE which could absorb this tail in a higher-dimension operator. What should we do?

The question was posed and the answer given in Refs. 15 and 16: Following the line of reasoning applied in OPE-based processes we should truncate the series at an optimal order and, *in addition*, introduce an infrared parameter δm_Q which will absorb the renormalon tail. Equation (45) implies that

$$n_* \sim \ln\frac{m_Q}{\Lambda}, \quad \frac{\delta m_Q}{m_Q} \sim e^{-n_*} \sim \frac{\Lambda}{m_Q}. \tag{46}$$

The perturbative expansion *per se* anticipates the onset of the nonperturbative regime (the impossibility of pinning down the would-be quark pole in perturbation theory to accuracy better than Λ).

Certainly, the renormalons do not represent the dominant component of the infrared dynamics. However, they provide the "renormalon guidance" playing a very important role of an indicator of the presence of the power-suppressed non-perturbative effects.

12. Renormalons in Weak Coupling Problems

In Sect. 4 it was mentioned that the study of renormalons at weak coupling can can shed new light on the general structure of field theory. According to the conjecture formulated in Refs. 34 and 35 at weak coupling any particular factorially divergent contribution, if Borel-nonsummable, must match a certain quasiclassical field configuration. Such configurations were identified[35] in two-dimensional $CP(N-1)$ models which present a close parallel[50] to four-dimensional Yang–Mills. However, subtle details are not yet satisfactory.

The instanton quarks supposedly matching the perturbative factorial divergence in cylindrical geometry have action $4\pi/(Ng^2)$ resulting in singularities at $4\pi k/N$ in the Borel plane;[n] here $k = 1, 2, \ldots, N$ is an integer. If we have a look at Fig. 3, we will see that the positions of the renormalon singularities are quantized in the units of $4\pi/\beta_0$. A matching relationship can only be achieved if $\beta_0 = \text{integer} \times N$. This is the case in $CP(N-1)$ models, but this is certainly *not* the case in Yang–Mills.[o]

Let us discuss this example — the two-dimensional $CP(N-1)$ model — in more detail. Assume it is considered on a cylinder $R_1 \times S_1(r)$ where r is the radius of the circle. At $r \to 0$ the problem reduces to quantum-mechanical, with the perturbative expansion being nonsummable *à la* Borel. The factorial divergence is not directly related to renormalons (which are absent in quantum mechanics), but it exists.

The corresponding singularity in the g^2 Borel plane lies at $8\pi/N$. This happens to be exactly the action of two instanton quarks.[35] Thus, one can (and does) achieve resurgence through construction of the corresponding trans-series. What is important for the following paragraph is the fact that action of two instanton quarks corresponds to the dimension of the lowest nontrivial operator in OPE, namely $(\partial S)^2$.

If $r \gtrsim \Lambda^{-1}$, we find ourselves in the strong coupling regime, the parameter r^{-1} plays the role of μ, and the perturbative factorial divergence is generated by renormalons, which should be treated, as usual, in the framework of OPE. Remarkably, the position of singularity in the Borel plane does *not* shift from $8\pi/N$. OPE explains the renormalon singularity at $8\pi/N$ since the first operator in OPE (after the trivial operator) has exactly the needed dimension and generates terms $\exp\left(-\frac{8\pi}{N}\right) \sim \Lambda^2$. Thus, it is not ruled out that the positions of the leading singularity in the Borel plane is r-independent in the interval $r \in (0, \text{const} \times \Lambda^{-1})$.

This r independence cannot survive in Yang–Mills theories, because of the β_0 factor mentioned above. What can happen in Yang–Mills, however, with luck, is a smooth r-dependence of the singularity positions in the Borel plane. This question remains open.

Cylindrical geometry exploited in Refs. 34 and 35 is not the only way to make Yang–Mills theory weakly coupled. Alternatively, one can Higgs the theory.

Assume we have SU(2) Yang–Mills theory fully Higgsed by an expectation value of the Higgs doublet field, just as in the standard model. The theory is at weak coupling. Assume we introduce $2n_f$ doublets of chiral (Weyl) fermions $(\chi_\alpha^i)^j$ and $(\psi_\alpha^i)^j$, where α and i are the Lorentz and SU(2) gauge indices, respectively, and

---

[n] In four dimensions it is convenient to consider the α plane, with $\alpha \equiv g^2/4\pi$. In two dimensions the g^2 plane is more convenient.

[o] There is an observation which, perhaps, gives hope for the future. In pure Yang–Mills $\beta_0 = \frac{11}{3}N$. As was noted by Khriplovich long ago[57] (see also Ref. 58, Sect. 25.1) the above value of β_0 has a distinct two-component structure. If one calculates β_0 in the physical gauge without ghosts (e.g. Coulomb), one will discover that in fact $\beta_0 = 4N - \frac{1}{3}N$ where the first term in the right-hand side presents *anti*screening inherent only to non-Abelian gauge theories, while the second term, with the fractional coefficient, is a conventional screening.

$j = 1, 2, \ldots, n_f$. For simplicity we will assume that both the fermion and the Higgs masses are the same as the mass of the W bosons M. After Higgsing this theory still has a global SU(2) symmetry. Three W bosons form a triplet under this global SU(2).

This SU(2) theory has no internal anomalies. In fact, it is vector-like. With the even number of doublets it avoids Witten's global anomaly too.[58,59]

In this fully IR regularized theory we can repeat the analysis outlined in Sect. 6. The renormalon diagram in Fig. 1 now yields an expression similar to that in (12) at $k^2 \gg M^2$. However, at $k^2 \lesssim M^2$ the integral over k^2 in (12) must be cut off from below at M^2 (with logarithmic accuracy).

Now, in perturbation theory the large-Q^2 expansion of the Adler function has the form

$$D(Q^2) = \sum c_{0,n} \left(\frac{1}{\ln Q^2/\Lambda^2}\right)^n + \sum c_{1,n} \left(\frac{1}{\ln Q^2/\Lambda^2}\right)^n \left(\frac{M}{Q}\right)^2$$
$$+ \sum c_{2,n} \left(\frac{1}{\ln Q^2/\Lambda^2}\right)^n \left(\frac{M}{Q}\right)^4 + \cdots . \tag{47}$$

It is not difficult to see that (although the critical value of n for the diagram in Fig. 1 changes compared to that in Sect. 6) the renormalon tail gives rise to a residual term in the second line proportional to $(\Lambda/Q)^4$, i.e. similar to what we have in the limit $M = 0$.

The weak coupling conjecture[34,35] assumes that (47) (more exactly, the $(\Lambda/Q)^4$ term representing its tail) must match a certain quasiclassical field configuration. At the moment the only candidate I see is the instanton–antiinstanton pair at a fixed (and small) separation. Is this the case? Can such a match be explicitly traced?

13. Conclusions

(1) Twenty years after its emergence,[15,16] the renormalon counting remains the only known method for evaluating nonperturbative corrections in the processes without OPE.

(2) Operator product expansion, with an explicit separation scale μ, conceptually solves the problem of factorial divergence of the perturbative series, at least at $N \to \infty$.

(3) Factorial divergence of the $(\Lambda/Q)^k$ series emerging in OPE at large k, as established in Ref. 56, needs further explorations and an appropriate theoretical description/understanding.

(4) The resurgence program put forward in Refs. 34 and 35 outlines a clear-cut parallel between factorial divergences at weak coupling on the one hand, which, being treated *à la* Ecalle, result in well-defined trans-series, and the OPE-based paradigm at strong coupling, on the other hand. More thinking is required to completely understand their relationship.

(5) There are theories with renormalon-induced factorial divergence but no instantons (e.g. two-dimensional $O(N)$ sigma model with $N > 3$). Construction of OPE in such models is not affected by the absence of instantons. If one considers them in cylindrical geometry, at weak coupling, finding substitutes for instanton quarks is a challenge. The first steps in this direction have been made, but more work is needed.

Acknowledgments

I am grateful to Martin Beneke, Ikaros Bigi, Alexei Cherman, Gerald Dunne, Sergei Monin, Mithat Ünsal, and Arkady Vainshtein for useful comments.

This work is supported in part by DOE grant DE-FG02- 94ER-40823.

Appendix. Definitions

We use the following convention for the β function:

$$\beta(\alpha_s) = \mu\frac{\partial\alpha_s}{\partial\mu} = -\frac{1}{2\pi}\beta_0\,\alpha_s^2 + \frac{1}{2(2\pi)^2}\beta_1\alpha_s^3 + \cdots. \tag{A.1}$$

where[60]

$$\beta_0 \equiv \beta_{0\text{gluon}} + \beta_0^f = \frac{11}{3}N_c - \frac{2}{3}N_f \tag{A.2}$$

and

$$\beta_1 = -2\left[\frac{17}{3}N_c^2 - \frac{N_f}{6N_c}(13N_c^2 - 3)\right]. \tag{A.3}$$

For three massless flavors $\beta_0 = 9$. The first two coefficients of the β function are scheme-independent.

References

1. V. A. Khoze, M. A. Shifman, N. G. Uraltsev and M. B. Voloshin, *Sov. J. Nucl. Phys.* **46**, 112 (1987).
2. I. I. Y. Bigi, V. A. Khoze, N. G. Uraltsev and A. I. Sanda, The question of CP noninvariance — As seen through the eyes of neutral beauty, *Adv. Ser. Direct. High Energy Phys.* **3**, 175 (1989).
3. I. Bigi, contribution to this volume, pp. 109–131.
4. M. A. Shifman, *Int. J. Mod. Phys. A* **14**, 5017 (1999), arXiv:hep-th/9906049.
5. M. A. Shifman and M. B. Voloshin, *Sov. J. Nucl. Phys.* **41**, 120 (1985); *Sov. Phys. JETP* **64**, 698 (1986); *Sov. J. Nucl. Phys.* **45**, 292 (1987).
6. H. Georgi, *Phys. Lett. B* **240**, 447 (1990).
7. N. Isgur and M. B. Wise, *Phys. Lett. B* **232**, 113 (1989).
8. I. I. Y. Bigi, N. G. Uraltsev and A. I. Vainshtein, *Phys. Lett. B* **293**, 430 (1992) [Erratum: *ibid.* **297**, 477 (1993)], arXiv:hep-ph/9207214.
9. J. Chay, H. Georgi and B. Grinstein, *Phys. Lett. B* **247**, 399 (1990).
10. M. A. Shifman, Lectures on heavy quarks in quantum chromodynamics, in *ITEP Lectures on Particle Physics and Field Theory*, Vol. 1, ed. M. Shifman (World Scientific, Singapore, 1999), p. 1, arXiv:hep-ph/9510377.

11. M. Beneke, *Phys. Rept.* **317**, 1 (1999), arXiv:hep-ph/9807443; M. Beneke and V. M. Braun, *Renormalons and power corrections*, in *At the Frontier of Particle Physics*, Vol. 3, ed. M. Shifman (World Scientific, Singapore, 2001), p. 1719, arXiv:hep-ph/0010208.

12. A. Manohar and M. Wise, *Heavy Quark Physics* (Cambridge University Press, 2000).

13. V. A. Novikov, M. A. Shifman, A. I. Vainshtein and V. I. Zakharov, *Nucl. Phys. B* **249**, 445 (1985).

14. M. A. Shifman, *Prog. Theor. Phys. Suppl.* **131**, 1 (1998), arXiv:hep-ph/9802214.

15. I. I. Y. Bigi, M. A. Shifman, N. G. Uraltsev and A. I. Vainshtein, *Phys. Rev. D* **50**, 2234 (1994), arXiv:hep-ph/9402360.

16. M. Beneke and V. M. Braun, *Nucl. Phys. B* **426**, 301 (1994), arXiv:hep-ph/9402364.

17. B. R. Webber, *Phys. Lett. B* **339**, 148 (1994), arXiv:hep-ph/9408222.

18. F. J. Dyson, *Phys. Rev.* **85**, 631 (1952).

19. J. C. Le Guillou and J. Zinn-Justin (eds.), *Large-Order Behavior of Perturbation Theory* (North-Holland, Amsterdam, 1990).

20. W. E. Thirring, *Helv. Phys. Acta* **26**, 33 (1953).

21. L. N. Lipatov, *Sov. Phys. JETP* **45**, 216 (1977).

22. E. B. Bogomolny and V. A. Fateev, *Phys. Lett. B* **76**, 210 (1978).

23. A. I. Vainshtein, *Decaying Systems and Divergence of the Series of Perturbation Theory*, Novosibirsk preprint, 1964. Its English translation was first published in Ref. 24.

24. A. I. Vainshtein, *Continuous Advances in QCD 2002, ArkadyFest*, eds. K. Olive, M. Shifman and M. Voloshin (World Scientific, Singapore, 2002), p. 619.

25. C. M. Bender and T. T. Wu, *Phys. Rev. D* **7**, 1620 (1973).

26. G. 't Hooft, Can we make sense out of quantum chromodynamics?, in *The Whys of Subnuclear Physics*, Erice, 1977, ed. A. Zichichi (Plenum, New York, 1979), p. 943; B. E. Lautrup, *Phys. Lett. B* **69**, 109 (1977); G. Parisi, *Phys. Lett. B* **76**, 65 (1978); *Nucl. Phys. B* **150**, 163 (1979); A. H. Mueller, *Nucl. Phys. B* **250**, 327 (1985).

27. J. Zinn-Justin, *Phys. Rept.* **70**, 109 (1981); *Quantum Field Theory and Critical Phenomena* (Clarendon Press, Oxford, 1999).

28. E. B. Bogomolny, *Phys. Lett. B* **91**, 431 (1980).

29. J. Zinn-Justin, *Nucl. Phys. B* **192**, 125 (1981).

30. G. V. Dunne and M. Ünsal, *JHEP* **1211**, 170 (2012), arXiv:1210.2423 [hep-th].

31. J. Zinn-Justin and U. D. Jentschura, *Phys. Lett. B* **596**, 138 (2004), arXiv:hep-ph/0405279; *Ann. Phys.* **313**, 197 (2004), arXiv:quant-ph/0501136; *Ann. Phys.* **313**, 269 (2004), arXiv:quant-ph/0501137.

32. G. V. Dunne and M. Ünsal, Generating energy eigenvalue trans-series from perturbation theory, arXiv:1306.4405 [hep-th].

33. I. Aniceto and R. Schiappa, Nonperturbative ambiguities and the reality of resurgent trans-series, arXiv:1308.1115 [hep-th].

34. P. Argyres and M. Ünsal, *Phys. Rev. Lett.* **109**, 121601 (2012), arXiv:1204.1661 [hep-th]; *JHEP* **1208**, 063 (2012), arXiv:1206.1890 [hep-th].

35. G. V. Dunne and M. Ünsal, *Phys. Rev. D* **87**, 025015 (2013), arXiv:1210.3646 [hep-th].

36. M. Ünsal, *Phys. Rev. Lett.* **100**, 032005 (2008), arXiv:0708.1772 [hep-th].

37. M. Ünsal, *Phys. Rev. D* **80**, 065001 (2009), arXiv:0709.3269 [hep-th].

38. G. 't Hooft, *Nucl. Phys. B* **72**, 461 (1974); see also G. 't Hooft, Planar diagram field theories, in *Under the Spell of the Gauge Principle* (World Scientific, Singapore, 1994), p. 378; *Commun. Math. Phys.* **86**, 449 (1982).

39. J. Koplik, A. Neveu and S. Nussinov, *Nucl. Phys. B* **123**, 109 (1977).

40. A. H. Mueller, *Proc. Int. Conf. QCD — 20 Years Later*, Aachen, 1992, eds. P. Zerwas and H. Kastrup (World Scientific, Singapore, 1993), Vol. 1, p. 162.
41. M. Neubert, *Phys. Rev. D* **51**, 5924 (1995), arXiv:hep-ph/9412265.
42. V. I. Zakharov, *Nucl. Phys. B* **385**, 452 (1992).
43. M. A. Shifman, A. I. Vainshtein and V. I. Zakharov, *Nucl. Phys. B* **147**, 385 (1979).
44. K. G. Wilson, *Phys. Rev.* **179**, 1499 (1969).
45. K. G. Wilson and J. B. Kogut, *Phys. Rept.* **12**, 75 (1974).
46. K. Symanzik, *Commun. Math. Phys.* **23**, 49 (1971); C. G. Callan, Jr., *Phys. Rev. D* **5**, 3202 (1972); W. Zimmermann, *Lectures on Elementary Particles and Quantum Field Theory*, Vol. 1, eds. S. Deser, M. Grisaru and H. Pendelton (MIT Press, Cambridge, MA, 1971).
47. F. David, *Nucl. Phys. B* **234**, 237 (1984).
48. F. David, *Nucl. Phys. B* **209**, 433 (1982).
49. W. A. Bardeen, B. W. Lee and R. E. Shrock, *Phys. Rev. D* **14**, 985 (1976).
50. V. A. Novikov, M. A. Shifman, A. I. Vainshtein and V. I. Zakharov, *Phys. Rept.* **116**, 103 (1984).
51. S. R. Coleman, *Commun. Math. Phys.* **31**, 259 (1973).
52. M. Beneke, V. M. Braun and N. Kivel, *Phys. Lett. B* **443**, 308 (1998), arXiv:hep-ph/9809287.
53. E. V. Shuryak, *The QCD Vacuum, Hadrons and Superdense Matter*, 2nd edn. (World Scientific, Singapore, 2004).
54. E. Witten, *Nucl. Phys. B* **160**, 57 (1979).
55. E. Witten, *Nucl. Phys. B* **149**, 285 (1979).
56. M. A. Shifman, Theory of preasymptotic effects in weak inclusive decays, in *Proc. Workshop on Continuous Advances in QCD*, ed. A. Smilga (World Scientific, Singapore, 1994), p. 249, arXiv:hep-ph/9405246; Recent progress in the heavy quark theory, in *Proc. V PASCOS Symp.*, Baltimore, March 1995, ed. J. Bagger (World Scientific, Singapore, 1996), p. 69, arXiv:hep-ph/9505289.
57. I. B. Khriplovich, *Sov. J. Nucl. Phys.* **10**, 235 (1969).
58. M. Shifman, *Advanced Topics in Quantum Field Theory* (Cambridge University Press, 2012), Chapter 5.
59. E. Witten, *Phys. Lett. B* **117**, 324 (1982) [reprinted in *Current Algebra and Anomalies*, eds. S. Treiman, R. Jackiw, B. Zumino and E. Witten (Princeton University Press, 1985), p. 429].
60. D. R. T. Jones, *Nucl. Phys. B* **75**, 531 (1974); W. E. Caswell, *Phys. Rev. Lett.* **33**, 244 (1974).

Inclusive Semileptonic B Decays and $|V_{cb}|$

Paolo Gambino

Università di Torino, Dipartimento di Fisica, and INFN,
Torino, Via Giuria 1, I-10125 Torino, Italy
gambino@to.infn.it

The CKM matrix elements $|V_{cb}|$ can be extracted from inclusive semileptonic B decays in a model independent way pioneered by Kolya Uraltsev and collaborators. I review here the present status and latest developments in this field.

1. A Semileptonic Collaboration

My continuing involvement with semileptonic B decays is mostly due to a fortuitous encounter with Kolya Uraltsev in 2002. A group of experimentalists of the DELPHI Collaboration, among whom Marco Battaglia and Achille Stocchi, had embarked in an analysis of semileptonic moments and asked Kolya and me to help them out. He was the expert, I was a novice in the field and things have stayed that way for a long time thereafter. The joint paper that appeared later that year[1] contained one of the first fits to semileptonic data to extract $|V_{cb}|$, the masses of the heavy quarks and some non-perturbative parameters, and it used Kolya's proposal to avoid any $1/m_c$ expansion.[2] The next step for us was to compute the moments with a cut on the lepton energy,[3] as measured by Cleo and Babar.[4,5] Impressed as I was by Kolya's deep physical insight and enthusiasm, I was glad that he asked me to continue our collaboration. Kolya thought that a global fit should be performed by experimentalists, but as theoretical issues kept arising he tirelessly discussed with them every single detail; the BaBar fit,[6] where the kinetic scheme analysis of Ref. 1 was extended to the BaBar dataset, and the global fit of Ref. 7 owe very much to his determination.

Kolya's patience in explaining was unlimited and admirable: countless times I took advantage of it and learned from him. Our semileptonic collaboration later covered perturbative corrections,[8] the extraction of $|V_{ub}|$,[9] and a reassessment of the zero-recoil heavy quark sum rule.[10,11] It was during one of his visits to Turin that he suffered a first heart attack, but it did not take long before he was back to his usual dynamism. Working with Kolya was sometimes complicated, but it was invariably rewarding. He was stubborn and we could passionately argue about a single point for hours. For him discussion, even heated discussion, was an essential part of doing physics. I will forever miss Kolya's passionate love of physics and his total dedication to science. They were the marks of a noble soul, a kind and discreet friend.

In the following I will review the present status of the inclusive $B \to X_c \ell \bar{\nu}$ decays, the subject of most of my work with Kolya, who was a pioneer of the field. Semileptonic B decays allow for a precise determination of the magnitude of the CKM matrix elements V_{cb} and V_{ub}, which are in turn crucial ingredients in the analysis of CP violation in the quark sector and in the prediction of flavour-changing neutral current transitions. In the case of inclusive decays, the Operator Product Expansion (OPE) allows us to describe the relevant non-perturbative physics in terms of a finite number of non-perturbative parameters that can be extracted from experiment, while in the case of exclusive decays like $B \to D^{(*)} \ell \bar{\nu}$ or $B \to \pi \ell \bar{\nu}$ the form factors have to be computed by non-perturbative methods, e.g. on the lattice. Presently, the most precise determinations of $|V_{cb}|$ (the inclusive one[12] and the one based on $B \to D^* \ell \nu$ at zero recoil and a lattice calculation of the form-factor[13]) show a $\sim 3\sigma$ discrepancy that does not seem to admit a new physics explanation, as I will explain later on. A similar discrepancy between the inclusive and exclusive determinations occurs in the case of $|V_{ub}|$.[14] It is a pity that Kolya will not witness how things eventually settle.

2. The Framework

Our understanding of inclusive semileptonic B decays is based on a simple idea: since inclusive decays sum over all possible hadronic final states, the quark in the final state hadronizes with unit probability and the transition amplitude is sensitive only to the long-distance dynamics of the initial B meson. Thanks to the large hierarchy between the typical energy release, of $O(m_b)$, and the hadronic scale Λ_{QCD}, and to asymptotic freedom, any residual sensitivity to non-perturbative effects is suppressed by powers of Λ_{QCD}/m_b.

The OPE allows us to express the nonperturbative physics in terms of B meson matrix elements of local operators of dimension $d \geq 5$, while the Wilson coefficients can be expressed as a perturbative series in α_s.[15-19] The OPE disentangles the physics associated with *soft* scales of order Λ_{QCD} (parameterized by the matrix elements of the local operators) from that associated with *hard* scales $\sim m_b$, which determine the Wilson coefficients. The total semileptonic width and the moments of the kinematic distributions are therefore double expansions in α_s and Λ_{QCD}/m_b, with a leading term that is given by the free b quark decay. Quite importantly, the power corrections start at $O(\Lambda_{\text{QCD}}^2/m_b^2)$ and are comparatively suppressed. At higher orders in the OPE, terms suppressed by powers of m_c also appear, starting with $O(\Lambda_{\rm QCD}^3/m_b^3 \times \Lambda_{\rm QCD}^2/m_c^2)$.[20] For instance, the expansion for the total semi-leptonic width is

$$
\Gamma_{sl} = \Gamma_0 \Bigg[1 + a^{(1)} \frac{\alpha_s(m_b)}{\pi} + a^{(2,\beta_0)} \beta_0 \left(\frac{\alpha_s}{\pi} \right)^2 + a^{(2)} \left(\frac{\alpha_s}{\pi} \right)^2
$$

$$
+ \left(-\frac{1}{2} + p^{(1)} \frac{\alpha_s}{\pi} \right) \frac{\mu_\pi^2}{m_b^2} + \left(g^{(0)} + g^{(1)} \frac{\alpha_s}{\pi} \right) \frac{\mu_G^2(m_b)}{m_b^2}
$$

$$
+ d^{(0)} \frac{\rho_D^3}{m_b^3} - g^{(0)} \frac{\rho_{LS}^3}{m_b^3} + \text{higher orders} \Bigg], \tag{1}
$$

where $\Gamma_0 = A_{\rm ew} |V_{cb}^2| G_F^2 m_b^5 (1 - 8\rho + 8\rho^3 - \rho^4 - 12\rho^2 \ln \rho)/192\pi^3$ is the tree level free quark decay width, $\rho = m_c^2/m_b^2$, and $A_{\rm ew} \simeq 1.014$ the leading electroweak correction. I have split the α_s^2 coefficient into a BLM piece proportional to $\beta_0 = 11 - 2/3 n_f$ and a remainder. The expansions for the moments have the same structure.

The relevant parameters in the double series of Eq. (1) are the heavy quark masses m_b and m_c, the strong coupling α_s, and the B meson expectation values of local operators of dimension 5 and 6, denoted by μ_π^2, μ_G^2, ρ_D^3, ρ_{LS}^3. As there are only two dimension five operators, two matrix elements appear at $O(1/m_b^2)$:

$$
\mu_\pi^2(\mu) = \frac{1}{2M_B} \langle B | \bar{b}_v \vec{\pi}^2 b_v | B \rangle_\mu, \tag{2}
$$

$$
\mu_G^2(\mu) = \frac{1}{2M_B} \langle B | \bar{b}_v \frac{i}{2} \sigma_{\mu\nu} G^{\mu\nu} b_v | B \rangle_\mu, \tag{3}
$$

where $\vec{\pi} = -i\vec{D}$, D^μ is the covariant derivative, $b_v(x) = e^{-im_b v \cdot x} b(x)$ is the b field deprived of its high-frequency modes, and $G^{\mu\nu}$ the gluon field tensor. The matrix element of the kinetic operator, μ_π^2, is naturally associated with the average kinetic energy of the b quark in the B meson, while that of the chromomagnetic operator, μ_G^2, is related to the B^*–B hyperfine mass splitting. They generally depend on a cutoff $\mu = O(1 \text{ GeV})$ chosen to separate soft and hard physics. The cutoff can be implemented in different ways. In the kinetic scheme,[21,22] a Wilson cutoff on the gluon momentum is employed in the b quark rest frame: all soft gluon contributions are attributed to the expectation values of the higher dimensional operators, while hard gluons with momentum $|\vec{k}| > \mu$ contribute to the perturbative corrections to the Wilson coefficients. Most current applications of the OPE involve $O(1/m_b^3)$ effects[23] as well, parameterized in terms of two additional parameters, generally denoted by ρ_D^3 and ρ_{LS}^3.[22] All of the OPE parameters describe universal properties of the B meson or of the quarks and are useful in several applications.

The interesting quantities to be measured are the total rate and some global shape parameters, such as the mean and variance of the lepton energy spectrum or of the hadronic invariant mass distribution. As most experiments can detect the leptons only above a certain threshold in energy, the lepton energy moments are defined as

$$\langle E_\ell^n \rangle = \frac{1}{\Gamma_{E_\ell > E_{\rm cut}}} \int_{E_\ell > E_{\rm cut}} E_\ell^n \frac{d\Gamma}{dE_\ell} dE_\ell , \tag{4}$$

where E_ℓ is the lepton energy in $B \to X_c \ell \nu$, $\Gamma_{E_\ell > E_{\rm cut}}$ is the semileptonic width above the energy threshold $E_{\rm cut}$ and $d\Gamma/dE_\ell$ is the differential semileptonic width as a function of E_ℓ. The hadronic mass moments are

$$\langle m_X^{2n} \rangle = \frac{1}{\Gamma_{E_\ell > E_{\rm cut}}} \int_{E_\ell > E_{\rm cut}} m_X^{2n} \frac{d\Gamma}{dm_X^2} dm_X^2 . \tag{5}$$

Here, $d\Gamma/dm_X^2$ is the differential width as a function of the squared mass of the hadronic system X. For both types of moments, n is the order of the moment. For $n > 1$, the moments can also be defined relative to $\langle E_\ell \rangle$ and $\langle m_X^2 \rangle$, respectively, in which case they are called central moments:

$$\ell_1(E_{\rm cut}) = \langle E_\ell \rangle_{E_\ell > E_{\rm cut}} , \quad \ell_{2,3}(E_{\rm cut}) = \langle (E_\ell - \langle E_\ell \rangle)^{2,3} \rangle_{E_\ell > E_{\rm cut}} ; \tag{6}$$

$$h_1(E_{\rm cut}) = \langle M_X^2 \rangle_{E_\ell > E_{\rm cut}} , \quad h_{2,3}(E_{\rm cut}) = \langle (M_X^2 - \langle M_X^2 \rangle)^{2,3} \rangle_{E_\ell > E_{\rm cut}} . \tag{7}$$

Since the physical information that can be extracted from the first three linear moments is highly correlated, it is more convenient to study the central moments ℓ_i and h_i, which correspond to the mean, variance, and asymmetry of the lepton energy and invariant mass distributions.

The OPE cannot be expected to converge in regions of phase space where the momentum of the final hadronic state is $O(\Lambda_{\rm QCD})$ and where perturbation theory has singularities. This is because what actually controls the expansion is not m_b but the energy release, which is $O(\Lambda_{\rm QCD})$ in those cases. The OPE is therefore valid only for sufficiently inclusive measurements and in general cannot describe differential distributions. The lepton energy moments can be measured very precisely, while the hadronic mass central moments are directly sensitive to higher dimensional matrix elements such as μ_π^2 and ρ_D^3. The leptonic and hadronic moments, which are independent of $|V_{cb}|$, give us constraints on the quark masses and on the non-perturbative OPE matrix elements, which can then be used, together with additional information, in the total semileptonic width to extract $|V_{cb}|$.

3. Higher Order Effects

The reliability of the inclusive method depends on our ability to control the higher order contributions in the double series and to constrain quark–hadron duality violation, i.e. effects beyond the OPE, which we know to exist but expect to be rather suppressed in semileptonic decays. The calculation of higher order effects allows us to verify the convergence of the double series and to reduce and properly estimate the residual theoretical uncertainty. Duality violation, see Ref. 24 for a review, is related to the analytic continuation of the OPE to Minkowski space-time. It can be constrained a posteriori, considering how well the OPE predictions fit the experimental data. This in turn depends on precise measurements and precise OPE predictions. As the experimental accuracy reached at the B factories is already better than the theoretical accuracy for most of the measured moments and will further improve at Belle-II, efforts to improve the latter are strongly motivated.

The main ingredients for an accurate analysis of the experimental data on the moments and the subsequent extraction of $|V_{cb}|$ have been known for some time. Let us consider first the purely perturbative contributions. The $O(\alpha_s)$ perturbative corrections to various kinematic distributions and to the rate have been computed long ago. In particular, the complete $O(\alpha_s)$ and $O(\alpha_s^2\beta_0)$ corrections to the charged leptonic spectrum have been first calculated in Refs. 25–27 and 28. The so-called BLM corrections,[29] of $O(\alpha_s^2\beta_0)$, are related to the running of the strong coupling inside the loops and are usually the dominant source of two-loop corrections in B decays. The first $O(\alpha_s)$ calculations of the hadronic spectra appeared in Refs. 30–32 and were later completed in Refs. 8, 33, 34, while the $O(\alpha_s^2\beta_0)$ contributions were studied in Refs. 31, 32, 34 and 8. The triple differential distribution was first computed at $O(\alpha_s)$ in Refs. 33 and 8; its $O(\alpha_s^n\beta_0^{n-1})$ corrections can be found in Ref. 8.

The complete two-loop perturbative corrections to the width and moments of the lepton energy and hadronic mass distributions have been computed in Refs. 35–37 by both numerical and analytic methods. The kinetic scheme implementation for actual observables can be found in Ref. 38. In general, using $\alpha_s(m_b)$ in the one-loop result and adopting the on-shell scheme for the quark masses, the non-BLM corrections amount to about -20% of the two-loop BLM corrections and give small contributions to normalized moments. In the kinetic scheme with cutoff $\mu = 1$ GeV, the perturbative expansion of the total width is

$$\Gamma[\bar{B} \to X_c e\bar{\nu}] \propto 1 - 0.96 \frac{\alpha_s(m_b)}{\pi} - 0.48\,\beta_0\left(\frac{\alpha_s}{\pi}\right)^2$$

$$+ 0.82\left(\frac{\alpha_s}{\pi}\right)^2 + O(\alpha_s^3) \approx 0.916\,. \qquad (8)$$

Higher order BLM corrections of $O(\alpha_s^n\beta_0^{n-1})$ to the width are also known[39,8] and can be resummed in the kinetic scheme: the resummed BLM result is numerically very close to that of from NNLO calculations.[39] The residual perturbative error in the total width is about 1%.

In the normalized leptonic moments the perturbative corrections cancel to a large extent, independently of the mass scheme, because hard gluon emission is comparatively suppressed. This pattern of cancelations, crucial for a correct estimate of the theoretical uncertainties, is confirmed by the complete $O(\alpha_s^2)$ calculation, although the numerical precision of the available results is not sufficient to improve the overall accuracy for the higher central leptonic moments.[38] The non-BLM corrections turn out to be more important for the hadronic moments. Even though it improves the overall theoretical uncertainty only moderately, the complete NNLO calculation leads to the meaningful inclusion of precise mass constraints in various perturbative schemes.

P. Gambino

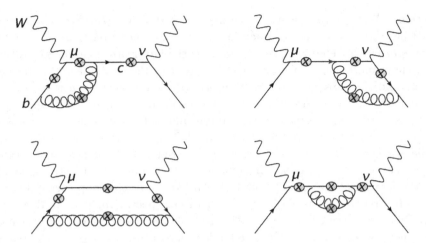

Fig. 1.   One-loop diagrams contributing to the current correlator. The background gluon can be attached wherever a cross is marked.

The coefficients of the non-perturbative corrections of $O(\Lambda_{QCD}^n/m_b^n)$ in the double series are Wilson coefficients of power-suppressed local operators and can be computed perturbatively. The calculation of the $O(\alpha_s \Lambda_{QCD}^2/m_b^2)$ corrections has been recently completed. The $O(\alpha_s)$ corrections to the coefficient of μ_π^2 have been computed numerically in Ref. 40 and analytically in Ref. 41. They can be also obtained from the parton level $O(\alpha_s)$ result using reparameterization invariance (RI) relations.[16,42,43] In fact, these RI relations have represented a useful check for the calculation of the remaining $O(\alpha_s \Lambda_{QCD}^2/m_b^2)$ corrections, those proportional to μ_G^2, which was completed in Ref. 44. The calculation consists in matching the one-loop diagrams in Fig. 1, representing the correlator of two axial-vector currents computed in an expansion around the mass-shell of the b quark, onto local HQET operators.

A recent independent calculation[45] of the semileptonic width at $m_c = 0$ seems to be in agreement with the $m_c \to 0$ limit of Ref. 44. References 41, 44 provide analytic results for the $O(\alpha_s \Lambda_{QCD}^2/m_b^2)$ corrections to the three relevant structure functions and hence to the triple differential semileptonic B decay width. The most general moment have now been computed to this order and employed to improve the precision of the fits to $|V_{cb}|$.[12]

Numerically, using for the heavy quark on-shell masses the values $m_b = 4.6$ GeV and $m_c = 1.15$ GeV, the total semileptonic width reads

$$\Gamma_{B \to X_c \ell \nu} = \Gamma_0 \left[\left(1 - 1.78 \frac{\alpha_s}{\pi} \right) \left(1 - \frac{\mu_\pi^2}{2m_b^2} \right) - \left(1.94 + 2.42 \frac{\alpha_s}{\pi} \right) \frac{\mu_G^2(m_b)}{m_b^2} \right],$$

where Γ_0 is the tree level width and we have omitted higher order terms of $O(\alpha_s^2)$ and $O(1/m_b^3)$. The coefficient of μ_π^2 is fixed by RI (or equivalently, by Lorentz invariance) at all orders. The parameter μ_G^2 is renormalized at the scale m_b. It is

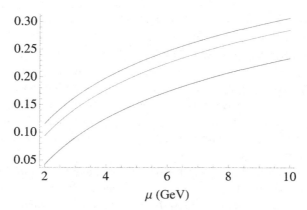

Fig. 2. Relative NLO correction to the μ_G^2 coefficients in the width (blue), first (red) and second central (yellow) leptonic moments as a function of the renormalization scale of μ_G^2.

advisable to evaluate the QCD coupling constant at a scale lower than m_b. If we adopt $\alpha_s = 0.25$ the $O(\alpha_s)$ correction increases the μ_G^2 coefficient by about 7%. In the kinetic scheme with cutoff $\mu = 1$ GeV and for the same values of the masses the width becomes

$$\Gamma_{B \to X_c \ell \nu} = \Gamma_0 \left[1 - 0.96 \frac{\alpha_s}{\pi} - \left(\frac{1}{2} - 0.99 \frac{\alpha_s}{\pi} \right) \frac{\mu_\pi^2}{m_b^2} - \left(1.94 + 3.46 \frac{\alpha_s}{\pi} \right) \frac{\mu_G^2(m_b)}{m_b^2} \right],$$

where the NLO corrections to the coefficients of μ_π^2, μ_G^2 are both close to 15% but have different signs. Overall, the $O(\alpha_s \Lambda_{\text{QCD}}^2/m_b^2)$ contributions decrease the total width by about 0.3%. However, NLO corrections also modify the coefficients of μ_π^2, μ_G^2 in the moments which are fitted to extract the non-perturbative parameters, and will ultimately shift the values of the OPE parameters to be employed in the width. Therefore, in order to quantify the eventual numerical impact of the new corrections on the semileptonic width and on $|V_{cb}|$, a new global fit has to be performed. The size of the $O(\alpha_s \mu_G^2/m_b^2)$ corrections depends on the renormalization scale μ of the chromomagnetic operator. This is illustrated in Fig. 2, where the size of the NLO correction relative to the tree level results is shown for the width and the first two leptonic central moments at different values of μ. The NLO corrections are quite small for $\mu \approx 2$ GeV and, as expected, increase with μ. For $\mu \gtrsim m_b$ the running of μ_G^2 appears to dominate the NLO corrections. In view of the importance of $O(1/m_b^3)$ corrections, if a theoretical precision of 1% in the decay rate is to be reached, the $O(\alpha_s/m_b^3)$ effects need to be calculated.

As to the higher power corrections, the $O(1/m_b^4)$ and $O(1/m_Q^5)$ effects were computed in Ref. 46. The main problem here is the proliferation of non-perturbative parameters: as many as nine new expectation values appear at $O(1/m_b^4)$ and more at the next order. Because they cannot all be extracted from experiment, in Ref. 46 they have been estimated in the ground state saturation approximation, thus reducing them to products of the known $O(1/m_b^{2,3})$ parameters, see also

Ref. 47. In this approximation, the total $O(1/m_Q^{4,5})$ correction to the width is about
+1.3%. The $O(1/m_Q^5)$ effects are dominated by $O(1/m_b^3 m_c^2)$ intrinsic charm con-
tributions, amounting to +0.7%.[20] The net effect on $|V_{cb}|$ also depends on the
corrections to the moments. Reference 46 estimate that the overall effect on $|V_{cb}|$
is a 0.4% increase. While this sets the scale of higher order power corrections, it
is as yet unclear how much the result depends on the assumptions made for the
expectation values. A new preliminary global fit[48] performed using different ansatz
for the new non-perturbative parameters seems to confirm that these corrections
lead to a small shift in $|V_{cb}|$.

Two implementations of the OPE calculation have been employed in global
analyses; they are based either on the kinetic scheme[21,22,39,3] or on the $1S$ mass
scheme for the b quark mass.[49,50] They both include power corrections up to and
including $O(1/m_b^3)$ and perturbative corrections of $O(\alpha_s^2 \beta_0)$. Beside differing in the
perturbative scheme adopted, the global fits may include a different choice of experimental data, employ specific assumptions, or estimate the theoretical uncertainties
in different ways. Recently, the kinetic scheme implementation has been upgraded
to include first the complete $O(\alpha_s^2)$ [38] and later the $(\alpha_s \Lambda^2/m_b^2)$ [12] contributions.

4. $|V_{cb}|$ and the Fit to Semileptonic Moments

The OPE parameters can be constrained by various moments of the lepton energy
and hadron mass distributions of $B \to X_c \ell \nu$ that have been measured with good
accuracy at the B-factories, as well as at CLEO, DELPHI, CDF.[51,6,52–56] The total
semileptonic width can then be employed to extract $|V_{cb}|$. The situation is less
favorable in the case of $|V_{ub}|$, where the total rate is much more difficult to access
experimentally because of the background from $B \to X_c \ell \nu$, but the results of the
semileptonic fits are crucial also in that case. This strategy has been rather successful and has allowed for a $\sim 2\%$ determination of V_{cb} and for a $\sim 5\%$ determination
of V_{ub} from inclusive decays.[14,82]

The first few moments of the charged lepton energy spectrum in $B \to X_c \ell \nu$
decays are experimentally measured with high precision — better than 0.2% in
the case of the first moment. At the B-factories a lower cut on the lepton energy,
$E_\ell \geq E_{\rm cut}$, is applied to suppress the background. Experiments measure the moments at different values of $E_{\rm cut}$, which provides additional information as the cut
dependence is also a function of the OPE parameters. The relevant quantities are
therefore $\ell_{1,2,3}$, $h_{1,2,3}$, as well as the ratio R^* between the rate with and without a
cut

$$R^*(E_{\rm cut}) = \frac{\int_{E_{\rm cut}}^{E_{\rm max}} dE_\ell \frac{d\Gamma}{dE_\ell}}{\int_0^{E_{\rm max}} dE_\ell \frac{d\Gamma}{dE_\ell}}. \tag{9}$$

This quantity is needed to relate the actual measurement of the rate with a cut to
the total rate, from which one conventionally extracts $|V_{cb}|$. All of these observables

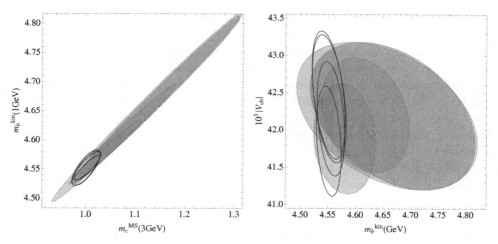

Fig. 3. Two-dimensional projections of the fits performed with different assumptions for the theoretical correlations. The orange, magenta, blue, light blue 1-sigma regions correspond to the four scenarios considered in Ref. 58. The black contours show the same regions when the m_c constraint of Ref. 59 is employed.

can be expressed as double expansions in α_s and inverse powers of m_b, schematically

$$M_i = M_i^{(0)} + \frac{\alpha_s(\mu)}{\pi} M_i^{(1)} + \left(\frac{\alpha_s}{\pi}\right)^2 M_i^{(2)} + \left(M_i^{(\pi,0)} + \frac{\alpha_s(\mu)}{\pi} M_i^{(\pi,1)}\right) \frac{\mu_\pi^2}{m_b^2}$$

$$+ \left(M_i^{(G,0)} + \frac{\alpha_s(\mu)}{\pi} M_i^{(G,1)}\right) \frac{\mu_G^2}{m_b^2} + M_i^{(D)} \frac{\rho_D^3}{m_b^3} + M_i^{(LS)} \frac{\rho_{LS}^3}{m_b^3} + \cdots, \qquad (10)$$

where all the coefficients $M_i^{(j)}$ depend on m_c, m_b, E_{cut}, and on various renormalization scales. The dots represent missing terms of $O(\alpha_s^3)$, $O(\alpha_s^2/m_b^2)$, $O(\alpha_s/m_b^3)$, and $O(1/m_b^4)$, which are either unknown or not yet included in the latest analysis.[12] It is worth stressing that according to the adopted definition the OPE parameters $\mu_\pi^2, \dots$ are matrix elements of local operators evaluated in the physical B meson, i.e. without taking the infinite mass limit.

The semileptonic moments are sensitive to a specific linear combination of m_c and m_b, $\approx m_b - 0.8m_c$,[57] see Fig. 3, which is close to the one needed for the extraction of $|V_{cb}|$, but they cannot resolve the individual masses with good accuracy. It is important to check the consistency of the constraints on m_c and m_b from semileptonic moments with precise determinations of these quark masses, as a step in the effort to improve our theoretical description of inclusive semileptonic decays. Moreover, the inclusion of these constraints in the semileptonic fits improves the accuracy of the $|V_{ub}|$ and $|V_{cb}|$ determinations. The heavy quark masses and the non-perturbative parameters obtained from the fits are also relevant for a precise calculation of other inclusive decay rates such as that of $B \to X_s\gamma$.[58]

In the past, the first two moments of the photon energy in $B \to X_s\gamma$ have generally been employed to improve the accuracy of the fit. Indeed, the first moment corresponds to a determination of m_b. However, in recent years rather precise determinations of the heavy quark masses (e^+e^- sum rules, lattice QCD, etc.) have become available, based on completely different methods, see e.g. Refs. 60, 59, 61–68 and Refs. 69, 70 for reviews. The charm mass determinations have a smaller absolute uncertainty and appear quite consistent with each other, providing a good external constraint for the semileptonic fits. Radiative moments remain interesting in their own respect, but they are not competitive with the charm mass determinations. Moreover, experiments place a lower cut on the photon energy, which introduces a sensitivity to the Fermi motion of the b-quark inside the B meson and tends to disrupt the OPE. One can still resum the higher-order terms into a non-local distribution function and parameterize it assuming different functional forms,[71–73] but the parameterization will depend on m_b, μ_π^2, etc., namely the same parameters one wants to extract. Another serious problem is that only the leading operator contributing to inclusive radiative decays can be described by an OPE. Therefore, radiative moments are in principle subject to additional $O(\Lambda_{QCD}/m_b)$ effects, which have not yet been estimated.[74] For all these reasons the most recent analyses[58,12] have relied solely on charm and possibly bottom mass determinations.

The global fits of Refs. 58, 12 are performed in the kinetic scheme with a cutoff $\mu = 1$ GeV and follow the implementation described in Refs. 3, 38. The two fits only differ in the inclusion of $O({\alpha_s}^2/m_b^2)$ corrections and in the consequent reduction of theoretical uncertainties. In order to use the high precision m_c determinations without introducing additional theoretical uncertainty due to the mass scheme conversion, it is convenient to employ the $\overline{\text{MS}}$ scheme for the charm mass, denoted by $\bar{m}_c(\mu_c)$, and to choose a normalization scale μ_c well above m_c, e.g. 3 GeV.

The experimental data for the moments are fitted to the theoretical expressions in order to constrain the non-perturbative parameters and the heavy quark masses. 43 measurements are included, see Ref. 58 for the list. The chromomagnetic expectation value μ_G^2 is also constrained by the hyperfine splitting

$$M_{B^*} - M_B = \frac{2}{3} \frac{\mu_G^2}{m_b} + O\left(\frac{\alpha_s \mu_G^2}{m_b}, \frac{1}{m_b^2}\right).$$

Unfortunately, little is known of the power corrections to the above relation and only a loose bound[75] can be set, see Ref. 11 for a recent discussion. For what concerns ρ_{LS}^3, it is somewhat constrained by the heavy quark sum rules.[75] References 58, 12 use the constraints

$$\mu_G^2 = (0.35 \pm 0.07)\ \text{GeV}^2, \quad \rho_{LS}^3 = (-0.15 \pm 0.10)\ \text{GeV}^3. \tag{11}$$

It should be stressed that ρ_{LS}^3 plays a minor role in the fits because its coefficients are generally suppressed with respect to the other parameters.

It is interesting to note that the fit without theoretical uncertainties is not good, with $\chi^2/\text{d.o.f.} \sim 2$, corresponding to a very small p-value and driven by a strong tension ($\sim 3.5\sigma$) between the constraints in Eq. (11) and the measured moments. On the other hand, the fit without the constraints (11) is not too bad. Indeed, theoretical uncertainties are not so much necessary for the OPE expressions to fit the moments — that would merely test Eq. (10) as a parameterization; they are instead needed to preserve the definition of the parameters as B expectation values of certain local operators, which in turn can be employed in the semileptonic widths and in other applications of the Heavy Quark Expansion.

As noted above, the OPE description of semileptonic moments is subject to two sources of theoretical uncertainty: missing higher order terms in Eq. (10) and terms that violate quark–hadron duality. Only of the first kind of uncertainty is usually considered: the violation of local quark–hadron duality would manifest itself as an inconsistency of the fit, which as we will see is certainly not present at the current level of theoretical and experimental accuracy.

In Ref. 12 we assume that missing perturbative corrections can affect the coefficients of μ_π^2 and μ_G^2 at the level of $\pm 7\%$, while missing perturbative and higher power corrections can effectively change the coefficients of ρ_D^3 and ρ_{LS}^3 by $\pm 30\%$. Moreover we assign an irreducible theoretical uncertainty of 8 MeV to the heavy quark masses, and vary $\alpha_s(m_b)$ by 0.018. The changes in M_i due to these variations of the fundamental parameters are added in quadrature and provide a theoretical uncertainty δM_i^{th}, to be subsequently added in quadrature with the experimental one, δM_i^{exp}. This method is consistent with the residual scale dependence observed at NNLO, and appears to be reliable: the NNLO corrections and the $O(1/m_b^{4,5})$ (using ground state saturation as in Ref. 46) have been found to be within the range of expectations based on the method in the original formulation of Ref. 3.

The correlation between theoretical errors assigned to different observables is much harder to estimate, but plays an important role in the semileptonic fits. Let us first consider moments computed at a fixed value of E_{cut}: as long as one deals with central higher moments, there is no argument of principle supporting a correlation between two different moments, for instance ℓ_1 and h_2. We also do not observe any clear pattern in the known corrections, and therefore regard the theoretical predictions for different central moments as completely uncorrelated. Let us now consider the calculation of a certain moment M_i for two close values of E_{cut}, say 1 GeV and 1.1 GeV. Clearly, the OPE expansion for $M_i(1 \text{ GeV})$ will be very similar to the one for $M_i(1.1 \text{ GeV})$, and we may expect this to be true at any order in α_s and $1/m_b$. The theoretical uncertainties we assign to $M_i(1 \text{ GeV})$ and $M_i(1.1 \text{ GeV})$ will therefore be very close to each other and *very highly correlated*. The degree of correlation between the theory uncertainty of $M_i(E_1)$ and $M_i(E_2)$ can intuitively be expected to decrease as $|E_1 - E_2|$ grows. Moreover, we know that higher power corrections are going to modify significantly the spectrum only close to the endpoint. Indeed, one observes that the $O(1/m_b^{4,5})$ contributions are equal for all cuts below about 1.2 GeV (see Fig. 2 of Ref. 46) and the same happens for the $O(\alpha_s \mu_\pi^2/m_b^2)$ corrections.[40] Therefore, the dominant sources of current theoretical uncertainty suggest very high correlations among the theoretical predictions of the moments for cuts below roughly 1.2 GeV.

Various assumptions on the theoretical correlations have been tried. A 100% correlation between a certain central moment computed at different values of E_{cut} has been assumed, e.g. in Ref. 7). This is too strong an assumption, which ends up distorting the fit because the dependence of M_i on E_{cut}, itself a function of the fit parameters, is then free of theoretical uncertainty. A fit performed in this way will underestimate the uncertainties. Another possibility has been proposed in Ref. 50, with the theoretical correlation matrix equal to the experimental one.

Four alternative approaches for the theoretical correlations are compared in Ref. 58. Figure 3 shows some of the results of the fits performed with the four options for the theoretical correlations. The fits include the two constraints of Eq. (11). In general, the results depend sensitively on the option adopted. In the case of the heavy quark masses, which are strongly correlated, we observe large errors differing significantly between the various options, although the central values are quite consistent. The results for the non-perturbative parameters depend even stronger on the option. The inclusion of precise mass constraints in the fit decreases the errors and neutralizes the ambiguity due to the ansatz for the theoretical correlations. It also allows us to check the consistency of the results with independent information. The effect of the inclusion of a precise charm mass constraint in the semileptonic fit is illustrated in Fig. 3. As expected, the uncertainty in the b mass becomes about 20–25 MeV in all scenarios, a marked improvement, also with respect to the precision resulting from the use of radiative moments.[14] The inclusion of the m_c constraint indeed stabilized the fits with respect to the ansatz for the theory correlations. On the other hand, there is hardly any improvement in the final precision of the non-perturbative parameters and of $|V_{cb}|$.

Figure 4 show two examples of leptonic and hadronic moments measurements compared with their theoretical prediction based on the results of the fit of Ref. 58 with theory uncertainty. As anticipated, theory errors are generally larger than experimental ones. The situation is similar also for the fits in Ref. 12.

The default fit of Ref. 12 uses $\bar{m}_c(3\ \text{GeV}) = 0.986(13)\ \text{GeV};$[60] the results are shown in Table 1, where the bottom mass is expressed in the kinetic scheme, $m_b^{\text{kin}}(1\ \text{GeV})$. Most available m_b determinations, however, use the $\overline{\text{MS}}$ mass $\bar{m}_b(\bar{m}_b)$ which is not well-suited to the description of semileptonic B decays as the calculation of the width and moments in terms of $\bar{m}_b(\bar{m}_b)$ involves large higher order corrections. Since the relation between the kinetic and the $\overline{\text{MS}}$ masses is known only to $O(\alpha_s^2)$, the ensuing uncertainty is not negligible. It has been estimated to be about 30 MeV,[38]

$$m_b^{\text{kin}}(1\ \text{GeV}) - \bar{m}_b(\bar{m}_b) = 0.37 \pm 0.03\ \text{GeV},$$

leading to a preferred value

$$\bar{m}_b(\bar{m}_b) = 4.183 \pm 0.037\ \text{GeV},$$

in good agreement with various recent m_b determinations,[60,76–81,68] as illustrated in Fig. 5. Of course, one can also include in the fit both m_c and m_b determinations, but because of the scheme translation error in m_b the gain in accuracy is limited.[58,12]

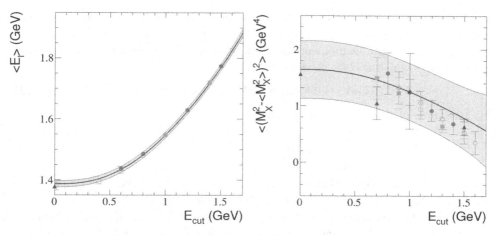

Fig. 4.   Default fit predictions for ℓ_1 and h_2 compared with measured values as functions of E_{cut}. The grey band is the theory prediction with total theory error. Filled symbols mean that the point was used in the fit. Open symbols are measurements that were not used in the fit. BaBar data are shown by circles, Belle by squares and other experiments (DELPHI, CDF, CLEO) by triangles.

Table 1.   Results of the global fit with m_c constraint in the default scenario of Ref. 12. All parameters are in GeV at the appropriate power and all, except m_c, in the kinetic scheme at $\mu = 1$ GeV. The last row gives the uncertainties.

m_b^{kin}	$\overline{m_c}\,(3\text{ GeV})$	μ_π^2	ρ_D^3	μ_G^2	ρ_{LS}^3	BR$_{c\ell\nu}$ (%)	$10^3|V_{cb}|$
4.553	0.987	0.465	0.170	0.332	−0.150	10.65	42.21
0.020	0.013	0.068	0.038	0.062	0.096	0.16	0.78

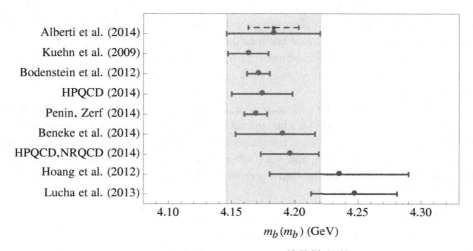

Fig. 5.   Comparison of different $\bar{m}_b(\bar{m}_b)$ determinations.[12,60,76–81,68] The dashed line denotes the error before scheme conversion.

As already noted, the semileptonic moments are highly sensitive to a linear combination of the heavy quark masses. When no external constraint is imposed on $m_{c,b}$, the semileptonic moments determine best a linear combination of the heavy quark masses which is close to their difference,

$$m_b^{\text{kin}}(1 \text{ GeV}) - 0.85\bar{m}_c(3 \text{ GeV}) = 3.714 \pm 0.018 \text{ GeV}. \tag{12}$$

The value of $|V_{cb}|$ is computed using

$$|V_{cb}| = \sqrt{\frac{|V_{cb}|^2 \, \text{BR}_{c\ell\nu}}{\tau_B \Gamma^{\text{OPE}}_{B \to X_c \ell\nu}}}, \tag{13}$$

with $\tau_B = 1.579(5)$ ps.[14] Its theoretical error is computed combining in quadrature the parametric uncertainty that results from the fit, and an additional 1.3% theoretical error to take into account missing higher order corrections in the expression for the semileptonic width. It turns out that using $\bar{m}_c(2 \text{ GeV})$ rather than $\bar{m}_c(2 \text{ GeV})$ leads to a better converging expansion for the width, with smaller theoretical error for $|V_{cb}|$, about 1%. However, the value of $|V_{cb}|$ extracted in this way, $42.01(68) \times 10^{-3}$ is compatible with that in Table 1. The same holds if the kinetic scheme is used also for m_c, in which case we obtain $|V_{cb}| = 42.04(67) \times 10^{-3}$.

The fits are generally good, with $\chi^2/\text{d.o.f.} \approx 0.4$ for the default fit. The low χ^2 of the default fit is due to the large theoretical uncertainties we have assumed. It may be tempting to interpret it as evidence that the theoretical errors have been overestimated. However, higher order corrections may effectively shift the parameters of the $O(1/m_b^2)$ and $O(1/m_b^3)$ contributions. If we want to maintain the formal definition of these parameters, and to be able to use them elsewhere, we therefore have to take into account the potential shift they may experience because of higher order effects.

The fits with a constraint on m_c are quite stable with respect to a change of inputs. In particular, small differences are found when experimental data at high E_{cut} are excluded, and when only hadronic or leptonic moments are considered. One may also wonder whether the inclusion of moments measured at different values of E_{cut} really benefits the final accuracy. It turns out that the benefit is minor but non-negligible.

In the kinetic scheme the inequalities $\mu_\pi^2(\mu) \geq \mu_G^2(\mu)$, $\rho_D^3(\mu) \geq -\rho_{LS}^3(\mu)$ hold at arbitrary values of the cutoff μ. The central values of the fit satisfy the inequalities. The dependence of the results on the scales of α_s and on the kinetic cutoff has been studied in Ref. 12. Changing the scale of α_s from m_b to 2 GeV the value of $|V_{cb}|$ increases by only 0.5%, well within its error, while m_b increases by less than 0.4%.

5. Conclusions

We have seen that the most recent value of $|V_{cb}|$ extracted from an analysis of inclusive semileptonic B decays is[12]

$$|V_{cb}| = (42.21 \pm 0.78) \times 10^{-3}. \tag{14}$$

A competitive determination of $|V_{cb}|$ derives from a comparison of the extrapolation of the $B \to D^*l\nu$ rate to the zero-recoil point with an unquenched lattice QCD calculation of the zero recoil form factor by the Fermilab-MILC collaboration,[13]

$$|V_{cb}| = (39.04 \pm 0.49_{\text{exp}} \pm 0.53_{\text{lat}} \pm 0.19_{\text{QED}}) \times 10^{-3}, \qquad (15)$$

where the error is split into experimental, lattice and QED components. The values in Eqs. (14), (15) disagree by 2.9σ. This is a long-standing tension, which has become stronger with recent improvements in the OPE and lattice calculations. A few comments are in order:

- The zero-recoil form factor of $B \to D^*l\nu$ can also be estimated using heavy quark sum rules, see Refs. 10, 11 for a recent reanalysis. Although this method is subject to larger uncertainties and it is difficult to improve its accuracy, it leads to a $|V_{cb}|$ compatible with the inclusive determination, $|V_{cb}| = 40.93(1.11) \times 10^{-3}$. There exist also less precise determinations of $|V_{cb}|$ based on the decay $B \to Dl\nu$, but they do not help resolving the issue at the moment, see Ref. 82 for a review.
- The extrapolation of the $B \to D^*l\nu$ experimental data to the zero-recoil point is performed by the experimental collaborations using the Caprini–Lellouch–Neubert parameterization,[83] based on HQET at next-to-leading order and expected to reproduce the form factor within 2% (not included in the present error budget). While this rigid parameterization with only two free parameters fits well the experimental data at $w \neq 1$, at the present level of precision its use to extrapolate the rate to zero recoil is questionable. Lattice calculations of the form factors at non-zero recoil are currently under way; they would allow us to avoid the extrapolation.
- It is also possible to determine $|V_{cb}|$ indirectly, using the CKM unitarity relations together with CP violation and flavor data, without the above direct information: SM analyses of this kind by the UTfit and CKMFitter collaborations give $(42.05\pm 0.65) \times 10^{-3}$ [84] and $(41.4^{+2.4}_{-1.4}) \times 10^{-3}$,[85] both closer to the inclusive value of Eq. (14).

In principle, the discrepancy between the values of $|V_{cb}|$ extracted from inclusive decays and from $B \to D^*l\nu$ could be ascribed to physics beyond the SM, as the $B \to D^*$ transition is sensitive only to the axial-vector component of the $V - A$ charged weak current. However, the new physics effect should be sizable (8%), and it would require new interactions which seem ruled out by electroweak constraints on the effective $Zb\bar{b}$ vertex.[86] The most likely explanation of the discrepancy between Eqs. (14), (15) is therefore a problem in the theoretical and/or experimental analyses of semileptonic decays.

I do not have space here to discuss the closely related determination of $|V_{ub}|$ from inclusive semileptonic B decays without charm, which shows a similar puzzling tension between inclusive and exclusive determinations. However, I cannot avoid stressing the central role played by Kolya from the beginning also in this field.[72,87,88,9] The interested reader is referred to Ref. 82 for a recent review.

Acknowledgments

This work is supported in part by MIUR under contract 2010YJ2NYW 006, by the EU Commission through the HiggsTools Initial Training Network PITN-GA-2012-316704, and by Compagnia di San Paolo under contract ORTO11TPXK.

References

1. M. Battaglia *et al.*, *Phys. Lett. B* **556**, 41–49 (2003), arXiv:hep-ph/0210319.
2. N. Uraltsev, *Nucl. Phys. B (Proc. Suppl.)* **117**, 554–557 (2002), arXiv:hep-ph/0210044.
3. P. Gambino and N. Uraltsev, *Eur. Phys. J. C* **34**, 181–189 (2004), arXiv:hep-ph/0401063.
4. A. Mahmood *et al.*, *Phys. Rev. D* **70**, 032003 (2004), arXiv:hep-ex/0403053.
5. B. Aubert *et al.*, *Phys. Rev. D* **69**, 111103 (2004), arXiv:hep-ex/0403031.
6. B. Aubert *et al.*, *Phys. Rev. Lett.* **93**, 011803 (2004), arXiv:hep-ex/0404017.
7. O. Büchmuller and H. Flächer, *Phys. Rev. D* **73**, 073008 (2006), arXiv:hep-ph/0507253.
8. V. Aquila, P. Gambino, G. Ridolfi and N. Uraltsev, *Nucl. Phys. B* **719**, 77–102 (2005), arXiv:hep-ph/0503083.
9. P. Gambino, P. Giordano, G. Ossola and N. Uraltsev, *JHEP* **0710**, 058 (2007), arXiv:0707.2493.
10. P. Gambino, T. Mannel and N. Uraltsev, *Phys. Rev. D* **81**, 113002 (2010), arXiv:1004.2859.
11. P. Gambino, T. Mannel and N. Uraltsev, *JHEP* **1210**, 169 (2012), arXiv:1206.2296.
12. A. Alberti, P. Gambino, K. J. Healey and S. Nandi, arXiv:1411.6560.
13. J. A. Bailey *et al.*, arXiv:1403.0635.
14. Y. Amhis *et al.*, arXiv:1207.1158.
15. J. Chay, H. Georgi and B. Grinstein, *Phys. Lett. B* **247**, 399 (1990).
16. I. I. Bigi, N. Uraltsev and A. Vainshtein, *Phys. Lett. B* **293**, 430–436 (1992), arXiv:hep-ph/9207214.
17. I. I. Bigi, M. A. Shifman, N. Uraltsev and A. I. Vainshtein, *Phys. Rev. Lett.* **71**, 496–499 (1993), arXiv:hep-ph/9304225.
18. B. Blok, L. Koyrakh, M. A. Shifman and A. Vainshtein, *Phys. Rev. D* **49**, 3356 (1994), arXiv:hep-ph/9307247.
19. A. V. Manohar and M. B. Wise, *Phys. Rev. D* **49**, 1310–1329 (1994), arXiv:hep-ph/9308246.
20. I. I. Bigi, T. Mannel, S. Turczyk and N. Uraltsev, *JHEP* **1004**, 073 (2010), arXiv:0911.3322.
21. I. I. Bigi *et al.*, *Phys. Rev. D* **56**, 4017–4030 (1997), arXiv:hep-ph/9704245.
22. I. I. Bigi *et al.*, *Phys. Rev. D* **52**, 196–235 (1995), arXiv:hep-ph/9405410.
23. M. Gremm and A. Kapustin, *Phys. Rev. D* **55**, 6924–6932 (1997), arXiv:hep-ph/9603448.
24. I. I. Bigi and N. Uraltsev, *Int. J. Mod. Phys. A* **16**, 5201–5248 (2001), arXiv:hep-ph/0106346.
25. M. Jezabek and J. H. Kuhn, *Nucl. Phys. B* **314**, 1 (1989).
26. M. Jezabek and J. H. Kuhn, *Nucl. Phys. B* **320**, 20 (1989).
27. A. Czarnecki and M. Jezabek, *Nucl. Phys. B* **427**, 3–21 (1994), arXiv:hep-ph/9402326.
28. M. Gremm and I. W. Stewart, *Phys. Rev. D* **55**, 1226–1232 (1997), arXiv:hep-ph/9609341.

29. S. J. Brodsky, G. P. Lepage and P. B. Mackenzie, *Phys. Rev. D* **28**, 228 (1983).
30. A. Czarnecki, M. Jezabek and J. H. Kuhn, *Acta Phys. Polon. B* **20**, 961 (1989).
31. A. F. Falk, M. E. Luke and M. J. Savage, *Phys. Rev. D* **53**, 2491–2505 (1996), arXiv:hep-ph/9507284.
32. A. F. Falk and M. E. Luke, *Phys. Rev. D* **57**, 424–430 (1998), arXiv:hep-ph/9708327.
33. M. Trott, *Phys. Rev. D* **70**, 073003 (2004), arXiv:hep-ph/0402120.
34. N. Uraltsev, *Int. J. Mod. Phys. A* **20**, 2099–2118 (2005), arXiv:hep-ph/0403166.
35. A. Pak and A. Czarnecki, *Phys. Rev. Lett.* **100**, 241807 (2008), arXiv:0803.0960.
36. K. Melnikov, *Phys. Lett. B* **666**, 336–339 (2008), arXiv:0803.0951.
37. S. Biswas and K. Melnikov, *JHEP* **1002**, 089 (2010), arXiv:0911.4142.
38. P. Gambino, *JHEP* **1109**, 055 (2011), arXiv:1107.3100.
39. D. Benson, I. I. Bigi, T. Mannel and N. Uraltsev, *Nucl. Phys. B* **665**, 367–401 (2003), arXiv:hep-ph/0302262.
40. T. Becher, H. Boos and E. Lunghi, *JHEP* **0712**, 062 (2007), arXiv:0708.0855.
41. A. Alberti, T. Ewerth, P. Gambino and S. Nandi, *Nucl. Phys. B* **870**, 16–29 (2013), arXiv:1212.5082.
42. M. E. Luke and A. V. Manohar, *Phys. Lett. B* **286**, 348–354 (1992), arXiv:hep-ph/9205228.
43. A. V. Manohar, *Phys. Rev. D* **82**, 014009 (2010), arXiv:1005.1952.
44. A. Alberti, P. Gambino and S. Nandi, *JHEP* **1401**, 1–16 (2014), arXiv:1311.7381.
45. T. Mannel, A. A. Pivovarov and D. Rosenthal, arXiv:1405.5072.
46. T. Mannel, S. Turczyk and N. Uraltsev, *JHEP* **1011**, 109 (2010), arXiv:1009.4622.
47. J. Heinonen and T. Mannel, *Nucl. Phys. B* **889**, 46–63 (2014), arXiv:1407.4384.
48. P. Gambino and S. Turczyk, work in progress.
49. A. H. Hoang, Z. Ligeti and A. V. Manohar, *Phys. Rev. D* **59**, 074017 (1999), arXiv:hep-ph/9811239.
50. C. W. Bauer, Z. Ligeti, M. Luke, A. V. Manohar and M. Trott, *Phys. Rev. D* **70**, 094017 (2004), arXiv:hep-ph/0408002.
51. S. Csorna *et al.*, *Phys. Rev. D* **70**, 032002 (2004), arXiv:hep-ex/0403052.
52. B. Aubert *et al.*, *Phys. Rev. D* **81**, 032003 (2010), arXiv:0908.0415.
53. P. Urquijo *et al.*, *Phys. Rev. D* **75**, 032001 (2007), arXiv:hep-ex/0610012.
54. C. Schwanda *et al.*, *Phys. Rev. D* **75**, 032005 (2007), arXiv:hep-ex/0611044.
55. D. Acosta *et al.*, *Phys. Rev. D* **71**, 051103 (2005), arXiv:hep-ex/0502003.
56. J. Abdallah *et al.*, *Eur. Phys. J. C* **45**, 35–59 (2006), arXiv:hep-ex/0510024.
57. M. Voloshin, *Phys. Rev. D* **51**, 4934–4938 (1995), arXiv:hep-ph/9411296.
58. P. Gambino and C. Schwanda, *Phys. Rev. D* **89**, 014022 (2014), arXiv:1307.4551.
59. B. Dehnadi, A. H. Hoang, V. Mateu and S. M. Zebarjad, *JHEP* **1309**, 103 (2013), arXiv:1102.2264.
60. K. G. Chetyrkin *et al.*, *Phys. Rev. D* **80**, 074010 (2009), arXiv:0907.2110.
61. S. Bodenstein *et al.*, *Phys. Rev. D* **83**, 074014 (2011), arXiv:1102.3835.
62. A. Signer, *Phys. Lett. B* **672**, 333–338 (2009), arXiv:0810.1152.
63. I. Allison *et al.*, *Phys. Rev. D* **78**, 054513 (2008), arXiv:0805.2999.
64. C. McNeile *et al.*, *Phys. Rev. D* **82**, 034512 (2010), arXiv:1004.4285.
65. B. Blossier *et al.*, *Phys. Rev. D* **82**, 114513 (2010), arXiv:1010.3659.
66. J. Heitger, G. M. von Hippel, S. Schaefer and F. Virotta, *PoS* **LATTICE2013**, 475 (2013), arXiv:1312.7693.
67. N. Carrasco *et al.*, *Nucl. Phys. B* **887**, 19–68 (2014), arXiv:1403.4504.
68. W. Lucha, D. Melikhov and S. Simula, *Phys. Rev. D* **88**(5), 056011 (2013), arXiv:1305.7099.
69. M. Antonelli *et al.*, *Phys. Rept.* **494**, 197–414 (2010), arXiv:0907.5386.

70. J. Beringer *et al.*, *Phys. Rev. D* **86**, 010001 (2012).

71. M. Neubert, *Phys. Rev. D* **49**, 4623–4633 (1994), arXiv:hep-ph/9312311.

72. I. I. Bigi, M. A. Shifman, N. Uraltsev and A. Vainshtein, *Int. J. Mod. Phys. A* **9**, 2467–2504 (1994), arXiv:hep-ph/9312359.

73. D. Benson, I. Bigi and N. Uraltsev, *Nucl. Phys. B* **710**, 371–401 (2005), arXiv:hep-ph/0410080.

74. G. Paz, arXiv:1011.4953.

75. N. Uraltsev, *Phys. Lett. B* **545**, 337–344 (2002), arXiv:hep-ph/0111166.

76. S. Bodenstein *et al.*, *Phys. Rev. D* **85**, 034003 (2012), arXiv:1111.5742.

77. M. Beneke *et al.*, arXiv:1411.3132.

78. B. Colquhoun *et al.*, arXiv:1408.5768.

79. B. Chakraborty *et al.*, arXiv:1408.4169.

80. A. Hoang, P. Ruiz-Femenia and M. Stahlhofen, *JHEP* **1210**, 188 (2012), arXiv:1209.0450.

81. A. A. Penin and N. Zerf, *JHEP* **1404**, 120 (2014), arXiv:1401.7035.

82. A. Bevan *et al.*, *Eur. Phys. J. C* **74**(11), 3026 (2014), arXiv:1406.6311.

83. I. Caprini, L. Lellouch and M. Neubert, *Nucl. Phys. B* **530**, 153–181 (1998), arXiv:hep-ph/9712417.

84. M. Bona *et al.*, *JHEP* **0610**, 081 (2006), arXiv:hep-ph/0606167, see http://www.utfit.org for the latest results.

85. J. Charles *et al.*, *Eur. Phys. J. C* **41**, 1–131 (2005), arXiv:hep-ph/0406184, see http://ckmfitter.in2p3.fr for recent results.

86. A. Crivellin and S. Pokorski, arXiv:1407.1320.

87. I. I. Bigi, R. Dikeman and N. Uraltsev, *Eur. Phys. J. C* **4**, 453–461 (1998), arXiv:hep-ph/9706520.

88. N. Uraltsev, *Int. J. Mod. Phys. A* **14**, 4641–4652 (1999), arXiv:hep-ph/9905520.

Moments in Inclusive Semileptonic B Meson Decays at the Belle Experiment

Christoph Schwanda

Institute for High Energy Physics, Austrian Academy of Sciences,
Nikolsdorfer Gasse 18, 1050 Vienna, Austria

Since my return to Austria in the year 2003, I have measured observables in inclusive B meson decays at the Belle experiment and worked together with theorists on the interpretation of these measurements in terms of the Cabibbo–Kobayashi–Maskawa matrix element $|V_{cb}|$. And in fact, only this memorial book project made me fully aware of Kolya Uraltsev's ground breaking theoretical contributions to this field. He was not a theorist who talked a lot to an experimentalist like me, and maybe this is not a bad thing for good science. I certainly remember his enthusiasm from conferences, e.g., when I was powerless to keep his presentation to the scheduled time as a session chair at the CKM2005 workshop in San Diego. Still I feel there is some amount of irony in the fact, that I know so little about a person whose work has been so decisive for my career in high energy physics.

To commemorate Kolya Uraltsev's pioneering work on inclusive semileptonic B meson decays $B \to X_c \ell \nu$ and on the Heavy Quark Expansion (HQE), which has already been paid tribute to in other articles in this volume, I will review the measurement of the electron energy and the hadronic mass moments in $B \to X_c \ell \nu$ decays performed at the Belle experiment. These measurements allow to both test his theoretical calculations and to extract $|V_{cb}|$ and non-perturbative quantities, such as the b-quark mass, from his formulae.

1. Introduction

The extraction of the Cabibbo–Kobayashi–Maskawa (CKM) matrix element $|V_{cb}|$ [1] from inclusive semileptonic B meson decays $B \to X_c \ell \nu$ is based on the Operator Production Expansion (OPE), which allows to calculate the semileptonic B decay width.[2,5] The leading term in this expansion in inverse powers of the b-quark m_b corresponds to the decay of a free b-quark while bound state corrections arise at $\mathcal{O}(1/m_b^2)$. These corrections contain perturbative coefficients (expressed as a series in powers of α_s) and non-perturbative quantities such as quark masses and expectation values of higher dimensional operators. While the latter cannot be calculated in perturbation theory, they can be determined in experiments by measuring inclusive quantities in $B \to X_c \ell \nu$ decays, such as the moments of the lepton energy E_ℓ and the hadronic mass squared M_X^2 spectra. OPEs for these quantities can be written

C. Schwanda

down[3–5] and, if measurements are sufficiently inclusive, the non-perturbative quantities in the OPE depend on the initial B meson state only.

Moments of the lepton energy spectrum in $B \to X_c \ell \nu$ for different minimum lepton energies have been obtained by the CLEO, DELPHI, BaBar and Belle experiments.[6–9] CLEO, CDF, DELPHI, BaBar and Belle have measured moments of the mass squared distribution of the hadronic system X_c for different threshold values of the lepton energy.[10,11,7,12,8] In this article, we review the Belle measurements which are — together with the measurements of BaBar — the most precise determinations of these quantites.

2. Belle Experiment and Data Sample

The Belle detector[13] is located in the interaction region of the KEKB asymmetric energy e^+e^- collider[14] of the KEK laboratory in Tsukuba, Japan. KEKB has operated between the years 1999 and 2010, mostly at the center-of-mass (c.m.) energy of the $\Upsilon(4S)$ resonance, where pairs of B mesons are produced by $e^+e^- \to \Upsilon(4S) \to B\bar{B}$. In total, Belle has accumulated an integrated luminosity of 711 fb^{-1} on the $\Upsilon(4S)$ resonance, corresponding to 771 million $B\bar{B}$ events. The Belle analyses of the lepton energy and hadronic mass moments use a data sample of only 140 fb^{-1} or 152 million $B\bar{B}$ events. Another 15 fb^{-1} taken at 60 MeV below the resonance are used to estimate the non-$B\bar{B}$ (continuum) background.

The Belle detector is a large-solid-angle magnetic spectrometer that consists of a silicon vertex detector (SVD), a 50-layer central drift chamber (CDC), an array of aerogel threshold Cherenkov counters (ACC), a barrel-like arrangement of time-of-flight scintillation counters (TOF), and an electromagnetic calorimeter comprised of CsI(Tl) crystals (ECL) located inside a super-conducting solenoid coil that provides a 1.5 T magnetic field. An iron flux-return located outside of the coil is instrumented to detect K_L^0 mesons and to identify muons (KLM).

Generic $B\bar{B}$ monte Carlo (MC) samples equivalent to about three times the integrated luminosity are also used. MC-simulated events are generated with EvtGen[15] and full detector simulation based on GEANT3[16] is applied. The decays $B \to D^* \ell \nu$ and $B \to D \ell \nu$ are generated assuming the form factor parameterization by Caprini et al.[17] The decays $B \to D^{**} \ell \nu$ [a] are simulated according to the Leibovich–Ligeti–Stewart–Wise (LLSW) model[18] (both relative abundance and form factor shape). The model for the $B \to X_u \ell \nu$ background is a hybrid mixture of exclusive modes and an inclusive component described by the De Fazio–Neubert model.[19] QED bremsstrahlung in $B \to X \ell \nu$ decays is included using the PHOTOS package.[20]

3. Experimental Procedure

3.1. *Hadronic tagging*

The first step in both Belle's electron energy and hadronic mass moment analysis[9,12] is the reconstruction of the hadronic decay of one B meson in $\Upsilon(4S) \to B\bar{B}$ (B_{tag}).

---

[a]The symbol D^{**} refers to the D_1, D_2^*, D_0^* and D_1' states.

We search the decay modes $B^+ \to \bar{D}^{(*)0}\pi^+$, $\bar{D}^{(*)0}\rho^+$, $\bar{D}^{(*)0}a_1^+$ and $B^0 \to D^{(*)-}\pi^+$, $D^{(*)-}\rho^+$, $D^{(*)-}a_1^+$.[b] Pairs of photons satisfying $E_\gamma > 50$ MeV in the laboratory-frame and 118 MeV/$c^2 < M(\gamma\gamma) < 150$ MeV/c^2 ($\pm 3.3\sigma$ around the π^0 mass) are combined to form π^0 candidates. K_S^0 mesons are reconstructed from pairs of oppositely charged tracks with invariant mass within ± 30 MeV/c^2 ($\pm 5.1\sigma$) of the nominal K_S^0 mass and a decay vertex displaced from the interaction point. Candidate ρ^+ and ρ^0 mesons are reconstructed in the $\pi^+\pi^0$ and $\pi^+\pi^-$ decay modes, requiring their invariant masses to be within ± 150 MeV/c^2 of the nominal ρ mass. Candidate a_1^+ mesons are obtained by combining a ρ^0 candidate with a charged pion and requiring an invariant mass between 1.0 and 1.6 GeV/c^2. D^0 candidates are searched for in the $K^-\pi^+$, $K^-\pi^+\pi^0$, $K^-\pi^+\pi^+\pi^-$, $K_S^0\pi^+\pi^-$ and $K_S^0\pi^0$ decay modes. The $K^-\pi^+\pi^+$ and $K_S^0\pi^+$ modes are used to reconstruct D^+ mesons. Charmed mesons are selected in a window corresponding to ± 3 times the mass resolution in the respective decay mode. D^{*+} mesons are reconstructed by pairing a charmed meson with a low momentum pion, $D^{*+} \to D^0\pi^+, D^+\pi^0$. The decay modes $D^{*0} \to D^0\pi^0$ and $D^{*0} \to D^0\gamma$ are used to search for neutral charmed vector mesons.

B_{tag} candidates are selected using the beam-energy constrained mass M_{bc} and the energy difference ΔE,

$$M_{\text{bc}} = \sqrt{(E_{\text{beam}}^*)^2 - (\vec{p}_B^*)^2}, \quad \Delta E = E_B^* - E_{\text{beam}}^*, \tag{1}$$

where E_{beam}^*, $\vec{p}_B^*$ and E_B^* are the beam energy, the 3-momentum and the energy of the B candidate in the $\Upsilon(4S)$ frame. In M_{bc} and ΔE, the signal peaks at the nominal B mass and zero, respectively. Defining a signal region around these values, about 60,000 charged and 40,000 neutral tags are properly reconstructed in the 140 fb^{-1} data sample.

3.2. *Lepton reconstruction*

Next, semileptonic decays of the other B meson (B_{signal}) are selected by searching for an identified charged lepton amongst the particles not used for the reconstruction of B_{tag}. Electron candidates are identified using the ratio of the energy detected in the ECL to the track momentum, the ECL shower shape, position matching between track and ECL cluster, the energy loss in the CDC and the response of the ACC counters. Muons are identified based on their penetration range and transverse scattering in the KLM detector.

The lepton energy moment analysis uses only electrons traversing the barrel region of the Belle detector ($35° < \theta < 125°$, where θ denotes the polar angle of the particle in the laboratory-frame with respect to the direction opposite to the positron beam). The study of hadronic mass moments uses both electrons and muons. Electron (muon) candidates are required to have a laboratory-frame

---

[b]The inclusion of the charge conjugate decay is implied.

momentum greater than 0.3 GeV/c (0.6 GeV/c) and satisfy $17° < θ < 150°$ ($25° < θ < 145°$). In both analyses, the effect of bremsstrahlung in electron events is partially recovered by searching for a photon with laboratory-frame energy $E_γ < 1$ GeV and a direction close to the electron track. If such a photon is found, it is merged with the electron and removed from the event.

In the momentum region relevant to these analyses, charged leptons are identified with an efficiency of about 90% and the probability to misidentify a pion as an electron (muon) is 0.25% (1.4%).[21,22]

3.3. *Hadronic mass reconstruction*

In the hadronic mass analysis, also the hadronic system X recoiling against $ℓν$ must be reconstructed. This is done by summing the 4-momenta of the remaining charged particles and photons in the event, denoting this sum by p_X. We exclude tracks passing very far away from the interaction point or compatible with a multiply reconstructed track generated by a low-momentum particle spiraling in the central drift chamber. Photons in the barrel region must have an energy greater than 50 MeV in the laboratory frame. Higher thresholds are applied in the endcap regions.

For a better resolution in the hadronic mass, we reject events with a missing mass larger than 3 GeV^2/c^4. Further improvement is obtained by recalculating the 4-momentum of the X system as

$$p'_X = (p_{\text{LER}} + p_{\text{HER}}) - p_{B_{\text{tag}}} - p_ℓ - p_ν , \tag{2}$$

where LER and HER refer to the colliding beams. The neutrino 4-momentum in this equation is taken to be $(|\vec{p}_{\text{miss}}|, \vec{p}_{\text{miss}})$, where $\vec{p}_{\text{miss}}$ is the missing 3-momentum.

The hadronic mass M_X^2 is the magnitude of the 4-momentum of the X system. The resolution in M_X^2 obtained from p'_X is about 0.8 GeV^2/c^4, compared to 1.4 GeV^2/c^4 in M_X^2 from p_X.

3.4. *Backgrounds*

Backgrounds in the spectrum of the electron momentum in the B meson rest frame (p_e^*) from the following sources are considered:

- *Continuum ($e^+e^- → q\bar{q}$ events with $q = u, d, s, c$)*: The shape of this component in p_e^* is derived from off-resonance data (Belle real data taken 60 MeV below the $Υ(4S)$ resonance), and normalized using the off- to on-resonance luminosity ratio and cross section difference.
- *Combinatorial*: This component arises from improperly reconstructed hadronic tags (B_{tag}). We derive the shape of this background from generic $B\bar{B}$ simulated events and normalize it to the side-band of the M_{bc} distribution (5.20 GeV/c^2 < M_{bc} < 5.25 GeV/c^2, Eq. (1)).

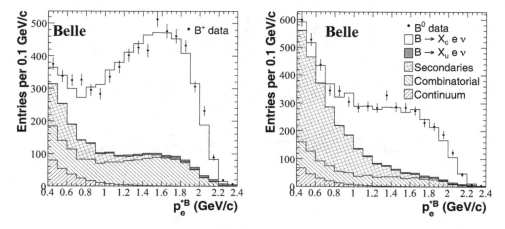

Fig. 1.   Measured electron momentum spectra in B^+ and B^0 tagged events with the different background components overlaid. Refer to the text for more details on these backgrounds.

- *Decays $B \to X_u \ell \nu$:* The shape of the p_e^* distribution of electrons from the inclusive $b \to u$ transition is estimated from simulation. This component is normalized to the number of reconstructed B^+ and B^0 tags in data, as determined from the M_{bc} distribution.
- *Secondaries and hadronic fakes:* There are also other background contributions in $B\bar{B}$ events such as cascade charm decays $b \to c \to \ell$, electrons from J/ψ, $\psi(2S)$, Dalitz decays and photon conversions. Hadrons (mostly pions) can be misidentified as electrons. The shapes of these contributions are determined from simulation and their collective normalization is determined by fitting the sum of the MC simulated signal and background contributions to the observed inclusive electron momentum spectrum to the observed inclusive electron momentum spectrum in the range $0.4 \text{ GeV}/c < p_e^* < 2.4 \text{ GeV}/c$. The values of the χ^2 per degree of freedom for the fits to B^+ and B^0 decay spectra are 1.3 and 1.1, respectively.

Figure 1 shows the measured electron momentum spectrum in the B meson rest frame (p_e^*) with all background contributions overlaid. Detector effects such as finite reconstruction efficiency and resolution are not yet corrected for. The signal and background yields in the range $0.4 \text{ GeV}/c < p_e^* < 2.4 \text{ GeV}/c$ are:

	B^+ tags	B^0 tags
Raw yield	6423 ± 80	5403 ± 74
Continuum background	249 ± 48	209 ± 39
Combinatorial background	1244 ± 20	696 ± 13
Secondaries (incl. hadronic fakes)	555 ± 11	1843 ± 22
$B \to X_u e \nu$ decays	74 ± 5	57 ± 6
Background subtracted yield	4300 ± 96	2597 ± 87

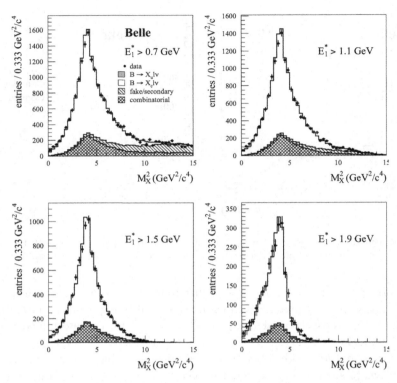

Fig. 2.   Measured hadronic mass spectrum for different lepton energy thresholds with the background components overlaid. The points with error bars are the experimental data after subtraction of the continuum background. Refer to the text for more details on the backgrounds.

In the analysis of the hadronic mass spectrum M_X^2 the same sources of background are considered: non-$B\bar{B}$ (continuum) events, combinatorial background from misreconstructed B_{tag}'s, contributions from secondary or fake leptons and $B \to X_u \ell \nu$ background. These contributions are also estimated in a very similar way as in the analysis of the electron spectrum.

The measured hadronic mass spectrum for different values of the lepton energy threshold in the B rest frame (events with energies exceeding the threshold E_{cut}^* are selected) is shown in Fig. 2. Again, detector effects are not yet corrected for. The signal yields and purities (in brackets) for different values of E_{cut}^* are:

E_{cut}^*	B^+ electron	B^+ muon	B^0 electron	B^0 muon
0.7	4105 ± 100 (70.5%)	3739 ± 108 (61.5%)	2491 ± 80 (65.9%)	2400 ± 86 (60.3%)
0.9	3855 ± 95 (73.2%)	3591 ± 104 (64.8%)	2353 ± 76 (73.4%)	2307 ± 83 (67.3%)
1.1	3466 ± 86 (74.9%)	3305 ± 96 (68.3%)	2098 ± 68 (77.1%)	2120 ± 76 (74.2%)
1.3	2894 ± 72 (75.8%)	2857 ± 84 (70.6%)	1749 ± 58 (80.4%)	1800 ± 66 (78.0%)
1.5	2195 ± 56 (74.6%)	2225 ± 66 (72.3%)	1322 ± 45 (84.2%)	1388 ± 52 (79.7%)
1.7	1384 ± 38 (77.2%)	1415 ± 44 (72.4%)	824 ± 30 (83.7%)	878 ± 34 (80.7%)
1.9	571 ± 19 (73.8%)	627 ± 22 (74.0%)	353 ± 15 (84.3%)	376 ± 17 (76.7%)

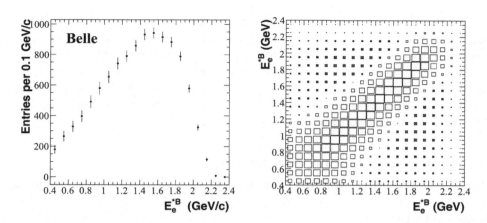

Fig. 3.   Electron energy distribution in the B meson rest frame (left), after correcting for the effects of QED bremsstrahlung, detector resolution and reconstruction efficiency. The bin contents are correlated and the statistical covariance matrix is shown on the right (filled boxes correspond to negative elements).

3.5. *Unfolding and moment calculation*

In both the electron energy and the hadronic mass spectrum analyses the measured distributions are corrected for the effects of QED bremsstrahlung, finite detector resolution and finite reconstruction efficiency by performing an unfolding procedure based on the Singular Value Decomposition (SVD) algorithm.[23]

In the electron energy moment analysis the true energy distribution in the B meson rest frame is obtained from the background subtracted spectra in Fig. 1 (assuming electrons to be massless, i.e., $E_e^* = p_e^*$) by combining contributions from B^+ and B^0 events and by unfolding the experimental distribution using a response matrix obtained from MC simulation (Fig. 3). Note that in the true electron energy spectrum the statistical uncertainties of the bin contents are correlated as a result of the unfolding.

Then, the partial branching fractions and the first four central moments of the electron energy spectrum are calculated from the unfolded spectrum with nine electron energy threshold values ($E_{\text{cut}}^* = 0.4, 0.6, 0.8, 1.0, 1.2, 1.4, 1.6, 1.8$ and 2.0 GeV). For example, the first moment and its statistical uncertainty are calculated as

$$\langle E_e^* \rangle = \frac{\sum_i (E_e^*)_i x_i'}{\sum_i x_i'}, \quad \sigma^2(\langle E_e^* \rangle) = \frac{\sum_{i,j} (E_e^*)_i X_{ij} (E_e^*)_j}{(\sum_i x_i')^2}. \tag{3}$$

Here, x' is the unfolded spectrum and X its covariance matrix. The sums run over the bins i above the energy threshold E_{cut}^* and $(E_e^*)_i$ is the central value of the ith bin.

In the hadronic mass moment analysis, the procedure is similar up to some important differences: Unfolding of the M_X^2 distribution must be repeated for every new energy threshold value E_{cut}^*. Also, the true hadronic mass spectrum is not continuous but receives significant contributions from four narrow states — D, D^*, D_1 and D_2^*. The binning of the unfolded spectrum is thus chosen such that these narrow states lie in the center of narrow bins. We measure the first, second and second central moments of the hadronic mass squared distribution M_X^2 for seven lepton energy thresholds ($E_{cut}^* = 0.7$, 0.9, 1.1, 1.3, 1.5, 1.7 and 1.9 GeV). Moments are measured separately in the four sub-samples, defined by the charge of B_{tag} (B^+, B^0) and the lepton type, and then averaged.

4. Results and Systematic Uncertainties

The first four central moments of the unfolded electron energy spectrum in the B meson rest frame and the partial branching fractions of the decay $B \to X_c e\nu$ measured by Belle are:

E_{cut}^* (GeV)	$\langle E_e^* \rangle$ (MeV)	$\langle (E_e^* - \langle E_e^* \rangle)^2 \rangle$ $(10^{-3}\ \text{GeV}^2)$	$\langle (E_e^* - \langle E_e^* \rangle)^3 \rangle$ $(10^{-3}\ \text{GeV}^3)$
0.4	$1393.92 \pm 6.73 \pm 3.02$	$168.77 \pm 3.68 \pm 1.53$	$-21.04 \pm 1.93 \pm 0.66$
0.6	$1427.82 \pm 5.82 \pm 2.55$	$146.15 \pm 2.88 \pm 1.08$	$-11.04 \pm 1.35 \pm 0.49$
0.8	$1480.04 \pm 4.81 \pm 2.13$	$117.97 \pm 2.05 \pm 0.55$	$-3.45 \pm 0.83 \pm 0.30$
1.0	$1547.76 \pm 3.96 \pm 1.45$	$88.17 \pm 1.42 \pm 0.36$	$0.83 \pm 0.49 \pm 0.20$
1.2	$1627.79 \pm 3.26 \pm 1.08$	$61.36 \pm 1.02 \pm 0.36$	$2.40 \pm 0.30 \pm 0.11$
1.4	$1719.96 \pm 2.58 \pm 1.10$	$38.99 \pm 0.71 \pm 0.24$	$2.33 \pm 0.16 \pm 0.07$
1.6	$1826.15 \pm 1.80 \pm 1.03$	$21.75 \pm 0.47 \pm 0.22$	$1.45 \pm 0.08 \pm 0.05$
1.8	$1943.18 \pm 0.93 \pm 1.16$	$10.14 \pm 0.28 \pm 0.18$	$0.68 \pm 0.03 \pm 0.04$
2.0	$2077.59 \pm 0.21 \pm 1.23$	$3.47 \pm 0.13 \pm 0.19$	$0.19 \pm 0.01 \pm 0.03$

E_{cut}^* (GeV)	$\langle (E_e^* - \langle E_e^* \rangle)^4 \rangle$ $(10^{-3}\ \text{GeV}^4)$	$\Delta\mathcal{B}$ (10^{-2})
0.4	$64.153 \pm 1.813 \pm 0.935$	$10.44 \pm 0.19 \pm 0.22$
0.6	$45.366 \pm 1.108 \pm 0.548$	$10.07 \pm 0.18 \pm 0.21$
0.8	$28.701 \pm 0.585 \pm 0.247$	$9.42 \pm 0.16 \pm 0.19$
1.0	$15.962 \pm 0.302 \pm 0.142$	$8.41 \pm 0.15 \pm 0.17$
1.2	$7.876 \pm 0.162 \pm 0.106$	$7.11 \pm 0.13 \pm 0.14$
1.4	$3.314 \pm 0.080 \pm 0.055$	$5.52 \pm 0.11 \pm 0.11$
1.6	$1.129 \pm 0.033 \pm 0.032$	$3.71 \pm 0.09 \pm 0.07$
1.8	$0.283 \pm 0.010 \pm 0.017$	$1.93 \pm 0.06 \pm 0.04$
2.0	$0.047 \pm 0.002 \pm 0.007$	$0.53 \pm 0.02 \pm 0.02$

Here, the first errors are statistical and the second uncertainties are systematic. For the latter, we consider the following contributions:

- *Detector related*: This category includes uncertainties related to the MC simulation of the event selection, the charged particle track reconstruction and the electron identification. For the partial branching fractions $\Delta\mathcal{B}$ this contribution is dominant.

- *Signal modeling*: Uncertainties in the form factor shapes and relative contributions of the decays $B \to D\ell\nu$, $B \to D^*\ell\nu$ and $B \to D^{**}\ell\nu$ to $B \to X_c\ell\nu$ contribute to this category. For $\langle E_e^* \rangle$ and $\langle (E_e^* - \langle E_e^* \rangle)^n \rangle$ $(n = 2, 3, 4)$ this is the dominant systematic error component.
- *Background and moment calculation*: Further contributions to the systematic uncertainty come from the subtraction on the non-$B\bar{B}$, wrongly reconstructed B_{tag} candidates, $b \to c \to e$ cascade and $B \to X_u\ell\nu$ backgrounds, and from the uncertainty related to unfolding.

These are the Belle results for the first, the second central and the second moment of M_X^2:

E_{cut}^* (GeV)	$\langle M_X^2 \rangle$ (GeV^2/c^4)	$\langle (M_X^2 - \langle M_X^2 \rangle)^2 \rangle$ (GeV^4/c^8)	$\langle M_X^4 \rangle$ (GeV^4/c^8)
] 0.7	$4.403 \pm 0.036 \pm 0.052$	$1.494 \pm 0.173 \pm 0.327$	$20.88 \pm 0.48 \pm 0.77$
0.9	$4.353 \pm 0.032 \pm 0.041$	$1.229 \pm 0.138 \pm 0.244$	$20.18 \pm 0.40 \pm 0.58$
1.1	$4.293 \pm 0.028 \pm 0.029$	$0.940 \pm 0.098 \pm 0.137$	$19.37 \pm 0.33 \pm 0.36$
1.3	$4.213 \pm 0.027 \pm 0.024$	$0.641 \pm 0.071 \pm 0.080$	$18.40 \pm 0.29 \pm 0.26$
1.5	$4.144 \pm 0.028 \pm 0.022$	$0.515 \pm 0.061 \pm 0.064$	$17.69 \pm 0.28 \pm 0.23$
1.7	$4.056 \pm 0.033 \pm 0.022$	$0.322 \pm 0.058 \pm 0.040$	$16.77 \pm 0.32 \pm 0.21$
1.9	$3.996 \pm 0.041 \pm 0.021$	$0.143 \pm 0.056 \pm 0.038$	$16.11 \pm 0.38 \pm 0.20$

Again, the first uncertainty is statistical and the second is systematic. Contributions to the systematic error come from signal modeling ($B \to D^{(*)}\ell\nu$, $B \to D^{**}\ell\nu$), background subtraction and detector-related uncertainties (reconstruction efficiency, unfolding). For most moments and E_{cut}^* values the signal modeling is the dominant contribution to the systematic error while at low E_{cut}^* values also background-related uncertainties are significant.

For both the electron energy and the hadronic mass moments the full covariance matrix of the measurements (including statistical and systematic contributions) has been estimated.[9,12] In the electron energy moment analysis the statistical correlations are derived directly from the covariance matrix X of the unfolded spectrum x' while the systematic correlations are constructed from the individual systematic error components assuming that they are either correlated, anti-correlated or uncorrelated for the given moments. For the hadronic mass moments, the total correlations are obtained using a toy MC approach.

5. Summary

We have reviewed the Belle measurements of the electron energy and the hadronic mass moments in $B \to X_c\ell\nu$. Belle measures the first four moments of the electron energy spectrum for threshold energies from 0.4 GeV to 2.0 GeV. In addition, the partial branching fractions for the same set of threshold energies are obtained. The analysis of the hadronic mass spectrum reports the first and the second central

C. Schwanda

and non-central moments of the hadronic mass squared spectrum for lepton energy thresholds ranging from 0.7 to 1.9 GeV. In both analyses, the full correlation matrix of measurements has been evaluated. These measurements have been used in the latest determination of $|V_{cb}|$ from inclusive semileptonic B decays.[24]

References

1. N. Cabibbo, *Phys. Rev. Lett.* **10**, 531 (1963); M. Kobayashi and T. Maskawa, *Prog. Theor. Phys.* **49**, 652 (1973).
2. D. Benson, I. I. Bigi, T. Mannel and N. Uraltsev, *Nucl. Phys.* B **665**, 367 (2003), arXiv:hep-ph/0302262.
3. P. Gambino and N. Uraltsev, *Eur. Phys. J.* C **34**, 181 (2004), arXiv:hep-ph/0401063.
4. P. Gambino, *JHEP* **1109**, 055 (2011), arXiv:1107.3100 [hep-ph].
5. C. W. Bauer, Z. Ligeti, M. Luke, A. V. Manohar and M. Trott, *Phys. Rev.* D **70**, 094017 (2004), arXiv:hep-ph/0408002.
6. CLEO Collab. (A. H. Mahmood *et al.*), *Phys. Rev.* D **70**, 032003 (2004), arXiv:hep-ex/0403053.
7. DELPHI Collab. (J. Abdallah *et al.*), *Eur. Phys. J.* C **45**, 35 (2006), arXiv:hep-ex/0510024.
8. BaBar Collab. (B. Aubert *et al.*), *Phys. Rev.* D **81**, 032003 (2010), arXiv:0908.0415 [hep-ex].
9. Belle Collab. (P. Urquijo *et al.*), *Phys. Rev.* D **75**, 032001 (2007), arXiv:hep-ex/0610012.
10. CLEO Collab. (S. E. Csorna *et al.*), *Phys. Rev.* D **70**, 032002 (2004), arXiv:hep-ex/0403052.
11. CDF Collab. (D. Acosta *et al.*), *Phys. Rev.* D **71**, 051103 (2005), arXiv:hep-ex/0502003.
12. BELLE Collab. (C. Schwanda *et al.*), *Phys. Rev.* D **75**, 032005 (2007), arXiv:hep-ex/0611044.
13. Belle Collab. (A. Abashian *et al.*), *Nucl. Instrum. Meth.* A **479**, 117 (2002).
14. S. Kurokawa, *Nucl. Instrum. Meth.* A **499**, 1 (2003), and other papers included in this volume.
15. D. J. Lange, *Nucl. Instrum. Meth.* A **462**, 152 (2001).
16. R. Brun, F. Bruyant, M. Maire, A. C. McPherson and P. Zanarini, CERN-DD/ EE/84-1.
17. I. Caprini, L. Lellouch and M. Neubert, *Nucl. Phys.* B **530**, 153 (1998), arXiv:hep-ph/9712417.
18. A. K. Leibovich, Z. Ligeti, I. W. Stewart and M. B. Wise, *Phys. Rev.* D **57**, 308 (1998), arXiv:hep-ph/9705467.
19. F. De Fazio and M. Neubert, *JHEP* **9906**, 017 (1999), arXiv:hep-ph/9905351.
20. E. Barberio and Z. Was, *Comput. Phys. Commun.* **79**, 291 (1994).
21. K. Hanagaki, H. Kakuno, H. Ikeda, T. Iijima and T. Tsukamoto, *Nucl. Instrum. Meth.* A **485**, 490 (2002), arXiv:hep-ex/0108044.
22. A. Abashian *et al.*, *Nucl. Instrum. Meth.* A **491**, 69 (2002).
23. A. Höcker and V. Kartvelishvili, *Nucl. Instrum. Meth.* A **372**, 469 (1996), arXiv:hep-ph/9509307.
24. P. Gambino and C. Schwanda, *Phys. Rev.* D **89**, 014022 (2014), arXiv:1307.4551 [hep-ph].

Memorial to Kolya Uraltsev: On the Way to $1/m_b^5$

Sascha Turczyk

PRISMA Cluster of Excellence & Mainz Institute for Theoretical Physics, Johannes Gutenberg University, 55099 Mainz, Germany

In this article I pay my tribute to the memory of Kolya Uraltsev. I look back on my experience with him, and focus especially on two joined projects with him on inclusive semi-leptonic $B \to X_c \ell \bar{\nu}_\ell$ decays. This is a field where he had a major impact over the past decades.

1. Comments on the Daily Life Experience

Let me start this Memorial article to honor the memory of Kolya Uraltsev with some private comments. I first met Kolya personally during a Workshop at CERN in the beginning of 2007, shortly after starting my PhD at University of Siegen. Some time later in 2008 Kolya joined the staff of Siegen. During this period we started our first project. Thus I am given the great opportunity to portray him from the perspective of a young student. He taught me a great part of what I know in heavy quark physics. Later on I will describe our work related to the Heavy Quark Expansion (HQE) and inclusive semi-leptonic $B \to X_c \ell \bar{\nu}_\ell$ decays, clearly a research area where Kolya was a great expert and a pioneer.[1,2]

First I want to portray Kolya in the way I got to know him personally during the short time we worked together. My advisor Thomas Mannel brought us together to discuss a possible project. Previously we had published a paper[3] on a topic similar to one Kolya had worked on with his collaborators.[4] It seemed there were, besides similarities and points of agreement, some differences which needed to be investigated. Indeed, we eventually clarified the subtleties[5] and found complete agreement. Hence we started a discussion, in which I explained the ideas behind our work. It utilized a systematic approach to calculate higher order nonperturbative (NP) corrections.[6] This was the starting point to explain everybody the details of both these works. The discussion started around 4 pm. On this occasion I really experienced his desire to clarify everything up to the tiniest subtlety. He was a very enthusiastic, discussion loving person, as I could realize during our first physics discussion. For a young researcher with naturally little knowledge it was a tough discussion. We stopped at 11pm due to my exhaustion. However it taught me the importance to think about and especially question all the little details, and this is already an important lesson. Besides, it really demonstrated the dedication of Kolya towards science and his persistence to clarify a topic.

In the following, we had several meetings where I learned a lot, but I also got to know him better personally. He was indeed very enthusiastic about physics. His verve also showed up in physics related, but more practical topics. For instance, he tried to fix and reuse various electronic components. Sometimes one would think to enter the office of an experimental physicist with all his equipment, demonstrating his versatile abilities and knowledge. I was responsible for computer administration. In case something broke, I asked him if he would need this before bringing to the electronics waste. It was admirable how he repaired not only electronic elements, but also mechanical devices. In some sense, his skills were legendary in our group. Once I brought a spare part for his car from the USA, because one could not buy this piece separately in Europe. That altogether shows how passionate he was with his life and his dedication. He truly pursued a goal until he finally succeeded. Kolya enjoyed playing with physics related gadgets, too. I remember, when he once flew — quite skilled — a small radio controlled helicopter in the corridor in front of his office. My recollection is that he really was fascinated by this.

Last but not least, let me mention his helpfulness. If I had a question, he answered it in a very detailed way, which proved very useful and I appreciated that a lot. Not only he answered the question completely including detailed explanations, but he also gave side remarks about validity and closely related topics for a complete explanation.

Finally I want to close my private resume, with a sentence once Kolya told me, because I always have self-doubts: "You sometimes underestimate what you can do, Sascha". Hopefully he got this right, as he always got his physics right. Kolya was not only a great physicist, he also was a great man.

2. Towards the Physics Projects

Both projects I will describe very briefly in the following sections, were devoted to NP corrections in the Heavy Quark Expansion (HQE). In semileptonic decays, one can factorize the leptonic part $L_{\mu\nu}$ and hadronic part $W_{\mu\nu}$. The decay rate is written as

$$d\Gamma = \frac{G_F^2 |V_{cb}|^2}{4M_B} \mathrm{Im}\, T_{\mu\nu} L^{\mu\nu} \, d\phi_{\mathrm{PS}}, \qquad (1)$$

where we have made use of the optical theorem $\mathrm{Im}\, T_{\mu\nu} \propto W_{\mu\nu}$. From this the CKM matrix element $|V_{cb}|$ can be extracted very cleanly to a high precision[7–10] of $\mathcal{O}(\lesssim 2\%)$.[11] The leptonic part can be calculated easily, while the hadronic part contains non-perturbative information of the b-quark bound to the B-Meson. In inclusive decays we make no reference to the final state, and hence the hadronic tensor is given by the discontinuity of the forward matrix element

$$T_{\mu\nu} = -\frac{i}{2m_B} \int d^4x\, e^{-iqx} \langle B(v)|T\left[J_\nu^\dagger(x)J_\mu(0)\right]|B(v)\rangle\,, \quad J_\mu = \bar{c}\gamma_\mu P_L\,. \quad (2)$$

The HQE allows us to calculate this discontinuity in terms of local forward matrix elements[1,12–14] because of an operator product expansion (OPE) and to order $1/m_b^n$ the relevant matrix elements are given by contractions of

$$\langle B(v)|\bar{b}_v(iD_{\mu_1})(iD_{\mu_2})\cdots(iD_{\mu_n})b_v|B(v)\rangle\,. \quad (3)$$

The first correction starts with the kinetic energy parameter μ_π^2 and the chromo-magnetic moment μ_G^2 at order $1/m_b^2$ and the Darwin term ρ_D^3 and the spin-orbit term ρ_{LS}^3 at order $1/m_b^3$, see Ref. 15 for additional details. Kolya made tremendous contributions to the theoretical foundations of this framework, a selection of important works are.[1,2,16–18,7,8]

2.1. The two roads to 'intrinsic charm' in B decays

In this first work together with Kolya, Ikaros Bigi and Thomas Mannel[5] we asked ourselves how to interpret higher order corrections that have been dubbed "Intrinsic Charm" in the literature. In particular, we investigated two complementary ways to describe these effects.[3,4] Apart from the lessons "Intrinsic Charm" effects can teach us about non-perturbative dynamics, their consideration is relevant for the precise extraction of $|V_{cb}|$. They complement the estimate of the potential impact of $1/m_b^4$ contributions in the total decay rate. In order to visualize these effects, we may cast the hadronic tensor into the form

$$2M_B W_{\mu\nu}(p) = \int d^4x\, e^{ip\cdot x} \langle B(v)|\bar{b}_v(x)\Gamma_\nu^\dagger c(x)\bar{c}(0)\Gamma_\mu b_v(0)|B(v)\rangle\,. \quad (4)$$

The appearance of the (local) "intrinsic charm" (IC) operator at order $1/m_b^3$

$$\mathcal{O}_{IC} = (\bar{b}_v\Gamma^\dagger c)\,(\bar{c}\Gamma b_v) \quad (5)$$

is seen from this by expanding Eq. (4) in an OPE. This operator mixes with the previously mentioned matrix elements of local operators built of light-fields at order $1/m_b^n$ in Eq. (3) via loop integrals depicted in the sample Feynman diagrams below.

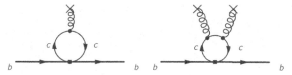

The IC operator induces new subtleties in the OPE at higher orders. Its contribution contains an infrared (IR) piece sensitive to the charm quark mass, proportional to $\ln m_c/m_b$ in leading order. This logarithm is generated either in the two-step procedure by the renormalization group running between m_b and m_c,[3] or it appears simply in the coefficient function when bottom and charm are treated on the same footing, i.e. integrated out at the same time. In fact, it has been shown[4] that at higher orders in the OPE even inverse powers of m_c appear. In order to investigate this IR sensitive behavior we write the decay rate as

$$\frac{d^2\Gamma}{dv \cdot p\,dp^2} = \frac{G_F^2 |V_{cb}|^2}{24\pi^3}\sqrt{v \cdot p^2 - p^2}\,\theta(v \cdot p^2 - p^2)\theta(m_b - v \cdot p)\theta(m_b^2 + p^2 - 2m_b v \cdot p)L_{\mu\nu}W^{\mu\nu}$$

with $p = p_b - q$ being the momentum of the final state charm quark. The expansion of the hadronic tensor in terms of the matrix elements (3) is then given by

$$2M_B W_{\mu\nu} = \sum_{n=0}^{\infty}\langle B(v)|\bar{b}_v\Gamma_\nu^\dagger(\slashed{p} + m_c)\frac{[(i\slashed{D})(\slashed{p} + m_c)]^n}{n!}\Gamma_\mu b_v|B(v)\rangle\delta_+^{(n)}(p^2 - m_c^2).$$
(6)

Hence an expansion IR sensitive to the charm quark mass appears with the form

$$\Gamma_n \propto \frac{1}{m_b^{3+k}}\left(\frac{1}{m_c^2}\right)^{(n-3)/2} \qquad \text{with } n = 5, 7, \ldots, \text{ and } k = 0, 2, \ldots .$$
(7)

This obviously starts at $1/m_b^3$ due to its connection to the IC operator entering at this order. The explicit calculation verifies the power-like diverging terms derived in Ref. 4. In fact only terms with an odd power in $1/m_b$ appear, consequently only at least one gluon matrix elements contribute. Note that with the realistic scaling behavior $m_c^2 \approx \Lambda_{\rm QCD}m_b$ these terms are formally of order $1/m_b^4$ and thus have to be taken into account to complete the $1/m_b^4$ calculation.

In the case of the decay into the light u-quark we are not allowed to integrate out this operator. As mentioned before, the Darwin term mixes with this operator. Thus we investigated a model to estimate the finite value of the IC matrix element

$$\frac{\rho_D^3}{m_b^3}\ln\frac{m_b^2}{m_c^2 + \Lambda^2}$$
(8)

to simulate the non-valence weak annihilation (WA) contribution. Expanding this to $1/m_c^2$ we make an educated guess for the scale $\Lambda \approx 0.7$ GeV by comparing with the $1/(m_b^3 m_c^2)$ terms. We used the vacuum saturation estimate for the operators.[4]

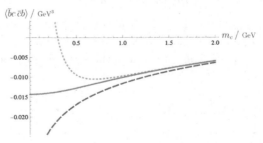

The blue dashed line corresponds to the logarithmic term only, the yellow dotted line including the power-like IR sensitive terms and the red solid line shows the model. Taking this value at face, we obtain a naive estimate of the non-valence WA

$$\frac{\delta\Gamma_{\rm sl}^{\rm n\,val}(b \to u)}{\Gamma_{\rm sl}(b \to u)} \approx -0.015.$$
(9)

2.2. *Higher order power corrections in inclusive B decays*

In a subsequent work,[19] we investigated the behavior of the OPE to higher orders in $1/m_b$. While we had the setup to include the full corrections to order $1/m_b^5$ available, our goal was to estimate the achievable precision and impact of the higher order corrections. The unknown higher order matrix elements are estimated by splitting the chain of derivatives in the matrix element into two parts A and C. Using an intermediate state representation suggested by Kolya, we derive the decomposition

$$\langle B|\bar{b}A(iD_0)^k C\Gamma b(0)|B\rangle$$
$$= \sum_n [(M_B - m_b) - (M_n - m_Q)]^k$$
$$\times (\langle B|\bar{b}AQ|n\rangle \cdot \langle n|\bar{Q}C\Gamma b|B\rangle + \langle B|\bar{b}A\gamma_5 Q|n\rangle \cdot \langle n|\bar{Q}C\gamma_5\Gamma b|B\rangle). \quad (10)$$

We assume the sum to be saturated by the lowest state, and each iD_0 corresponds to an average excitation energy $-\bar{\epsilon} \approx -0.4$ GeV, effectively expressing higher order contributions through known ME and $\bar{\epsilon}$. With this information at hand, we can numerically investigate the impact on the extraction of $|V_{cb}|$ due to direct and indirect effects. The direct effects on the extraction of $|V_{cb}|$ are given by the additional terms due to higher orders in the total decay rate and are estimated[19] to be

$$\frac{\delta\Gamma_{1/m^2}}{\Gamma_{\text{tree}}} = -0.043, \quad \frac{\delta\Gamma_{1/m^4}}{\Gamma_{\text{tree}}} = 0.0075,$$
$$\frac{\delta\Gamma^{\text{IC}}}{\Gamma_{\text{tree}}} = 0.007, \quad \frac{\delta\Gamma_{1/m^5}}{\Gamma_{\text{tree}}} = 0.006. \quad (11)$$

This result suggests that the power series for $\Gamma_{\text{sl}}(b \to c)$ is well behaved and under good control provided the NP expectation values are known. The estimated overall shift due to $1/m_b^{4+5}$ higher order terms would yield a 0.65% *direct* reduction in $|V_{cb}|$. The NP parameters are extracted by measuring moments of kinematic variables

$$\frac{\int dE_\ell\, dq_0\, dq^2\, K(E_\ell, q_0, q^2) C(E_\ell, q_0, q^2) \frac{d^3\Gamma_{\text{sl}}}{dE_\ell\, dq_0\, dq^2}}{\int dE_\ell\, dq_0\, dq^2\, C(E_\ell, q_0, q^2) \frac{d^3\Gamma_{\text{sl}}}{dE_\ell\, dq_0\, dq^2}}. \quad (12)$$

Hence adding higher order terms will shift the extracted values. C is a possible cut due to experimental constraints, and K denotes the kinematical variable in question. This will consequently change *indirectly* the value of the $|V_{cb}|$ extraction from the normalization. To assess these effects, we account for the change of the moment $\mathcal{M}$ with respect to only one HQE parameter as a simplifying assumption

$$\delta m_b = -\frac{\delta\mathcal{M}}{\frac{\partial\mathcal{M}}{\partial m_b}}, \quad \delta\mu_\pi^2 = -\frac{\delta\mathcal{M}}{\frac{\partial\mathcal{M}}{\partial\mu_\pi^2}}, \quad \delta\rho_D^3 = -\frac{\delta\mathcal{M}}{\frac{\partial\mathcal{M}}{\partial\rho_D^3}}, \quad \frac{\delta|V_{cb}|}{|V_{cb}|} = -\frac{1}{2}\frac{1}{\Gamma_{\text{sl}}}\frac{\partial\Gamma_{\text{sl}}}{\partial \text{HQP}}\delta\,\text{HQP}.$$

In reality a shift in the moments is compensated by a simultaneous shift of all parameters. We judge changes of the HQE parameters in a sense, that a very small shift indicates a well constrained parameter, while a large shift hints to an observable insensitive to the particular HQE parameter. We deem changes of $|\delta m_b| \gtrsim 10$ MeV, $|\delta \mu_\pi^2| \gtrsim 0.1$ GeV^2 and $|\delta \rho_D^3| \gtrsim 0.1$ GeV^3 as significant bearing in mind the estimated accuracy of the existing OPE predictions.[8] This would result in

$$\frac{\delta|V_{cb}|}{|V_{cb}|} \simeq \begin{cases} -0.0066 & \text{at} \quad \delta m_b = 10 \text{ MeV} \\ 0.0013 & \text{at} \quad \delta \mu_\pi^2 = 0.1 \text{ GeV}^2 \\ 0.009 & \text{at} \quad \delta \rho_D^3 = 0.1 \text{ GeV}^3 \end{cases} \quad . \tag{13}$$

We find a mild cut dependence for $E_e \lesssim 1$ GeV, the plots can be found in Ref. 19. For a typical experimental cut of $E_e \geq 1$ GeV we find the shifts listed in the table below.

	$\langle E_\ell \rangle$	$\langle (E_\ell - \langle E_\ell \rangle)^2 \rangle$	$\langle (E_\ell - \langle E_\ell \rangle)^3 \rangle$	$\langle M_X^2 \rangle$	$\langle (M_X^2 - \langle M_X^2 \rangle)^2 \rangle$	$\langle (M_X^2 - \langle M_X^2 \rangle)^3 \rangle$
δm_b, MeV	-39	-60		-21		
$(\delta|V_{cb}|/|V_{cb}|)$	(0.026)	(0.040)	—	(0.014)	—	—
$\delta \mu_\pi^2$, GeV^2	-0.30	-0.12	-0.04	-0.13	-0.08	0.33
$(\delta|V_{cb}|/|V_{cb}|)$	(-0.004)	(-0.0016)	(-0.0005)	(-0.0017)	(-0.0010)	(0.0043)
$\delta \rho_D^3$, GeV^3	0.16	0.09	0.02	0.09	0.05	0.10
$(\delta|V_{cb}|/|V_{cb}|)$	(0.014)	(0.008)	(0.020)	(0.008)	(0.005)	(0.009)

The pattern suggests, that the main effect is coming from an increase of 0.1 GeV^3 in ρ_D^3, possibly combined with a slight shift of μ_π^2 of around ± 0.05 GeV^2, leading to an increase of $\sim 1\%$ in $|V_{cb}|$. Further discussion can be found in Ref. 19. Adding this *indirect* to the *direct* effect results in a shift, which could also be interpreted as the achievable precision due to NP effects, of

$$\frac{\delta|V_{cb}|}{|V_{cb}|} \approx +(0.003\text{--}0.005) . \tag{14}$$

3. Summary

In this short memorial I have revisited my collaboration with Kolya Uraltsev. Especially it is devoted to subtle effects in higher orders of inclusive semileptonic decays. He was a pioneer in these calculations, and we have clarified the role of the charm quark in higher order corrections. Using the terms up to $1/m_b^5$ and an educated estimate of the unknown matrix elements, we could infer an expected size of the non-valence weak annihilation contribution. Its contribution to the $b \to u$ transition is estimated to be of the order of -1.5%.

Furthermore we could evaluate the influence on $|V_{cb}|$ to be a net increase of approximately 0.3–0.5% due to incorporating these higher order terms. This demonstrates the very good convergence of the heavy quark expansion.

References

1. I. I. Bigi, M. A. Shifman, N. G. Uraltsev and A. I. Vainshtein, *Phys. Rev. Lett.* **71**, 496 (1993), arXiv:hep-ph/9304225.
2. I. I. Bigi, M. A. Shifman, N. G. Uraltsev and A. I. Vainshtein, *Int. J. Mod. Phys. A* **9**, 2467 (1994), arXiv:hep-ph/9312359.
3. C. Breidenbach, T. Feldmann, T. Mannel and S. Turczyk, *Phys. Rev. D* **78**, 014022 (2008), arXiv:0805.0971 [hep-ph].
4. I. I. Bigi, N. Uraltsev and R. Zwicky, *Eur. Phys. J. C* **50**, 539 (2007), arXiv:hep-ph/0511158.
5. I. Bigi, T. Mannel, S. Turczyk and N. Uraltsev, *JHEP* **1004**, 073 (2010), arXiv:0911.3322 [hep-ph].
6. B. M. Dassinger, T. Mannel and S. Turczyk, *JHEP* **0703**, 087 (2007), arXiv:hep-ph/0611168.
7. D. Benson, I. I. Bigi, T. Mannel and N. Uraltsev, *Nucl. Phys. B* **665**, 367 (2003), arXiv:hep-ph/0302262.
8. P. Gambino and N. Uraltsev, *Eur. Phys. J. C* **34**, 181 (2004), arXiv:hep-ph/0401063.
9. C. W. Bauer, Z. Ligeti, M. Luke and A. V. Manohar, *Phys. Rev. D* **67**, 054012 (2003), arXiv:hep-ph/0210027.
10. A. H. Hoang, Z. Ligeti and A. V. Manohar, *Phys. Rev. D* **59**, 074017 (1999), arXiv:hep-ph/9811239.
11. Particle Data Group Collab. (J. Beringer *et al.*), *Phys. Rev. D* **86**, 010001 (2012).
12. J. Chay, H. Georgi and B. Grinstein, *Phys. Lett. B* **247**, 399 (1990).
13. A. V. Manohar and M. B. Wise, *Phys. Rev. D* **49**, 1310 (1994), arXiv:hep-ph/9308246.
14. T. Mannel, *Nucl. Phys. B* **413**, 396 (1994), arXiv:hep-ph/9308262.
15. P. Gambino, contribution to this volume, pp. 27–44.
16. I. I. Bigi, N. G. Uraltsev and A. I. Vainshtein, *Phys. Lett. B* **293**, 430 (1992) [Erratum: *ibid.* **297**, 477 (1993)], arXiv:hep-ph/9207214.
17. M. A. Shifman, N. G. Uraltsev and A. I. Vainshtein, *Phys. Rev. D* **51**, 2217 (1995) [Erratum: *ibid.* **52**, 3149 (1995)], arXiv:hep-ph/9405207.
18. I. I. Bigi, M. A. Shifman, N. Uraltsev and A. I. Vainshtein, *Phys. Rev. D* **56**, 4017 (1997), arXiv:hep-ph/9704245.
19. T. Mannel, S. Turczyk and N. Uraltsev, *JHEP* **1011**, 109 (2010), arXiv:1009.4622 [hep-ph].

Lifetimes and Heavy Quark Expansion

Alexander Lenz

Institute for Particle Physics Phenomenology,
Durham University, DH1 3LE Durham, UK
alexander.lenz@durham.ac.uk

Kolya Uraltsev was one of the inventors of the Heavy Quark Expansion (HQE), that describes inclusive weak decays of hadrons containing heavy quarks and in particular lifetimes. Besides giving a pedagogic introduction to the subject, we review the development and the current status of the HQE, which just recently passed several non-trivial experimental tests with an unprecedented precision. In view of many new experimental results for lifetimes of heavy hadrons, we also update several theory predictions: $\tau(B^+)/\tau(B_d) = 1.04^{+0.05}_{-0.01} \pm 0.02 \pm 0.01$, $\tau(B_s)/\tau(B_d) = 1.001 \pm 0.002$, $\tau(\Lambda_b)/\tau(B_d) = 0.935 \pm 0.054$ and $\bar{\tau}(\Xi^0_b)/\bar{\tau}(\Xi^+_b) = 0.95 \pm 0.06$. The theoretical precision is currently strongly limited by the unknown size of the non-perturbative matrix elements of four-quark operators, which could be determined with lattice simulations.

1. Introduction

Lifetimes are among the most fundamental properties of elementary particles. In this work we consider lifetimes of hadrons containing heavy quarks, which decay via the weak interaction. Their masses and lifetimes read (according to PDG[1] and HFAG[2])

B-mesons

	$B_d = (\bar{b}d)$	$B^+ = (\bar{b}u)$	$B_s = (\bar{b}s)$	$B_c^+ = (\bar{b}c)$	
Mass (GeV)	5.27955(26)	5.27925(26)	5.3667(4)	6.2745(18)	(1)
Lifetime (ps)	1.519(7)	1.641(8)	1.516(11)	0.452(33)	
$\tau(X)/\tau(B_d)$	1	1.079 ± 0.007	0.998 ± 0.009	0.30 ± 0.02	

b-baryons

	$\Lambda_b = (udb)$	$\Xi^0_b = (usb)$	$\Xi^-_b = (dsb)$	$\Omega^-_b = (ssb)$	
Mass (GeV)	5.6194(6)	5.7918(5)	5.79772(55)	6.071(40)	(2)
Lifetime (ps)	1.451(13)	1.477(32)	1.599(46)	$1.54\left(^{+26}_{-22}\right)$	
$\tau(X)/\tau(B_d)$	0.955 ± 0.009	0.972 ± 0.021	1.053 ± 0.030	$1.01\left(^{+17}_{-14}\right)$	

The masses and the lifetimes of the Ξ_b^0, Ξ_b^- and the Ω_b^- have been measured by the LHCb Collaboration[3-5] just after the first version of this article appeared on the arXiv. We have given above these new values instead of the HFAG and PDG averages. Alternative lifetime averages were, e.g., obtained in Ref. 6.

D-mesons

	$D^0 = (\bar{u}c)$	$D^+ = (\bar{d}c)$	$D_s^+ = (\bar{s}c)$
Mass (GeV)	1.86491(17)	1.8695(4)	1.9690(14)
Lifetime (ps)	0.4101(15)	1.040(7)	0.500(7)
$\tau(X)/\tau(D^0)$	1	2.536 ± 0.017	1.219 ± 0.017

$$(3)$$

c-baryons

	$\Lambda_c = (udc)$	$\Xi_c^+ = (usc)$	$\Xi_c^0 = (dsc)$	$\Omega_c = (ssc)$
Mass (GeV)	2.28646(14)	$2.4676\left(^{+4}_{-10}\right)$	$2.47109\left(^{+35}_{-100}\right)$	$2.6952\left(^{+18}_{-16}\right)$
Lifetime (ps)	0.200(6)	0.442(26)	$0.112\left(^{+13}_{-10}\right)$	0.069(12)
$\tau(X)/\tau(D^0)$	0.488 ± 0.015	1.08(6)	0.27(3)	0.17 ± 0.03

$$(4)$$

One of the first observations to make is the fact that all lifetimes are of the same order of magnitude, they are all in the pico-second range and they differ at most by a factor of 25. Looking exclusively at hadrons containing one *b*-quark (and no *c*-quark), one even finds that all lifetimes are equal within about 10%. This clearly calls for a theoretical explanation.

In this review we will discuss the theoretical framework describing decay rates of inclusive decays of hadrons containing a heavy quark, the so-called **Heavy Quark Expansion**. A special case of such observables are the lifetimes of hadrons, which are given by the inverse of the total decay rates. Kolya Uraltsev was one of the main pioneers in the development of the HQE, which has its roots back in the 1970s. When I began my career, Kolya's theory was already a kind of textbook knowledge and my PhD and my first scientific papers were devoted to the calculation of higher order QCD corrections within the framework of the HQE. Thus it was very inspiring to meet Kolya personally at one of my first international conferences, which was held in 2000 in Durham, where we discussed the so-called "missing charm puzzle"[7] and the decay rate difference $\Delta\Gamma_s$ of B_s-mesons.[8] I benefited a lot from many follow-up encounters with Kolya, e.g., at CERN, in Portoroz and in Siegen. At the end of 2012 I was working with a student from Munich[9] on *D*-meson lifetimes and in that respect Kolya was sending me several long emails regarding the history of lifetime predictions, which clearly influenced this review.

Many of the most convincing precision tests of the HQE have just been performed recently. In the beginning of 2012 $\Delta\Gamma_s$ has been measured for the first time, i.e., with a statistical significance of five standard deviations, in accordance with the HQE prediction. The long standing puzzle concerning the lifetime of the Λ_b-baryon — for many years a very strong challenge for the HQE — seems to have been settled experimentally, with the latest results just appearing in 2014. It is a real tragedy that Kolya did not have more time to celebrate the successes of the theory, to which he contributed so much.

We start in Section 2 with a basic introduction into lifetimes of weakly decaying particles. In Section 2.2 we discuss the structure of the HQE in detail and in Section 2.3 we give a brief review of the discussed observables. In Section 3 we investigate the history of the HQE and we highlight Kolya's contribution, while we discuss the status quo in experiment and theory in Section 4. Here we also give some numerical updates of theory predictions for lifetime ratios. In Section 5 we give an outlook on what has to be done in order to improve further the theoretical accuracy and we conclude.

2. Basic Considerations About Lifetimes

2.1. *Naive estimates*

2.1.1. *Charm-quark decay*

Before trying to investigate the complicated meson decays, let us look at the decay of free c- and b-quarks. Later on we will show that the free quark decay is the leading term in a systematic expansion in the inverse of the heavy (decaying) quark mass — the HQE.

A charm quark can decay weakly into a strange- or a down-quark and a W^+-boson, which then further decays either into leptons (semi-leptonic decay) or into quarks (non-leptonic decay). Calculating the total inclusive decay rate of a charm-quark we get

$$\Gamma_c = \frac{G_F^2 m_c^5}{192\pi^3} |V_{cs}|^2 c_{3,c}, \tag{5}$$

with

$$
\begin{aligned}
c_{3,c} = {} & g\left(\frac{m_s}{m_c}, \frac{m_e}{m_c}\right) + g\left(\frac{m_s}{m_c}, \frac{m_\mu}{m_c}\right) \\
& + N_c|V_{ud}|^2 h\left(\frac{m_s}{m_c}, \frac{m_u}{m_c}, \frac{m_d}{m_c}\right) + N_c|V_{us}|^2 h\left(\frac{m_s}{m_c}, \frac{m_u}{m_c}, \frac{m_s}{m_c}\right) \\
& + \left|\frac{V_{cd}}{V_{cs}}\right|^2 \left\{ g\left(\frac{m_d}{m_c}, \frac{m_e}{m_c}\right) + g\left(\frac{m_d}{m_c}, \frac{m_\mu}{m_c}\right) \right. \\
& \left. + N_c|V_{ud}|^2 h\left(\frac{m_d}{m_c}, \frac{m_u}{m_c}, \frac{m_d}{m_c}\right) + N_c|V_{us}|^2 h\left(\frac{m_d}{m_c}, \frac{m_u}{m_c}, \frac{m_s}{m_c}\right) \right\}.
\end{aligned}
\tag{6}
$$

h denotes a new phase space function, when there are three massive particles in the final state. If we set all phase space factors to one ($f(m_s/m_c) = f(0.0935/1.471) = 1 - 0.03, \ldots$ with $m_s = 93.5(2.5)$ MeV [1]) and use $|V_{ud}|^2 + |V_{us}|^2 \approx 1 \approx |V_{cd}|^2 + |V_{cs}|^2$, then we get $|V_{cs}|^2 c_{3,c} = 5$, similar to the τ decay. In that case we predict a charm lifetime of

$$\tau_c = \begin{cases} 0.84 \text{ ps} \\ 1.70 \text{ ps} \end{cases} \quad \text{for} \quad m_c = \begin{cases} 1.471 \text{ GeV} & \text{(Pole-scheme)} \\ 1.277(26) \text{ GeV} & (\overline{\text{MS}}\text{-scheme}) \end{cases}. \tag{7}$$

These predictions lie roughly in the ball-bark of the experimental numbers for D-meson lifetimes, but at this stage some comments are appropriate:

- Predictions of the lifetimes of free quarks have a huge parametric dependence on the definition of the quark mass ($\propto m_q^5$). This is the reason, why typically only lifetime ratios (the dominant m_q^5 dependence as well as CKM factors and some sub-leading non-perturbative corrections cancel) are determined theoretically. We show in this introduction for pedagogical reasons the numerical results of the theory predictions of lifetimes and not only ratios. In our case the value obtained with the $\overline{\text{MS}}$-scheme for the charm quark mass is about a factor of 2 larger than the one obtained with the pole-scheme. In LO-QCD the definition of the quark mass is completely arbitrary and we have these huge uncertainties. If we calculate everything consistently in NLO-QCD, the treatment of the quark masses has to be defined within the calculation, leading to a considerably weaker dependence of the final result on the quark mass definition.

    Bigi, Shifman, Uraltsev and Vainshtein have shown in 1994[10] that the pole mass scheme is always affected by infra-red renormalons, see also the paper of Beneke and Braun[11] that appeared on the same day on the arXiv and the review in this issue.[12] Thus short-distance definitions of the quark mass, like the $\overline{\text{MS}}$-mass[13] seem to be better suited than the pole mass. More recent suggestions for quark mass concepts are the kinetic mass from Bigi, Shifman, Uraltsev and Vainshtein[14,15] introduced in 1994, the potential subtracted mass from Beneke[16] and the $\Upsilon(1s)$-scheme from Hoang, Ligeti and Manohar,[17,18] both introduced in 1998. In Ref. 19 we compared the above quark mass schemes for inclusive non-leptonic decay rates and found similar numerical results for the different short distance masses. Thus we rely in this review — for simplicity — on predictions based on the $\overline{\text{MS}}$-mass scheme and we discard the pole mass, even if we give several times predictions based on this mass scheme for comparison.

    Concerning the concrete numerical values for the quark masses we also take the same numbers as in Ref. 19. In that work relations between different quark mass schemes were strictly used at NLO-QCD accuracy (higher terms were discarded), therefore the numbers differ slightly from the PDG[1]-values, which would result in

$$\tau_c = \begin{cases} 0.44 \text{ ps} \\ 1.71 \text{ ps} \end{cases} \quad \text{for} \quad m_c = \begin{cases} 1.67(7) \text{ GeV} & \text{(Pole-scheme)} \\ 1.275(25) \text{ GeV} & (\overline{MS}\text{-scheme}) \end{cases}. \tag{8}$$

Since our final lifetime predictions are only known up to NLO accuracy and we expand every expression consistently up to order α_s, we will stay with the parameters used in Ref. 19.

- Taking only the decay of the c-quark into account, one obtains the same lifetimes for all charm-mesons, which is clearly a very bad approximation, taking the large spread of lifetimes of different D-mesons into account, see Eq. (3). Below we will see that in the case of charmed mesons a very sizable contribution comes from non-spectator effects where also the valence quark of the D-meson is involved in the decay.
- Perturbative QCD corrections will turn out to be very important, because $\alpha_s(m_c)$ is quite large.
- In the above expressions we neglected, e.g., annihilation decays like $D^+ \to l^+\nu_l$, which have very small branching ratios[1] (the corresponding Feynman diagrams have the same topology as the decay $B^- \to \tau^-\bar\nu_\tau$, that was mentioned earlier). In the case of D_s^+ meson the branching ratio into $\tau^+\nu_\tau$ will, however, be sizable[1] and has to be taken into account

$$\text{Br}(D_s^+ \to \tau^+\nu_\tau) = (5.43 \pm 0.31)\%. \tag{9}$$

In the framework of the HQE the non-spectator effects will turn out to be suppressed by $1/m_c$ and since m_c is not very large, the suppression is also not expected to be very pronounced. This will change in the case of B-mesons. Because of the larger value of the b-quark mass, one expects a better description of the meson decay in terms of the simple b-quark decay.

2.1.2. *Bottom-quark decay*

Calculating the total inclusive decay rate of a b-quark we get

$$\Gamma_b = \frac{G_F^2 m_b^5}{192\pi^3}|V_{cb}|^2 c_{3,b}. \tag{10}$$

Neglecting the masses of all final state particles, except for the charm-quark and for the tau lepton, as well as the contributions proportional to $|V_{ub}|^2$ and using further $|V_{ud}|^2 + |V_{us}|^2 \approx 1 \approx |V_{cd}|^2 + |V_{cs}|^2$, we get the following simplified formula

$$c_{3,b} = \left[(N_c + 2)f\left(\frac{m_c}{m_b}\right) + g\left(\frac{m_c}{m_b}, \frac{m_\tau}{m_b}\right) + N_c g\left(\frac{m_c}{m_b}, \frac{m_c}{m_b}\right)\right]. \tag{11}$$

If we have charm quarks in the final states, then the phase space functions show a huge dependence on the numerical value of the charm quark mass (values taken from Ref. 19)

$$f\left(\frac{m_c}{m_b}\right) = \begin{cases} 0.484 \\ 0.518 \quad \text{for} \\ 0.666 \end{cases} \begin{cases} m_c^{\text{Pole}} = 1.471 \text{ GeV}, & m_b^{\text{Pole}} = 4.650 \text{ GeV}, \\ \bar m_c(\bar m_c) = 1.277 \text{ GeV}, & \bar m_b(\bar m_b) = 4.248 \text{ GeV}, \\ \bar m_c(\bar m_c) = 0.997 \text{ GeV}, & \bar m_b(\bar m_b) = 4.248 \text{ GeV}. \end{cases} \tag{12}$$

The big spread in the values for the space functions clearly shows again that the definition of the quark mass is a critical issue for a precise determination of lifetimes. The value for the pole quark mass is only shown to visualise the strong mass dependence. As discussed above short-distance masses like the $\overline{\text{MS}}$-mass are theoretically better suited. Later on we will argue further for using $\bar{m}_c(\bar{m}_b)$ and $\bar{m}_b(\bar{m}_b)$ — so both masses at the scale m_b —, which was suggested in Ref. 20, in order to sum up large logarithms of the form $\alpha_s^n (m_c/m_b)^2 \log^n (m_c/m_b)^2$ to all orders. Thus only the result using $\bar{m}_c(\bar{m}_b)$ and $\bar{m}_b(\bar{m}_b)$ should be considered as the theory prediction, while the additional numbers are just given for completeness.

The phase space function for two identical particles in the final states reads[21–24] (see Ref. 25 for the general case of three different masses)

$$g(x) = \sqrt{1 - 4x^2}\,\left(1 - 14x^2 - 2x^4 - 12x^6\right) + 24x^4 \left(1 - x^4\right) \log \frac{1 + \sqrt{1 - 4x^2}}{1 - \sqrt{1 - 4x^2}}, \quad (13)$$

with $x = m_c/m_b$. Thus we get in total for all the phase space contributions

$$c_{3,b} = \begin{cases} 9 \\ 2.97 \\ 3.25 \\ 4.66 \end{cases} \text{for} \quad \begin{cases} m_c = 0\,, \\ m_c^{\text{Pole}},\ m_b^{\text{Pole}}\,, \\ \bar{m}_c(\bar{m}_c),\ \bar{m}_b(\bar{m}_b)\,, \\ \bar{m}_c(\bar{m}_b),\ \bar{m}_b(\bar{m}_b)\,. \end{cases} \quad (14)$$

The phase space effects are now quite dramatic. For the total b-quark lifetime we predict (with $V_{cb} = 0.04151^{+0.00056}_{-0.00115}$ from Ref. 26, for similar results see Ref. 27)

$$\tau_b = 2.60 \text{ ps} \quad \text{for} \quad \bar{m}_c(\bar{m}_b),\ \bar{m}_b(\bar{m}_b)\,. \quad (15)$$

This number is about 70% larger than the experimental number for the B-meson lifetimes. There are in principle two sources for that discrepancy: first we neglected several CKM-suppressed decays, which are however not phase space suppressed as well as penguin decays. An inclusion of these decays will enhance the total decay rate roughly by about 10% and thus reduce the lifetime prediction by about 10%. Second, there are large QCD effects, that will be discussed in the next subsection; including them will bring our theory prediction very close to the experimental number.

Next we introduce the missing, but necessary concepts for making reliable predictions for the lifetimes of heavy hadrons.

2.2. The structure of the HQE

2.2.1. The effective Hamiltonian

Above we tried to make clear, that for any numerical reliable quantitative estimate of meson decays, QCD effects have to be taken properly into account. To do so, weak decays of heavy quarks are not described within the full standard model, but with the help of an effective Hamiltonian. We start here simply with the explicit form of the effective Hamiltonian and refer the interested reader to some excellent reviews by Buchalla, Buras and Lautenbacher,[28] by Buras,[29] by Buchalla[30] and a recent one by Grozin.[31] The effective Hamiltonian reads

$$\mathcal{H}_{\text{eff}} = \frac{G_F}{\sqrt{2}} \left[\sum_{q=u,c} V_c^q (C_1 Q_1^q + C_2 Q_2^q) - V_p \sum_{j=3} C_j Q_j \right]. \quad (16)$$

Without QCD corrections only the operator Q_2 arises and the *Wilson coefficient* C_2 is equal to one, $C_2 = 1$. Q_2 has a current–current structure:

$$Q_2 = c_\alpha \gamma_\mu (1 - \gamma_5) \bar{b}_\alpha \times d_\beta \gamma^\mu (1 - \gamma_5) \bar{u}_\beta \, . \tag{17}$$

α and β denote colour indices. The Vs describe different combinations of CKM elements. With the inclusion of QCD one gets additional operators. Q_1 has the same Dirac structure as Q_2, but it has a different colour structure, $Q_3, \ldots, Q_6$ arise from QCD penguin diagrams, etc. Due to renormalisation all Wilson coefficients become scale dependent functions. In LO-QCD we get[a] $C_2(4.248 \text{ GeV}) = 1.1$ and $C_1(4.248 \text{ GeV}) = -0.24$ and the penguin coefficients are below 5%, with the exception of C_8, the coefficient of the chromo-magnetic operator. With this operator product expansion (OPE) a separation of the scales was achieved. The high energy physics is described by the Wilson coefficients, they can be calculated in perturbation theory. The low energy physics is described by the matrix elements of the operators Q_i. Moreover large logarithms of the form $\alpha_s(m_b) \ln(m_b^2/M_W^2)$, which spoil the perturbative expansion in the full standard model, are now summed up to all orders. For semi-leptonic decays like $b \to c l^- \bar{\nu}_l$ the Wilson coefficient C_2 is simply 1, while the remaining ones vanish.

2.2.2. *The free quark decay with the effective Hamiltonian*

Now we can calculate the free quark decay starting from the effective Hamiltonian instead of the full standard model. If we again neglect penguins, we get in leading logarithmic approximation,

$$c_{3,b}^{\text{LO-QCD}} = c_{3,b}^{ce\bar{\nu}_e} + c_{3,b}^{c\mu\bar{\nu}_\mu} + c_{3,b}^{c\bar{u}d} + c_{3,b}^{c\bar{u}s} + c_{3,b}^{c\tau\bar{\nu}_\tau} + c_{3,b}^{c\bar{c}s} + c_{3,b}^{c\bar{c}d} \cdots \tag{18}$$

$$= \left[(2 + \mathcal{N}_a(\mu)) f\left(\frac{m_c}{m_b}\right) + g\left(\frac{m_c}{m_b}, \frac{m_\tau}{m_b}\right) + \mathcal{N}_a(\mu) g\left(\frac{m_c}{m_b}, \frac{m_c}{m_b}\right) \right], \tag{19}$$

with changing the colour factor $N_c = 3$ — stemming from QCD — into

$$\mathcal{N}_a(\mu) = 3C_1^2(\mu) + 3C_2^2(\mu) + 2C_1(\mu)C_2(\mu) \approx 3.3 \quad (\text{LO, } \mu = 4.248 \text{ GeV}) \, . \tag{20}$$

This effect enhances the total decay rate by about 10% and thus brings down (if also the sub-leading decays are included) the prediction for the lifetime of the b-quark to about

$$\tau_b \approx 2.10 \text{ ps} \quad \text{for} \quad \bar{m}_c(\bar{m}_b), \, \bar{m}_b(\bar{m}_b) \, . \tag{21}$$

Going to next-to-leading logarithmic accuracy we have to use the Wilson coefficients of the effective Hamiltonian to NLO accuracy and we have to determine one-loop QCD corrections within the effective theory. These NLO-QCD corrections turned

---

[a]We use as an input for the strong coupling $a_s(M_Z) = 0.1184$.

out to be very important for the inclusive b-quark decays. For massless final state quarks the calculation was done in 1991:[32]

$$c_{3,b} = c_{3,b}^{\text{LO-QCD}} + 8\frac{\alpha_s}{4\pi}\left[\left(\frac{25}{4} - \pi^2\right) + 2\left(C_1^2 + C_2^2\right)\left(\frac{31}{4} - \pi^2\right) - \frac{4}{3}C_1C_2\left(\frac{7}{4} + \pi^2\right)\right].$$
(22)

The first QCD corrections in Eq. (22) stems from semi-leptonic decays, the second and the third term in Eq. (22) stem from non-leptonic decays. It turned out, however, that effects of the charm quark mass are crucial, see, e.g., the estimate in Ref. 33. NLO-QCD corrections with full mass dependence were determined for $b \to c l^- \bar{\nu}$ already in 1983,[34] for $b \to c \bar{u} d$ in 1994,[35] for $b \to c \bar{c} s$ in 1995,[36] for $b \to$ no charm in 1997[37] and for $b \to sg$ in 2000.[38,39] Since there were several misprints in Ref. 36 — leading to IR divergent expressions —, the corresponding calculation was redone in Ref. 19 and the numerical result was updated.[b] With the results in Ref. 19 we predict (using $\bar{m}_c(\bar{m}_b)$ and $\bar{m}_b(\bar{m}_b)$)

$$c_{3,b} = \begin{cases} 9 & (m_c = 0 = \alpha_s) \\ 5.29 \pm 0.35 & \text{(LO-QCD)} \\ 6.88 \pm 0.74 & \text{(NLO-QCD)} \end{cases}.$$
(23)

Comparing this result with Eq. (14) one finds a huge phase space suppression, which reduces the value of $C_{3,b}$ from 9 in the mass less case to about 4.7 when including charm quark mass effect. Switching on in addition QCD effects $c_{3,b}$ is enhanced back to a value of about 6.9. The LO $b \to c$ transitions contribute about 70% to this value, the full NLO-QCD corrections about 24% and the $b \to u$ and penguin contributions about 6%.[19]

For the total lifetime we predict thus

$$\tau_b = (1.65 \pm 0.24)\,\text{ps},$$
(24)

which is our final number for the lifetime of a free b-quark. This number is now very close to the experimental numbers in Eq. (1), unfortunately the uncertainty is still quite large. To reduce this, a calculation at the NNL order would be necessary. Such an endeavour seems to be doable nowadays. The dominant Wilson coefficients C_1 and C_2 are known at NNLO accuracy[40] and the two loop corrections in the effective theory have been determined e.g. in Refs. 41–46 for semi-leptonic decays and partly in Ref. 47 for non-leptonic decays.

With this input we predict the semi leptonic branching ratio (following Ref. 19) to be

$$B_{sl} = (11.6 \pm 0.8)\%,$$
(25)

which agrees well with recent measurements[1,48]

$$B_{sl}(B_d) = (10.33 \pm 0.28)\%.$$
(26)

---

[b]The authors of Ref. 36 left particle physics and it was not possible to obtain the correct analytic expressions. The numerical results in Ref. 36 were, however, correct.

2.2.3. *The HQE*

Now we are ready to derive the heavy quark expansion for inclusive decays.[c] The decay rate of the transition of a B-meson to an inclusive final state X can be expressed as a phase space integral over the square of the matrix element of the effective Hamiltonian sandwiched between the initial B-meson[d] state and the final state X. Summing over all final states X with the same quark quantum numbers we obtain

$$\Gamma(B \to X) = \frac{1}{2m_B} \sum_X \int_{PS} (2\pi)^4 \delta^{(4)}(p_B - p_X)|\langle X|\mathcal{H}_{\text{eff}}|B\rangle|^2. \tag{27}$$

If we consider, e.g., a decay into three particles, i.e. $B \to 1 + 2 + 3$, then the phase space integral reads

$$\int_{PS} = \prod_{i=1}^{3} \left[\frac{d^3 p_i}{(2\pi)^3 2E_i} \right] \tag{28}$$

and $p_X = p_1 + p_2 + p_3$. With the help of the optical theorem the total decay rate in Eq. (27) can be rewritten as

$$\Gamma(B \to X) = \frac{1}{2m_B} \langle B|\mathcal{T}|B\rangle, \tag{29}$$

with the transition operator

$$\mathcal{T} = \text{Im } i \int d^4 x \, T[\mathcal{H}_{\text{eff}}(x)\mathcal{H}_{\text{eff}}(0)], \tag{30}$$

consisting of a non-local double insertion of the effective Hamiltonian.

A second operator-product-expansion, exploiting the large value of the b-quark mass m_b, yields for $\mathcal{T}$

$$\mathcal{T} = \frac{G_F^2 m_b^5}{192\pi^3} |V_{cb}|^2 \left[c_{3,b}\bar{b}b + \frac{c_{5,b}}{m_b^2} \bar{b} g_s \sigma_{\mu\nu} G^{\mu\nu} b + 2\frac{c_{6,b}}{m_b^3} (\bar{b}q)_\Gamma (\bar{q}b)_\Gamma + \cdots \right] \tag{31}$$

and thus for the decay rate

$$\Gamma = \frac{G_F^2 m_b^5}{192\pi^3} |V_{cb}|^2 \left[c_{3,b} \frac{\langle B|\bar{b}b|B\rangle}{2M_B} + \frac{c_{5,b}}{m_b^2} \frac{\langle B|\bar{b}g_s\sigma_{\mu\nu}G^{\mu\nu}b|B\rangle}{2M_B} \right.$$

$$\left. + \frac{c_{6,b}}{m_b^3} \frac{\langle B|(\bar{b}q)_\Gamma(\bar{q}b)_\Gamma|B\rangle}{M_B} + \cdots \right]. \tag{32}$$

The individual contributions in Eq. (32) have the following origin and interpretation:

Leading term in Eq. (32):
To get the first term we contracted all quark lines, except the beauty-quark lines, in the product of the two effective Hamiltonians. This leads to the following two-loop

---

[c]We delay almost all referencing related to the creation of the HQE to Section 3.
[d]The replacements one has to do when considering a D-meson decay are either trivial or we explicitly comment on them.

diagram on the l.h.s., where the circles with the crosses denote the $\Delta B = 1$-operators from the effective Hamiltonian.

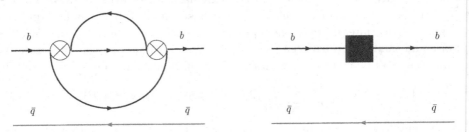

Performing the loop integrations in this diagram we get the Wilson coefficient $c_{3,b}$ that contains all the loop functions and the dimension-three operator $\bar{b}b$, which is denoted by the black square in the diagram on the r.h.s.. This has been done already in Eqs. (19), (22) and (23).

A crucial finding for the HQE was the fact, that the matrix element of the dimension-three operator $\bar{b}b$ can also be expanded in the inverse of the b-quark mass. According to the Heavy Quark Effective Theory (HQET) we get[e]

$$\frac{\langle B|\bar{b}b|B\rangle}{2M_B} = 1 - \frac{\mu_\pi^2 - \mu_G^2}{2m_b^2} + \mathcal{O}\left(\frac{1}{m_b^3}\right), \tag{33}$$

with the matrix element of the kinetic operator μ_π^2 and the matrix element of the chromo-magnetic operator μ_G^2, defined in the B-rest frame as[f]

$$\mu_\pi^2 = \frac{\langle B|\bar{b}(i\vec{D})^2 b|B\rangle}{2M_B} + \mathcal{O}\left(\frac{1}{m_b}\right),$$

$$\mu_G^2 = \frac{\langle B|\bar{b}\frac{g_s}{2}\sigma_{\mu\nu}G^{\mu\nu}b|B\rangle}{2M_B} + \mathcal{O}\left(\frac{1}{m_b}\right). \tag{34}$$

With the above definitions for the non-perturbative matrix-elements the expression for the total decay rate in Eq. (32) becomes

$$\Gamma = \frac{G_F^2 m_b^5}{192\pi^3} V_{cb}^2 \left\{ c_{3,b}\left[1 - \frac{\mu_\pi^2 - \mu_G^2}{2m_b^2} + \mathcal{O}\left(\frac{1}{m_b^3}\right)\right] \right.$$

$$\left. + 2c_{5,b}\left[\frac{\mu_G^2}{m_b^2} + \mathcal{O}\left(\frac{1}{m_b^3}\right)\right] + \frac{c_{6,b}}{m_b^3}\frac{\langle B|(\bar{b}q)\Gamma(\bar{q}b)\Gamma|B\rangle}{M_B} + \cdots \right\}. \tag{35}$$

The leading term in Eq. (35) describes simply the decay of a free quark. Since here the spectator-quark (red) is not involved in the decay process at all, this

---

[e]We use here the conventional relativistic normalisation $\langle B|B\rangle = 2EV$, where E denotes the energy of the meson and V the space volume. In the original literature sometimes different normalisations have been used, which can lead to confusion.
[f]We use here $\sigma_{\mu\nu} = \frac{i}{2}[\gamma_\mu, \gamma_\nu]$. In the original literature sometimes the notation $i\sigma G := i\gamma_\mu\gamma_\nu G^{\mu\nu}$ was used, which differs by a factor of i from our definition of σ.

contribution will be the same for all different b-hadrons, thus predicting the same lifetime for all b-hadrons.

The first corrections are already suppressed by two powers of the heavy b-quark mass — we have no corrections of order $1/m_b$! This non-trivial result explains, why our description in terms of the free b-quark decay was so close to the experimental values of the lifetimes of B-mesons.

In the case of D-mesons the expansion parameter $1/m_c$ is not small and the higher order terms of the HQE will lead to sizable corrections. The leading term $c_{3,c}$ for charm-quark decays gives at the scale $\mu = M_W$ for vanishing quark mass $c_{3,c} = 5$. At the scale $\mu = \bar{m}_c(\bar{m}_c)$ and realistic values of final states masses we get

$$c_{3,c} = \begin{cases} 5 & (m_s = 0 = \alpha_s) \\ 6.29 \pm 0.72 & \text{(LO-QCD)} \\ 11.61 \pm 1.55 & \text{(NLO-QCD)} \end{cases} . \tag{36}$$

Here we have a large QCD enhancement of more than a factor of two, while phase space effects seem to be negligible.

The $1/m_b^2$-corrections in Eq. (35) have two sources: first the expansion in Eq. (33) and the second one — denoted by the term proportional to $c_{5,b}$ — will be discussed below.

Concerning the different $1/m_b^3$-corrections, indicated in Eq. (35), we will see that the first two terms of the expansion in Eq. (32) are triggered by a two-loop diagram, while the third term is given by a one-loop diagram. This will motivate, why the $1/m_b^3$-corrections proportional to $c_{3,b}$ and $c_{5,b}$ can be neglected in comparison to the $1/m_b^3$-corrections proportional to $c_{6,b}$; the former ones will, however, be important for precision determination of semi-leptonic decay rates.[g]

Second term in Eq. (32):

To get the second term in Eq. (32) we couple in addition a gluon to the vacuum. This is denoted by the diagram below, where a gluon is emitted from one of the internal quarks of the two-loop diagram. Doing so, we obtain the so-called chromo-magnetic operator $\bar{b}g_s\sigma_{\mu\nu}G^{\mu\nu}b$, which already appeared in the expansion in Eq. (33).

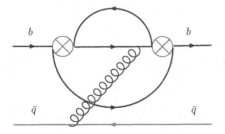
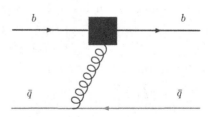

Since this operator is of dimension five, the corresponding contribution is — as seen before — suppressed by two powers of the heavy quark mass, compared to the leading term. The corresponding Wilson coefficient $c_{5,b}$ reads, e.g., for the semi-leptonic decay $b \to c e^- \bar{\nu}_e$ [h] and the non-leptonic decays $b \to c \bar{u} d$ and $b \to c \bar{c} s$

$$c_{5,b}^{ce\bar{\nu}_e} = -(1-z)^4 \left[1 + \frac{\alpha_s}{4\pi} \cdots \right], \tag{37}$$

$$c_{5,b}^{c\bar{u}d} = -|V_{ud}|^2 (1-z)^3 \left[\mathcal{N}_a(\mu)(1-z) + 8 C_1 C_2 + \frac{\alpha_s}{4\pi} \cdots \right], \tag{38}$$

$$c_{5,b}^{c\bar{c}s} = -|V_{cs}|^2 \left\{ \mathcal{N}_a(\mu) \left[\sqrt{1-4z}(1-2z)(1-4z-6z^2) \right. \right.$$

$$\left. + 24 z^4 \log \left(\frac{1+\sqrt{1-4z}}{1-\sqrt{1-4z}} \right) \right] + 8 C_1 C_2 \left[\sqrt{1-4z} \left(1 + \frac{z}{2} + 3z^2 \right) \right.$$

$$\left. \left. - 3z(1-2z^2) \log \left(\frac{1+\sqrt{1-4z}}{1-\sqrt{1-4z}} \right) \right] + \frac{\alpha_s}{4\pi} \cdots \right\}, \tag{39}$$

with the quark mass ratio $z = (m_c/m_b)^2$. For vanishing charm-quark masses and $V_{ud} \approx 1$ we get $c_{5,b}^{c\bar{u}d} = -3$ at the scale $\mu = M_W$, which reduces in LO-QCD to about -1.2 at the scale $\mu = m_b$.

For the total decay rate we have to sum up all possible quark level-decays

$$c_{5,b} = c_{5,b}^{ce\bar{\nu}_e} + c_{5,b}^{c\mu\bar{\nu}_\mu} + c_{5,b}^{c\tau\bar{\nu}_\tau} + c_{5,b}^{c\bar{u}d} + c_{5,b}^{c\bar{c}s} + \cdots. \tag{40}$$

Neglecting penguin contributions we get numerically

$$c_{5,b} = \begin{cases} \approx -9 & (m_c = 0 = \alpha_s) \\ -3.8 \pm 0.3 & (\bar{m}_c(\bar{m}_c), \alpha_s(m_b)) \end{cases}. \tag{41}$$

For $c_{5,b}$ both QCD effects as well as phase space effects are quite pronounced. The overall coefficient of the matrix element of the chromo-magnetic operator μ_G^2 normalised to $2m_b^2$ in Eq. (35) is given by $c_{3,b} + 4 c_{5,b}$, which is sometimes denoted as $c_{G,b}$. For semi-leptonic decays like $b \to c e^- \bar{\nu}_e$, it reads [i]

$$c_{G,b}^{ce\bar{\nu}_e} = c_{3,b}^{ce\bar{\nu}_e} + 4 c_{5,b}^{ce\bar{\nu}_e} = (-3) \left[1 - \frac{8}{3} z + 8z^2 - 8z^3 + \frac{5}{3} z^4 + 4z^2 \ln(z) \right]. \tag{42}$$

For the sum of all inclusive decays we get

$$c_{G,b} = \begin{cases} -27 = -3 c_3 & (m_c = 0 = \alpha_s) \\ -7.9 \approx -1.1 c_3 & (\bar{m}_c(\bar{m}_c), \alpha_s(m_b)) \end{cases}, \tag{43}$$

leading to the following form of the total decay rate

$$\Gamma = \frac{G_F^2 m_b^5}{192 \pi^3} V_{cb}^2 \left[c_{3,b} - c_{3,b} \frac{\mu_\pi^2}{2m_b^2} + c_{G,b} \frac{\mu_G^2}{2m_b^2} + \frac{c_{6,b}}{m_b^3} \frac{\langle B|(\bar{b}q)_\Gamma (\bar{q}b)_\Gamma|B \rangle}{M_B} + \cdots \right]. \tag{44}$$

[h] The result in Eq. (94) of the review[49] has an additional factor 6 in $c_5^{ce\bar{\nu}_e}$.
[i] We differ here slightly from Eq. (7) of Ref. 50, who have a different sign in the coefficients of z^2 and z^3. We agree, however, with the corresponding result in Ref. 25.

Both $1/m_b^2$-corrections are reducing the decay rate and their overall coefficients are of similar size as $c_{3,b}$. To estimate more precisely the numerical effect of the $1/m_b^2$ corrections, we still need the values of μ_π^2 and μ_G^2. Current values[51,52] of these parameters read for the case of B_d and B^+-mesons

$$\mu_\pi^2(B) = (0.414 \pm 0.078) \text{ GeV}^2 \,, \tag{45}$$

$$\mu_G^2(B) \approx \frac{3}{4} \left(M_{B^*}^2 - M_B^2\right) \approx (0.35 \pm 0.07) \text{ GeV}^2 \,. \tag{46}$$

For B_s-mesons only small differences compared to B_d and B^+-mesons are predicted[53]

$$\mu_\pi^2(B_s) - \mu_\pi^2(B_d) \approx (0.08 \ldots 0.10) \text{ GeV}^2 \,, \tag{47}$$

$$\frac{\mu_G^2(B_s)}{\mu_G^2(B_d)} \approx 1.07 \pm 0.03 \,, \tag{48}$$

while sizable differences are expected[53] for Λ_b-baryons.

$$\mu_\pi^2(\Lambda_b) - \mu_\pi^2(B_d) \approx (0.1 \pm 0.1) \text{ GeV}^2 \,, \quad \mu_G^2(\Lambda_b) = 0 \,. \tag{49}$$

Inserting these values in Eq. (44) we find that the $1/m_b^2$-corrections are decreasing the decay rate slightly ($m_b = \bar{m}_b(\bar{m}_b) = 4.248$ GeV):

	B_d	B^+	B_s	Λ_d
$-\dfrac{\mu_\pi^2}{2m_b^2}$	-0.011	-0.011	-0.014	-0.014
$\dfrac{c_{G,b}}{c_{3,b}} \dfrac{\mu_\pi^2}{2m_b^2}$	-0.011	-0.011	-0.011	0.00

$$\tag{50}$$

The kinetic and the chromo-magnetic operator each reduce the decay rate by about 1%, except for the case of the Λ_b-baryon, where the chromo-magnetic operator vanishes. The $1/m_b^2$-corrections exhibit now also a small sensitivity to the spectator-quark. Different values for the lifetimes of b-hadrons can arise due to different values of the non-perturbative parameters μ_G^2 and μ_π^2, the corresponding numerical effect will, however, be small.

$X:$	B^+	B_s	Λ_d
$\dfrac{\mu_\pi^2(X) - \mu_\pi^2(B_d)}{2m_b^2}$	0.000 ± 0.000	0.002 ± 0.000	0.003 ± 0.003
$\dfrac{c_{G,b}}{c_{3,b}} \dfrac{\mu_G^2(X) - \mu_G^2(B_d)}{2m_b^2}$	0.000 ± 0.000	$0.000 \ldots 0.001$	-0.011 ± 0.003

$$\tag{51}$$

Thus we find that the $1/m_b^2$-corrections give no difference in the lifetimes of B^+- and B_d-mesons, they enhance the B_s-lifetime by about 3 per mille, compared to the B_d-lifetime and they reduce the Λ_b-lifetime by about 1% compared to the B_d-lifetime.

To get an idea of the size of these corrections in the charm-system, we first investigate the Wilson coefficient c_5.

$$c_{5,c} = \begin{cases} \approx -5 & (m_c = 0 = \alpha_s) \\ -1.7 \pm 0.3 & (\bar{m}_c(\bar{m}_c), \alpha_s(m_b)) \end{cases}, \tag{52}$$

At the scale $\mu = m_c$ the non-leptonic contribution to c_5 is getting smaller than in the bottom case and it even changes sign. For the coefficient c_G we find

$$c_{G,c} = \begin{cases} \approx -15 = -3c_{3,c} & (m_c = 0 = \alpha_s) \\ 4.15 \pm 1.48 = (0.37 \pm 0.13)c_{3,c} & (\bar{m}_c(\bar{m}_c), \alpha_s(m_b)) \end{cases}. \tag{53}$$

We see for that for the charm case the overall coefficient of the chromo-magnetic operator has now a positive sign and the relative size is less than in the bottom case. For D^0- and D^+-mesons the value of the chromo-magnetic operator reads

$$\mu_G^2(D) \approx \frac{3}{4}\left(M_{D^*}^2 - M_D^2\right) \approx 0.41 \text{ GeV}^2\,, \tag{54}$$

which is of similar size as in the B-system. Normalising this value to the charm quark mass $m_c = \bar{m}_c(\bar{m}_c) = 1.277$ GeV, we get however a bigger contribution compared to the bottom case and also a different sign

$$c_{G,c}\frac{\mu_G^2(D)}{2m_c^2} \approx +0.05c_{3,c}\,. \tag{55}$$

Now the second order corrections are non-negligible, with a typical size of about + 5% of the total decay rate. Concerning lifetime differences of D-mesons, we find no visible effect due to the chromo-magnetic operator[9]

$$\frac{\mu_G^2(D^+)}{\mu_G^2(D^0)} \approx 0.993\,, \quad \frac{\mu_G^2(D_s^+)}{\mu_G^2(D^0)} \approx 1.012 \pm 0.003\,. \tag{56}$$

For the kinetic operator a sizable SU(3) flavour breaking was found by Bigi, Mannel and Uraltsev[53]

$$\mu_\pi^2(D_s^+) - \mu_\pi^2(D^0) \approx 0.1 \text{ GeV}^2\,, \tag{57}$$

leading to an reduction of the D_s^+-lifetime of the order of 3% compared to the D^0-lifetime

$$\frac{\mu_\pi^2(D_s^+) - \mu_\pi^2(D^0)}{2m_c^2} \approx 0.03\,. \tag{58}$$

Third term in Eq. (32):

The next term is obtained by only contracting two quark lines in the product of the two effective Hamiltonian in Eq. (30). The b-quark and the spectator quark of the considered hadron are not contracted. For B_d-mesons ($q = d$) and B_s-mesons ($q = s$) we get the following so-called *weak annihilation* diagram.

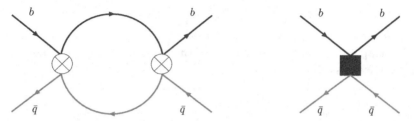

Performing the loop integration on the diagram on the l.h.s. we get the Wilson coefficient c_6 and dimension six four-quark operators $(\bar{b}q)_\Gamma(\bar{q}b)_\Gamma$, with Dirac structures Γ. The corresponding matrix elements of these $\Delta B = 0$ operators are typically written as

$$\langle B|(\bar{b}q)_\Gamma(\bar{q}b)_\Gamma|B\rangle = c_\Gamma f_B^2 M_B B_\Gamma \,, \tag{59}$$

with the bag parameter B_Γ, the decay constant f_B and a numerical factor c_Γ that contains some colour factors and sometimes also ratios of masses.

For the case of the B^+-meson we get a similar diagram, with the only difference that now the external spectator-quark lines are crossed, this is the so-called *Pauli interference* diagram.

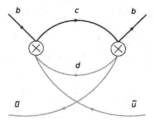

There are two very interesting things to note. First this is now a one-loop diagram. Although being suppressed by three powers of the b-quark mass it is enhanced by a phase space factor of $16\pi^2$ compared to the leading two-loop diagrams. Second, now we are really sensitive to the flavour of the spectator-quark, because in principle, each different spectator quark gives a different contribution.[j] These observations are responsible for the fact that lifetime differences in the system of heavy hadrons are almost entirely due to the contribution of weak annihilation and Pauli interference diagrams.

---

[j]This difference is, however, negligible, if one considers, e.g., B_s vs. B_d.

In the case of the B_d meson four different four-quark operators arise

$$Q^q = \bar{b}\gamma_\mu(1 - \gamma_5)q \times \bar{q}\gamma^\mu(1 - \gamma_5)b\,,$$

$$Q_S^q = \bar{b}(1 - \gamma_5)q \times \bar{q}(1 - \gamma_5)b\,,$$ \hfill (60)

$$T^q = \bar{b}\gamma_\mu(1 - \gamma_5)T^a q \times \bar{q}\gamma^\mu(1 - \gamma_5)T^a b\,,$$

$$T_S^q = \bar{b}(1 - \gamma_5)T^a q \times \bar{q}(1 - \gamma_5)T^a b\,,$$ \hfill (61)

with $q = d$ for the case of B_d-mesons. Q denotes colour singlet operators and T colour octet operators. For historic reasons the matrix elements of these operator are typically expressed as

$$\frac{\langle B_d|Q^d|B_d\rangle}{M_{B_d}} = f_B^2 B_1 M_{B_d}\,, \qquad \frac{\langle B_d|Q_S^d|B_d\rangle}{M_{B_d}} = f_B^2 B_2 M_{B_d}\,, \qquad (62)$$

$$\frac{\langle B_d|T^d|B_d\rangle}{M_{B_d}} = f_B^2 \epsilon_1 M_{B_d}\,, \qquad \frac{\langle B_d|T_S^d|B_d\rangle}{M_{B_d}} = f_B^2 \epsilon_2 M_{B_d}\,. \qquad (63)$$

The bag parameters $B_{1,2}$ are expected to be of order one in vacuum insertion approximation, while the $\epsilon_{1,2}$ vanish in that limit. We will discuss below several estimates of B_i and ϵ_i. Decay constants can be determined with lattice-QCD, see, e.g., the reviews of FLAG[54] or with QCD sum rules, see, e.g., the recent determination in Ref. 55. Later on, we will see, however, that the Wilson coefficients of B_1 and B_2 are affected by sizable numerical cancellations, enhancing hence the relative contribution of the colour suppressed ϵ_1 and ϵ_2. The corresponding Wilson coefficients of the four operators can be written as

$$c_6^{Q^d} = 16\pi^2\left[|V_{ud}|^2 F^u + |V_{cd}|^2 F^c\right]\,, \quad c_6^{Q_S^d} = 16\pi^2\left[|V_{ud}|^2 F_S^u + |V_{cd}|^2 F_S^c\right]\,, \quad (64)$$

$$c_6^{T^d} = 16\pi^2\left[|V_{ud}|^2 G^u + |V_{cd}|^2 G^c\right]\,, \quad c_6^{T_S^d} = 16\pi^2\left[|V_{ud}|^2 G_S^u + |V_{cd}|^2 G_S^c\right]\,. \quad (65)$$

F^q describes an internal $c\bar{q}$ loop in the above weak annihilation diagram. The functions F and G are typically split up in contributions proportional to C_2^2, $C_1 C_2$ and C_1^2:

$$F^u = C_1^2 F_{11}^u + C_1 C_2 F_{12}^u + C_2^2 F_{22}^u\,, \qquad (66)$$

$$F_S^u = \cdots\,. \qquad (67)$$

Next, each of the F_{ij}^q can be expanded in the strong coupling

$$F_{ij}^u = F_{ij}^{u,(0)} + \frac{\alpha_s}{4\pi} F_{ij}^{u,(1)} + \cdots\,, \qquad (68)$$

$$F_{S,ij}^u = \cdots\,. \qquad (69)$$

As an example we give the following LO results

$$F_{11}^{u,(0)} = -3(1 - z)^2\left(1 + \frac{z}{2}\right)\,, \qquad F_{S,11}^{u,(0)} = 3(1 - z)^2(1 + 2z)\,, \qquad (70)$$

$$F_{12}^{u,(0)} = -2(1 - z)^2\left(1 + \frac{z}{2}\right)\,, \qquad F_{S,12}^{u,(0)} = 2(1 - z)^2(1 + 2z)\,, \qquad (71)$$

$$F_{22}^{u,(0)} = -\frac{1}{3}(1-z)^2\left(1+\frac{z}{2}\right), \quad F_{S,22}^{u,(0)} = \frac{1}{3}(1-z)^2(1+2z), \tag{72}$$

$$G_{22}^{u,(0)} = -2(1-z)^2\left(1+\frac{z}{2}\right), \quad G_{S,22}^{u,(0)} = 2(1-z)^2(1+2z), \tag{73}$$

with $z = m_c^2/m_b^2$.

Putting everything together we arrive at the following expression for the decay rate of a B_d-meson

$$\Gamma_{B_d} = \frac{G_F^2 m_b^5}{192\pi^3} V_{cb}^2 \left[c_3 - c_3 \frac{\mu_\pi^2}{2m_b^2} + c_G \frac{\mu_G^2}{2m_b^2} + \frac{16\pi^2 f_B^2 M_{B_d}}{m_b^3} \tilde{c}_6^{B_d} + \mathcal{O}\left(\frac{1}{m_b^3}, \frac{16\pi^2}{m_b^4}\right) \right]$$

$$\approx \frac{G_F^2 m_b^5}{192\pi^3} V_{cb}^2 \left[c_3 - 0.01c_3 - 0.01c_3 + \frac{16\pi^2 f_B^2 M_{B_d}}{m_b^3} \tilde{c}_6^{B_d} + \mathcal{O}\left(\frac{1}{m_b^3}, \frac{16\pi^2}{m_b^4}\right) \right], \tag{74}$$

with

$$\tilde{c}_6^{B_d} = |V_{ud}|^2 \left(F^u B_1 + F_S^u B_2 + G^u \epsilon_1 + G_S^u \epsilon_2 \right)$$

$$+ |V_{cd}|^2 \left(F^c B_1 + F_S^c B_2 + G^c \epsilon_1 + G_S^c \epsilon_2 \right). \tag{75}$$

The size of the third contribution in Eq. (74) is governed by size of $\tilde{c}_6$ and its pre-factor. The pre-factor gives

$$\frac{16\pi^2 f_{B_d}^2 M_{B_d}}{m_b^3} \approx 0.395 \approx 0.05c_3, \tag{76}$$

where we used $f_{B_d} = (190.5 \pm 4.2)$ MeV[54] for the decay constant. If $\tilde{c}_6$ is of order 1, we would expect corrections of the order of 5% to the total decay rate, which are larger than the formally leading $1/m_b^2$-corrections. The LO-QCD expression for $\tilde{c}_6^{B_d}$ can be written as

$$\tilde{c}_6^{B_d} = |V_{ud}|^2(1-z)^2 \left\{ \left(3C_1^2 + 2C_1C_2 + \frac{1}{3}C_2^2 \right) \left[(B_2 - B_1) + \frac{z}{2}(4B_2 - B_1) \right] \right.$$

$$\left. + 2C_2^2 \left[(\epsilon_2 - \epsilon_1) + \frac{z}{2}(4\epsilon_2 - \epsilon_1) \right] \right\}. \tag{77}$$

However, in Eq. (77) several cancellations are arising. In the first line there is a strong cancellation among the bag parameters B_1 and B_2. In vacuum insertion approximation $B_1 - B_2$ is zero and the next term proportional to $4B_2 - B_1$ is suppressed by $z \approx 0.055$. Using the latest lattice determination of these parameters,[56] dating back to 2001,

$$B_1 = 1.10 \pm 0.20, \quad B_2 = 0.79 \pm 0.10,$$

$$\epsilon_1 = -0.02 \pm 0.02, \quad \epsilon_2 = 0.03 \pm 0.01 \tag{78}$$

one finds $B_1 - B_2 \in [0.01, 0.61]$ and $(4B_2 - B_1)z/2 \in [0.07, 0.12]$, so the second contribution is slightly suppressed compared to the first one. Moreover there is an additional cancellation among the $\Delta B = 1$ Wilson coefficients. Without QCD the combination $3C_1^2 + 2C_1 C_2 + \frac{1}{3} C_2^2$ is equal to $1/3$, in LO-QCD this combination is reduced to about 0.05 ± 0.05 at the scale of m_b (varying the renormalisation scale between $m_b/2$ and $2m_b$). Hence B_1 and B_2 give a contribution between 0 and 0.07 to $\tilde{c}_6^{B_d}$, leading thus at most to a correction of about 4 per mille to the total decay rate. This statement depends, however, crucially on the numerical values of the bag parameters, where we are lacking a state-of-the-art determination.

There is no corresponding cancellation in the coefficients related to the colour-suppressed bag parameters $\epsilon_{1,2}$. According to Ref. 56 $\epsilon_2 - \epsilon_1 \in [0.02, 0.08]$, leading to a correction of at most 1.0% to the decay rate. Relying on the lattice determination in Ref. 56 we find that the colour-suppressed operators can be numerical more important than the colour allowed operators and the total decay rate of the B_d-meson can be enhanced by the weak annihilation at most by about 1.4%. The status at NLO-QCD will be discussed below.

The Pauli interference contribution to the B^+-decay rate gives

$$\tilde{c}_6^{B^+} = (1 - z)^2 \left[\left(C_1^2 + 6C_1 C_2 + C_2^2 \right) B_1 + 6 \left(C_1^2 + C_2^2 \right) \epsilon_1 \right] . \tag{79}$$

The contribution of the colour-allowed operator is slightly suppressed by the $\Delta B = 1$ Wilson coefficients. Without QCD the bag parameter B_1 has a pre-factor of one, which changes in LO-QCD to about -0.3. Taking again the lattice values for the bag parameter from Ref. 56, we expect Pauli interference contributions proportional to B_1 to be of the order of about -1.8% of the total decay rate. In the coefficient of ϵ_1 no cancellation is arising and we expect (using again Ref. 56) this contribution to be between 0 and -1.5% of the total decay rate. All in all Pauli interference seems to reduce the total B^+-decay rate by about 1.8% to 3.3%. The status at NLO-QCD will again be discussed below.

In the charm system the pre-factor of the coefficient c_6 reads

$$\frac{16\pi^2 f_D^2 M_D}{m_c^3} \approx \begin{cases} 6.2 \approx 0.6 \, c_3 & \text{for} \quad D^0, \, D^+ \\ 9.2 \approx 0.8 \, c_3 & \text{for} \quad D_s^+ \end{cases} , \tag{80}$$

where we used $f_{D^0} = (209.2 \pm 3.3)$ MeV and $f_{D_s^+} = (248.3 \pm 2.7)$ MeV [54] for the decay constants. Depending on the strength of the cancellation among the $\Delta C = 1$ Wilson coefficients and the bag parameters, large corrections seem to be possible now: In the case of the weak annihilation the cancellation of the $\Delta C = 1$ Wilson coefficients seems to be even more pronounced than at the scale m_b. Thus a knowledge of the colour-suppressed operators is inalienable. In the case of Pauli interference no cancellation occurs and we get values for the coefficient of B_1, that are smaller than -1 and we get a sizable, but smaller contribution from the colour-suppressed operators. Unfortunately there is no lattice determination of the $\Delta C = 0$ matrix elements available, so we cannot make any final, profound statements about the status in the charm system. Numerical results for the NLO-QCD case will also be discussed below.

Fourth term in Eq. (32):

If one takes in the calculation of the weak annihilation and Pauli interference diagrams also small momenta and masses of the spectator quark into account, one gets corrections that are suppressed by four powers of m_b compared to the free-quark decay. These dimension seven terms are either given by four-quark operators times the small mass of the spectator quark or by a four quark operator with an additional derivative. An example is the following $\Delta B = 0$ operator

$$P_3 = \frac{1}{m_b^2} \bar{b}_i \overleftarrow{D}_\rho \gamma_\mu (1 - \gamma_5) D^\rho d_i \times \bar{d}_j \gamma^\mu (1 - \gamma_5) b_j. \tag{81}$$

These operators have currently only been estimated within vacuum insertion approximation. However, for the corresponding operators appearing in the decay rate difference of neutral B-meson first studies with QCD sum rules have been performed.[57,58]

Putting everything together we arrive at the *Heavy-Quark Expansion* of decay rates of heavy hadrons

$$\Gamma = \Gamma_0 + \frac{\Lambda^2}{m_b^2}\Gamma_2 + \frac{\Lambda^3}{m_b^3}\Gamma_3 + \frac{\Lambda^4}{m_b^4}\Gamma_4 + \cdots, \tag{82}$$

where the expansion parameter is denoted by Λ/m_b. From the above explanations it is clear that Λ is not simply given by Λ_{QCD} — the pole of the strong coupling constant — as stated often in the literature. Very naively one expects Λ to be of the order of Λ_{QCD}, because both denote non-perturbative effects. The actual value of Λ, has, however, to be determined by an explicit calculation for each order of the expansion separately. At order $1/m_b^2$ one finds that Λ is of the order of μ_π or μ_G, so roughly below 1 GeV. For the third order Λ^3 is given by $16\pi^2 f_B^2 M_B$ times a numerical suppression factor, leading to values of Λ larger than 1 GeV. Moreover, each of the coefficients Γ_j, which is a product of a perturbatively calculable Wilson coefficient and a non-perturbative matrix element, can be expanded in the strong coupling

$$\Gamma_j = \Gamma_j^{(0)} + \frac{\alpha_s(\mu)}{4\pi}\Gamma_j^{(1)} + \frac{\alpha_s^2(\mu)}{(4\pi)^2}\Gamma_j^{(2)} + \cdots. \tag{83}$$

Before we apply this framework to experimental observables, we would like to make some comments of caution.

A possible drawback of this approach might be that the expansion in the inverse heavy quark mass does not converge well enough — advocated under the labelling *violation of quark hadron duality*. There is a considerable amount of literature about theoretical attempts to prove or to disprove duality, but all of these attempts have to rely on strong model assumptions.

Kolya published some general investigations of quark hadron duality violation in Refs. 59, 60 and some investigations within the two dimensional 't Hooft model,[61,62] that indicated the validity of quark hadron duality. Other investigations in that direction were e.g. performed by Grinstein and Lebed in 1997[63] and 1998[64] and by Grinstein in 2001.[65,66] In our opinion the best way of tackling this question is to confront precise HQE-based predictions with precise experimental data. An especially well suited candidate for this problem is the decay $b \to c\bar{c}s$, which is CKM dominant, but phase space suppressed. The actual expansion parameter of the HQE is in this case not $1/m_b$ but $1/(m_b\sqrt{1-4z})$; so violations of duality should be more pronounced. Thus a perfect observable for testing the HQE is the decay rate difference $\Delta\Gamma_s$ of the neutral B_s mesons, which is governed by the $b \to c\bar{c}s$ transition. The first measurement of this quantity in 2012 and several follow-up measurements are in perfect agreement with the HQE prediction and exclude thus huge violations of quark hadron duality, see Ref. 67 and the discussion below.

2.3. *Overview of observables*

In this section we give a brief overview of observables, whose experimental values can be compared with HQE predictions. As we have discussed above, the general expression for the lifetime ratio of two heavy hadrons H_1 and H_2 reads

$$\frac{\tau(H_1)}{\tau(H_2)} = 1 + \frac{\mu_\pi^2(H_1) - \mu_\pi^2(H_2)}{2m_b^2} + \frac{c_G}{c_3}\frac{\mu_G^2(H_2) - \mu_G^2(H_1)}{2m_b^2}$$

$$+ \frac{c_6(H_2)}{c_3}\frac{\langle H_2|Q|H_2\rangle}{m_b^3 M_B} - \frac{c_6(H_1)}{c_3}\frac{\langle H_1|Q|H_1\rangle}{m_b^3 M_B} + \mathcal{O}\left(\frac{\Lambda^4}{m_b^4}\right), \qquad (84)$$

where we have used the HQE expression for Γ_1 and expanded the ratio consistently in $1/m_b$. Another possibility would be to use the experimental value for the lifetime τ_1 of the hadron H_1 and the relation $\Gamma_1 = 1/\tau_1$ to express the decay rate Γ_1. This gives

$$\frac{\tau(H_1)}{\tau(H_2)} = 1 + \frac{G_F^2 m_b^3}{384\pi^3}V_{cb}^2\tau_1\left[c_3\left(\mu_\pi^2(H_1) - \mu_\pi^2(H_2)\right) + c_G\left(\mu_G^2(H_2) - \mu_G^2(H_1)\right)\right]$$

$$+ \frac{G_F^2 m_b^2}{192\pi^3}V_{cb}^2\tau_1\left[\frac{c_6(H_2)\langle H_2|Q|H_2\rangle - c_6(H_1)\langle H_1|Q|H_1\rangle}{M_B} + \mathcal{O}\left(\frac{\Lambda}{m_b}\right)\right].$$

$$(85)$$

Both methods yield similar numerical results. The relative difference of them is given by the deviation of the b-lifetime prediction in Eq. (24) from the measured lifetime:

$$\delta = \frac{1.65 \text{ ps}}{1.519 \text{ ps}} = 1.086. \qquad (86)$$

Switching between the two methods will change the relative size of the HQE-corrections by 9%. This intrinsic uncertainty has to be kept in mind for error estimates; it could be reduced by an NNLO-QCD calculation of c_3.

We will discuss the following classes of lifetime ratios:

- In the case of B-mesons, there are two well-measured ratios $\tau(B_s)/\tau(B_d)$ and $\tau(B^+)/\tau(B_d)$. We have an almost perfect cancellation in the first ratio, therefore this clean ratio can be used to search for new physics effects, see, e.g., Refs. 68, 69. The second ratio is dominated by Pauli interference.
- Concerning b-baryons, we expect some visible $1/m_b^2$- and $1/m_b^3$-corrections. Until recently only the Λ_b lifetime was studied experimentally. In 2014 also more precise numbers for the Ξ_b-baryons became available[3,4] and we can study now ratios like $\tau(\Lambda_b)/\tau(B_d)$ and $\tau(\Xi_b^+)\tau(\Xi_b^0)$.
- The B_c-meson is quite different from the above discussion, because now both constituent quarks have large decay rates and we have simultaneously an expansion in $1/m_b$ and in $1/m_c$.
- The ratio of D-meson lifetimes is similar to the ones of B-mesons. The big issue is here simply if the HQE shows any convergence at all in $\tau(D_s^+)/\tau(D^0)$ and $\tau(D^+)/\tau(D^0)$.

Decay rate differences $\Delta\Gamma$ of neutral mesons can determined by a very similar HQE approach as discussed above, see, e.g., Ref. 67 for an introduction into mixing. The general expressions for the mixing contribution Γ_{12} starts at order $1/m_b^3$ and it can be written as

$$\Gamma_{12}^q = \left(\frac{\Lambda}{m_b}\right)^3 \Gamma_3 + \left(\frac{\Lambda}{m_b}\right)^4 \Gamma_4 + \cdots . \qquad (87)$$

In the mixing sector we get the following observables:

- In the neutral B-meson system $\Delta\Gamma_q$ denotes the difference of the total decay rates of the heavy (H) mesons eigenstate and the light (L) eigenstate. They are extracted from Γ_{12} via the relations

$$\Delta\Gamma_d = \Gamma_L^d - \Gamma_H^d = 2|\Gamma_{12}^d|\cos\phi_d, \quad \Delta\Gamma_s = \Gamma_L^s - \Gamma_H^s = 2|\Gamma_{12}^s|\cos\phi_s, \qquad (88)$$

with the mixing phase defined as $\phi_q = \arg(-M_{12}^q/\Gamma_{12}^q)$. Related quantities, that also rely on the HQE for Γ_{12} are the so-called semi-leptonic asymmetries

$$a_{sl}^d = \left|\frac{\Gamma_{12}^d}{M_{12}^d}\right|\sin\phi_d, \quad a_{sl}^s = \left|\frac{\Gamma_{12}^s}{M_{12}^s}\right|\sin\phi_d, \qquad (89)$$

that were already discussed in 1987 by Bigi, Khoze, Uraltsev and Sanda[70] and even earlier in Refs. 71–73.

- In the case of neutral D-mesons the expression of the decay rate difference $\Delta\Gamma_D$ in terms of Γ_{12} and M_{12} is more complicated, than in the case of B-mesons. Here, typically the quantity y is discussed

$$y = \frac{\Delta\Gamma_D}{2\Gamma_D} . \qquad (90)$$

Before comparing recent data with HQE predictions, we will do some historical investigations of the origin of the HQE.

3. A Brief History of Lifetimes and the HQE

We give here a brief history of the theoretical investigations of lifetimes of heavy hadrons and the heavy quark expansion. We do not discuss the development of the Heavy Quark Effective Theory (HQET), which happened in the late 1980s and early 1990s. We also concentrate on total decay rates, thus leaving out many of the important contributions to the theory of semi-leptonic decays.

Heavy hadrons were discovered as J/ψ-states in 1974.[74,75] At about that time the first investigations of weak decays of heavy hadrons started. We structure the theoretical development in three periods: pioneering studies, systematic studies and precision studies. It is of course quite arbitrary, where the exact borders between these periods are drawn.

3.1. *Pioneering studies*

Here we summarise the first investigations of heavy meson decays, without having a systematic expansion at hand.

- According to Kolya (see, e.g., Ref. 53)[k] the first time, that heavy flavour hadrons have been described asymptotically by a free quark decay was in 1973 by Nikolaev.[76] The charm-quark decay as the dominant contribution to D-meson decays was considered, e.g., in 1974/5 by Gaillard, Lee and Rosner,[77] by Kingsley, Treiman, Wilczek and Zee,[78] by Ellis, Gaillard and Nanopoulos[79] and by Altarelli, Cabibbo and Maiani.[80] In Ref. 79 the total lifetime of the charm meson was calculated to be about 0.5 ps^{-1}, by taking only the LO-QCD value of c_3 with vanishing internal quark masses into account.

- Pauli interference was introduced in 1979 by Guberina, Nussinov, Peccei and Rückl.[81] Without having any systematic expansion at hand these authors found

$$\frac{\tau(D^+)}{\tau(D^0)}^{\text{PI 1979}} = \frac{c_-^2 + 2c_+^2 + 2}{4c_+^4 + 2} = \frac{\mathcal{N}_a + 2}{\mathcal{N}_a + 2 + (C_1^2 + 6C_1 C_2 + C_2^2)}. \qquad (91)$$

This result can be obtained from our formulae by the following modifications:

— For the D^0 decay rate only Γ_0, i.e., only the free quark decay, is taken into account in LO-QCD and with vanishing internal quark masses, i.e., no $1/m_c^2$- and $1/m_c^3$-corrections are considered.

— For the D^+ decay rate only Γ_0 and the Pauli interference in Γ_3 are taken into account in LO-QCD and with vanishing internal quark masses. Since at that time no systematic expansion was available, the contributions were simply

---

[k]In an email from 4.11.2012 Kolya wrote to me: *The present generation may not appreciate how nontrivial (or even heretic) such a proposition could sound that time! It was the era of traditional hadron physics where descriptions like Veneziano model or Regge theory were assumed to underlie hadrons, and their common (indisputable) feature was soft interactions leading to exponential suppression of any form factor . . . .*

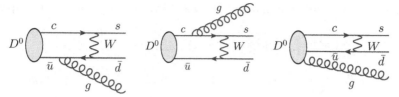

Fig. 1.   Gluon emission from the weak annihilation diagram.

added. This corresponds to making the following replacements in our formulae: $(4\pi f_D)^2 \approx (2.63 \text{ GeV})^2 \to m_c^2$ and $M_D \approx m_c$, which is of course very crude and more importantly not really justified. In addition the bag parameters were used in vacuum insertion approximation, i.e., $B = 1$ and $\epsilon = 0$.

With modern inputs Eq. (91) gives a value of about 1.5, while the authors obtained with input parameters from 1979 and without using the renormalisation group for the $\Delta C = 1$ Wilson coefficients a ratio of about 10. It is also quite interesting to note Fig. 1 of Ref. 81, which presents the leading $\bar{c}c$-term, weak annihilation and Pauli interference. Further studies of Pauli interference were done slightly later in, e.g., Ref. 82.

• Weak annihilation suffers from chirality suppression, thus it was proposed in 1979 by Bander, Silverman and Soni[83] and also by Fritzsch and Minkowski[84] and by Bernreuther and Nachtmann and Stech[85] to consider gluon emission from the ingoing quark lines in order to explain the large lifetime ratio in the charm system, see Fig. 1. This yields a large contribution proportional to $f_D^2/\langle E_{\bar{q}}^2 \rangle$, where $f_D \approx 200 \text{MeV}$ is the D meson decay constant and $\langle E_{\bar{q}} \rangle$ denotes the average energy of the initial anti-quark. Thus the one-gluon emission weak annihilation seems to be not suppressed at all, compared to the leading free-quark decay. In Ref. 85, the authors additionally included the Cabibbo-suppressed weak annihilation of D^+ and obtained for the effects of weak annihilation in D^0 and D^+

$$\frac{\tau(D^+)}{\tau(D^0)}^{\text{WA 1980}} \approx 5.6 - 6.9 \,. \tag{92}$$

One should keep in mind, that Pauli interference, which is now known to be the dominant effect, is still neglected here. Comparing with the experimental numbers in the Introduction, one sees what a severe overestimation these early analyses, that did not allow for any power-counting, were. If the arguments of Refs. 83–85 were correct, then no systematic HQE would be possible — we come back to this point below.

• More systematic studies and further investigations of the Pauli interference effect can be found in Refs. 86–88. The following formula — Eq. (93) — was first derived by Shifman and Voloshin and presented several years later in the review of Khoze and Shifman from 1983.[86] It was pointed out in February 1984 by Bilic that in the original version there was a sign error, which was corrected in the same year[87] by

Shifman and Voloshin and shortly afterwards by Bilic, Guberina, Trampetic,[88]

$$\Gamma(D^+) = \frac{G_F^2}{2M_D} \langle D^+ | \frac{m_c^5}{64\pi^3} \frac{2C_+^2 + C_-^2}{3} \bar{c}c$$

$$+ \frac{m_c^2}{2\pi} \left[\left(C_+^2 + C_-^2 \right) (\bar{c}\Gamma_\mu T^A d)(\bar{d}\Gamma_\mu T^A c) + \frac{2C_+^2 - C_-^2}{3} (\bar{c}\Gamma_\mu d)(\bar{d}\Gamma_\mu c) \right] |D^+\rangle.$$

$$(93)$$

We have rewritten the original expression in the colour-singlet and colour-octet basis commonly used today for $\Delta C = 0$ operators. In order to compare easier with our formulae we can switch from the C_+, C_--basis to the C_1, C_2-basis

$$2C_+^2 + C_-^2 = \mathcal{N}_a, \tag{94}$$

$$C_+^2 + C_-^2 = 2(C_1^2 + C_2^2), \tag{95}$$

$$2C_+^2 - C_-^2 = C_1^2 + 6C_1C_2 + C_2^2. \tag{96}$$

Neglecting weak annihilation, the total decay rate for D^0 is given by the first term in Eq. (93). For the bag parameters vacuum insertion approximation is used. In early analyses the lifetime ratios were generally underestimated

$$\frac{\tau(D^+)}{\tau(D^0)}^{\text{HQE 1984}} \approx 1.5, \tag{97}$$

which was mainly due to a too small estimate for the decay constant $f_D \approx 160 - 170$ MeV. The present value[54] of $f_D = 209.2$ MeV yields $\tau(D^+)/\tau(D^0) \approx 2.2$, which drastically improves the consistency with experiments. To some extent Eq. (93) given in Refs. 86, 87 can be seen as a starting point for a systematic expansion in the inverse of the heavy quark mass.

• In 1986[89] Shifman and Voloshin considered for the first time the effects of hybrid renormalisation, coming thus much closer to the present state of theory predictions for the ratio of D^+ and D^0 lifetimes. Moreover they predicted[89]

$$\frac{\tau(B_s)}{\tau(B_d)}^{\text{HQE 1986}} \approx 1, \quad \frac{\tau(B^+)}{\tau(B_d)}^{\text{HQE 1986}} \approx 1.1, \quad \frac{\tau(\Lambda_b)}{\tau(B_d)}^{\text{HQE 1986}} \approx 0.96, \tag{98}$$

which is amazingly close to current experimental values.

In Refs. 87, 89, it was also argued, that $\tau(D_s^+) \approx \tau(D^0)$, which contradicted the experimental situation at that time. In 1986 it was further shown by Guberina, Rückl and Trampetic[90] that the HQE was able to correctly reproduce the hierarchy of lifetimes in the charm sector

$$\text{HQE 1986}: \quad \tau(D^+) > \tau(D^0) > \tau(\Xi_c^+) > \tau(\Lambda_c^+) > \tau(\Xi_c^0) > \tau(\Omega_c^0). \tag{99}$$

In 1986 Khoze, Shifman, Uraltsev and Voloshin[91] refined the analysis of Refs. 86, 87, by taking into account weak annihilation and B-mixing. In particular they

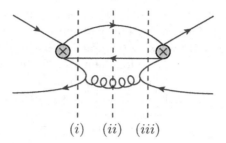

$$(i) \quad (ii) \quad (iii)$$

Fig. 2. Different cuts contributing to the weak annihilation. The $f_D^2/\langle E_{\bar{u}}\rangle^2$ enhanced term due to the cut (ii) considered in Refs. 83–85 is cancelled by interference effects (i) and (iii), such that the fully inclusive rate experiences the correct $1/m_c^3$ scaling behaviour predicted by the HQE.

found that the decay rate difference in the neutral B_s-system may be sizable

$$\frac{\Delta \Gamma_s}{\Gamma_s}^{\text{HQE 1986}} \approx 0.07 \left(\frac{f_{B_s}}{(130 \text{ MeV})}\right)^2 \approx 0.22 , \qquad (100)$$

where we inserted the most recent FLAG-average[54] for the decay constant. The authors of Ref. 91 emphasised also that the weak annihilation effects suggested in Refs. 83–85 formally leads to huge corrections in the $1/m_q$-expansion, which spoils a systematic expansion. This problem somehow stopped[l] further work in that direction until the issue was settled in January 1992 by Bigi and Uraltsev.[93]

3.2. *Systematic studies*

Here we describe the development of the HQE in its current form.

- For inclusive semi-leptonic decays, where the above issue was not severe, it was shown already in 1990 by Chay, Georgi and Grinstein,[94] that in an expansion in inverse powers of the heavy quark mass no $1/m_q$-corrections are appearing and therefore a consistent, systematic expansion seemed to be in reach for these decays.[m]
- In 1992 Bigi and Uraltsev[93] explained the apparent contradiction between the $1/m_q$ scaling of the HQE and the $f_D^2/\langle E_{\bar{q}}\rangle^2$ enhanced gluon bremsstrahlung of Refs. 83–85. They showed, that these power-enhanced terms cancel in fully inclusive rates between different cuts as indicated in Fig. 2 and pre-asymptotic effects hence scale with $1/m_c^3$, consistently with the HQE. This seminal work opened now the way for the HQE in its current form. The explicit proof of the cancellation of all power-enhanced terms was done in 1998 by Beneke, Buchalla, Greub, Lenz and Nierste,[97] in the context of the calculation of $\Gamma_3^{(1)}$ for $\Delta \Gamma_s$.

---

[l]Blok and Shifman stated in 1992:[92] *"Probably for this reason the problem of pre-asymptotic corrections in the inclusive widths has been abandoned for many years."*
[m]The famous Luke's Theorem[95] was proven in the context of the HQET. This theorem can be considered as a generalisation of the Ademollo-Gatto theorem from 1964.[96]

- The HQE in its current form was written down in July 1992 in Ref. 98 by Bigi, Uraltsev and Vainshtein for semi-leptonic and non-leptonic decays with one heavy quark in the final state. By working out the expansion in Eq. (33) the absence of $1/m_q$-corrections was shown also for the non-leptonic decays. In addition $\Gamma_2^{(0)}$ was determined with the inclusion of charm mass effects, i.e., the values of the Wilson coefficient c_5 given in Eq. (37) and Eq. (38). In the original paper there are some misprints, that were partly corrected[n] in an erratum. The full set of correct formulae was given end of 1992 by Bigi, Blok, Shifman, Uraltsev and Vainshtein in Ref. 99. In these two papers[98,99] a different normalisation was used for the physical meson states than we did in Eq. (33).

  At about the same time Blok and Shifman investigated the rule of discarding terms of order $1/N_c$ in inclusive $b \to c\bar{u}d$- and $c \to s\bar{d}u$-decays,[92] as well as in the $b \to c\bar{c}s$-decay.[22] In that respect they also determined the $1/m_b^2$-corrections for inclusive non-leptonic decays. More precisely they determined the contribution of $c_{5,b}^{c\bar{u}d}$ proportional to $C_1 C_2$ — see Eq. (38) — in Ref. 92 and the contribution of $c_{5,b}^{c\bar{u}d}$ proportional to $C_1 C_2$ — see Eq. (39) — in Ref. 22.

  The complete formulae for the case of two heavy particles in the final state with identical masses, e.g., $b \to c\bar{c}s$ — see Eq. (39) — are given in December 1993 by Bigi, Blok, Shifman and Vainshtein and in January 1994 in a book contribution of Bigi, Blok, Shifman, Uraltsev and Vainshtein from 1994.[24] In these papers now the same normalisation for the meson states as in Eq. (33) is used. The case for two arbitrary masses was studied by Falk, Ligeti, Neubert and Nir[25] in 1994.

- Now the door was open for many phenomenological investigations, which led also to several challenges for the new theory tool:

  — Inclusive non-leptonic decays were considered by Palmer and Stech in May 1993.[100] It turned out that the theory prediction for the decay $b \to c\bar{c}s$ did not fit to the data. Related investigations of the *missing charm puzzle* and the inclusive semi-leptonic branching ratio were done in November 1993 by Bigi, Blok, Shifman and Vainshtein[23] (*The baffling semi-leptonic branching ratio.*) In May 1994 it was suggested by Dunietz, Falk and Wise[101] (*Inconclusive inclusive nonleptonic B decays*) that this discrepancy points towards a violation of local quark hadron duality in the decay $b \to c\bar{c}s$ — a suggestion, which is now ruled out by the 2012 measurement of $\Delta\Gamma_s$, which is in perfect agreement with the HQE prediction, see below. Moreover the current theory prediction for the semi-leptonic branching ratio in Eq. (25) agrees well with the experimental numbers given in Eq. (26), although there is still some space for deviations.

  — An early extraction of V_{cb}, m_c and m_b was done in 1993 by Luke and Savage[50] and by Bigi and Uraltsev[102] and further in 1995 by Falk, Luke and Savage.[103] This kind of studies form a big industry now, see, e.g., the review about the determination of V_{cb} and V_{ub} by Kowalewski and Mannel in the PDG.[1]

---

[n]In Eq. (4) of the erratum the factors m_Q^2 should be in the denominator instead of the numerator.

— Bigi and Uraltsev applied the HQE to charm lifetimes in Ref. 102 and also some aspects in Ref. 104. For the D_s^+ meson they found

$$\frac{\tau(D_s^+)}{\tau(D^0)}^{\text{HQE1994}} = 0.9 - 1.3 \,, \tag{101}$$

where the uncertainty dominantly arises from the weak annihilation.

— Lifetimes of b-hadrons were also further studied. In that respect the expressions for $\Gamma_3^{(0)}$ with charm quark mass dependence were presented by Kolya Uraltsev[105] in 1996° and slightly later by Neubert and Sachrajda.[106] This mass dependence turned out to be important. Moreover the inclusion of colour-suppressed four quark operators was found to be crucial. Neubert and Sachrajda[106] gave a very nice and comprehensive review of the status quo in 1996 for the different b-hadron lifetime predictions in LO-QCD. At that time the measured Λ_b-lifetime was in conflict with early HQE predictions, that predicted a value of around 1.5 ps, see Eq. (98). The old data[107–110] pointed, however, more to values around $1.0 - 1.3$ ps.

Year	Exp	Decay	$\tau(\Lambda_b)$ [ps]	$\tau(\Lambda_b)/\tau(B_d)$
1998	OPAL	$\Lambda_c l$	1.29 ± 0.25	0.85 ± 0.16
1997	ALEPH	$\Lambda_c l$	1.21 ± 0.11	0.80 ± 0.07
1995	ALEPH	$\Lambda_c l$	1.02 ± 0.24	0.67 ± 0.16
1992	ALEPH	$\Lambda_c l$	1.12 ± 0.37	0.74 ± 0.24

$$\tag{102}$$

Neubert and Sachrajda concluded that this points — if the experimental values stay — either to anomalously large matrix elements (they will be discussed below) or to a violation of quark-hadron duality. The latter attitude was quite popular at that time, see, e.g., the paper by Altarelli, Martinelli, Petrarca and Rapuano from 1996[111] or the paper from Cheng from 1997[112] or the work from Ito, Matsuda and Matsui also from 1997.[113] Nowadays we know that the Λ_b-lifetime was a purely experimental problem and the measured values are in good agreement with the HQE estimates. These estimates suffer, however, from sizable hadronic uncertainties, which could be reduced by a state of the art lattice calculation.

— The lifetime of the B_c-meson, where both the b- and the c-quark decay weakly was studied systematically in 1996 by Beneke and Buchalla.[114]

°In an email dated from 4.11.2012, Kolya claimed that these results were known since a long time: "*Effects of the internal quark masses were in fact considered; the expressions were at hand, and plugging numbers were so a simple matter that this was not even noted specially. The expressions (they are given, for instance, in arXiv:hep-ph/9602324) were taken from the same mid-1980s notes I mentioned above.*"

3.3. *Precision studies*

Some motivation and some topics of precision studies can be found already in the recommendations given in the seminal 1992 paper:[98] "*The general procedure outlined above can be improved in four respects*:

(i) *Some of the numerical predictions stated above were tentative since not all the relevant calculations have been performed yet. Since the "missing" computations involve perturbation theory this presents "merely" a technical delay.*

(ii) *The real accuracy obtainable in this approach can be determined by calculating terms of order $1/m_Q^4$ and estimating the size of the relevant matrix elements.*

(iii) *..."*

Item (i) concerns the field of determining higher order QCD-corrections. After being involved in several NLO-QCD calculations within the HQE, I of course disagree with the use of the word "*merely*" above. Besides being a tedious task, these efforts had also a conceptual value, since they provided an explicit proof of the arguments for a cancellation of singularities due to quark thresholds given by Bigi and Uraltsev.[93] Item (ii) suggests the discussion of higher order terms in HQE, which has been done currently for many observables, see below. Another crucial topic, that was, however, not emphasised in Ref. 98, is the non-perturbative determination of the arising matrix elements.

3.3.1. *NLO-QCD*

For semi-leptonic decays the NLO QCD corrections in $\Gamma_2^{(1)}$ proportional to μ_π^2 were determined in 2007 by Becher, Boos and Lunghi[115] and confirmed in 2012.[116] The corresponding corrections proportional to μ_G^2 were calculated very recently by Alberti, Gambino and Nandi.[117]

As discussed above, the NLO QCD corrections in $\Gamma_3^{(1)}$ were crucial for proofing the consistency of the HQE. They were determined for $\Delta\Gamma_s$ in 1998 by Beneke, Buchalla, Greub, Lenz and Nierste.[97] In this case the diagrams in Fig. 3 appear. The same authors as well as the Rome group — Franco, Lubicz, Mescia and Tarantino — calculated the NLO-QCD corrections for $\tau(B+)/\tau(B_d)$ in 2002.[20,118] The Rome group included also the NLO-corrections for $\tau(B_s)/\tau(B_d)$ and partly for $\tau(\Lambda_d)/\tau(B_d)$[118] — here some penguin diagrams are still missing. Some dominant NLO-corrections for $\tau(B_s)/\tau(B_d)$ have already been determined in 1998 by Keum and Nierste.[119] In Ref. 20 it was also shown that the use of $\bar{m}_c(\bar{m}_b)$ automatically sums up logarithms of the form $\alpha_s^n z \log^n z$ to all orders. For $\Delta\Gamma_d$ and the semi-leptonic asymmetries a_{sl}^d and a_{sl}^s, $\Gamma_3^{(1)}$ was determined in 2003 by Beneke, Buchalla, Lenz and Nierste[120] and by Ciuchini, Franco, Lubicz, Mescia and Tarantino.[121] For the decay rate difference of neutral D-mesons the above formulae were rewritten in 2010 by Bobrowski, Lenz, Riedl and Rohrwild[122] and for D-meson lifetime ratios some missing contributions were calculated in 2013 by Lenz and Rauh.[9]

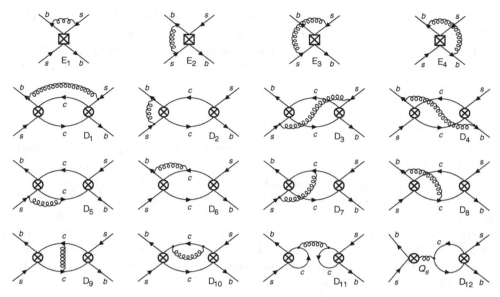

Fig. 3. NLO-QCD diagrams contributing to $\Delta\Gamma_s$.

A pioneering studies of some integrals that appear in NNLO-QCD for $\Delta\Gamma_s$ has been performed in 2012 by Asatrian, Hovhannisyan and Yeghiazaryan.[123]

3.3.2. *Higher order terms in the OPE*

For semi-leptonic decays $1/m_b$ corrections to the kinetic and chromo-magnetic operator were studied in 1994 by Bigi, Shifman, Uraltsev and Vainshtein.[14] Similar contributions to semi-leptonic decays were studied in 1995 by Blok, Dikeman and Shifman[124] and 1996 by Kremm and Kapustin.[125] Even higher corrections — $\Gamma_4^{(0)}$ and $\Gamma_5^{(0)}$ — to the semi-leptonic decay width were investigated in 2010 by Mannel, Turczyk and Uraltsev.[126]

$1/m_q$-corrections to weak annihilation and Pauli interference, i.e., $\Gamma_4^{(0)}$ were determined for $\Delta\Gamma_s$ in 1996 by Beneke, Buchalla and Dunietz[127] and they turned out to be sizable. The corresponding corrections for $\Delta\Gamma_d$ were calculated in 2001 by Dighe, Hurth, Kim and Yoshikawa[128] and for b-lifetimes in by Gabbiani, Onishchenko and Petrov in 2003[129] and 2004[130] (in the latter one also $1/m_q^2$-corrections were investigated) and by Lenz and Nierste in 2003.[131] Badin, Gabbiani and Petrov studied also $\Gamma_5^{(0)}$ for $\Delta\Gamma_s$ in 2007.[132] In $\Gamma_5^{(0)}$ several completely unknown matrix elements are arising. Moreover the Wilson coefficients have very small numerical values. Thus we are not including these corrections in our estimates.

One can also try to apply the methods of the HQE to D-mixing. First efforts in that direction were made in 1992 by Georgi[133] and by Ohl, Ricciardi and Simmons.[134] It turns out that the leading term, Γ_3, suffers from a severe GIM cancellation and thus the HQE leads to very small predictions for D-mixing. One idea to circumvent this severe cancellation was to consider higher orders in the HQE, in particular Γ_6 and Γ_9. Bigi and Uraltsev have shown in 2000[135] how in Γ_6 and Γ_9 the $1/m_c$-suppression could be overcompensated by a lifting of the GIM-suppression. They concluded that values of x and y of up to 1% are not excluded within the HQE.

3.3.3. *Non-perturbative parameters*

Early studies of μ_π^2 have been done, e.g., in 1993 by Bigi, Shifman, Uraltsev and Vainshtein.[136] In 1994[137] some ideas how to extract this quantity from experiment were developed by Bigi, Grozin, Shifman, Uraltsev and Vainshtein. The same quantity has also been determined with QCD sum rules in 1993 by Ball and Braun.[138] A kind of contradicting result was obtained in 1996 by Neubert[139] with the same method. Calculations within lattice QCD were, e.g., performed by Kronfeld and Simone in 2000.[140] The most recent value for μ_π^2 for B-mesons comes from a fit of semi-leptonic decays by Gambino and Schwanda in 2013,[51] somehow confirming the first QCD sum rule calculation:

$$
\begin{array}{|l|l|}
\hline
\mu_\pi^2 & \\
\hline
(0.52 \pm 0.12)\ \mathrm{GeV}^2 & \text{QCD-SR1993} \\
\hline
(0.10 \pm 0.05)\ \mathrm{GeV}^2 & \text{QCD-SR 1996} \\
\hline
(0.45 \pm 0.12)\ \mathrm{GeV}^2 & \text{Lattice 2000} \\
\hline
(0.414 \pm 0.078)\ \mathrm{GeV}^2 & \text{Fit 2013} \\
\hline
\end{array}
\tag{103}
$$

μ_G^2 can in principle be determined from experiment — see Eq. (46) —, for B-mesons it was further investigated by Kolya in 2001.[52] The differences of μ_G^2 and μ_π^2, if one considers instead of the lightest B-mesons, B_s-mesons or Λ_b-baryons were studied by Bigi, Mannel and Uraltsev in 2011.[53]

The new results seem also to confirm the bound

$$
\mu_\pi^2 > \mu_G^2
\tag{104}
$$

that was derived by Bigi, Shifman, Uraltsev and Vainshtein[136,14] and by Voloshin.[141] Kapustin, Ligeti, Wise and Grinstein[142] claimed that the above bound will be weakened due to perturbative corrections. A study of Kolya Uraltsev[143] came, however, to a different conclusion.

Matrix elements of four-quark operators relevant for lifetime ratios are almost unknown. For $\tau(B^+)/\tau(B_d)$ the latest lattice calculation was performed by Becirevic in 2001 and only published as proceedings.[56] Unfortunately these parameters, see Eq. (78) have never been updated. The same matrix elements can also be used for the case of $\tau(B_s)/\tau(B_d)$. There is also an earlier lattice study from Di Pierro and Sachrajda from 1999,[144] as well as two QCD sum rule studies from Baek, Lee,

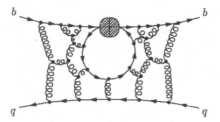

Fig. 4.   Penguin contractions that cancel in lifetime ratios like $\tau(B^+)/\tau(B_d)$ and $\tau(\Xi_b^0)/\tau(\Xi_b^+)$. They will, however, give a contribution to $\tau(\Lambda)/\tau(B_d)$. Since these contributions introduce a mixing of operators of different dimensionality, they are difficult to handle.

Liu and Song in 1997[145] and one year later from Cheng and Yang.[146]

B_1	B_2	ϵ_1	ϵ_2	
1.01 ± 0.01	0.99 ± 0.01	-0.08 ± 0.02	-0.01 ± 0.03	1997 QCD-SR
1.06 ± 0.08	1.01 ± 0.06	-0.01 ± 0.03	-0.01 ± 0.02	1998 Lattice
0.96 ± 0.04	0.95 ± 0.02	-0.14 ± 0.01	-0.08 ± 0.01	1998 QCD-SR
1.10 ± 0.20	0.79 ± 0.10	-0.02 ± 0.02	0.03 ± 0.01	2001 Lattice

$$(105)$$

Comparing these numbers, the authors[145,146] of the QCD sum rule evaluation seem to have very aggressive error estimates. Because of the very pronounced cancellations in Eq. (77) precise values for these bag parameters are crucial for an investigation of the lifetime ratio $\tau(B^+)/\tau(B_d)$. To some extent our definition of the bag parameters given in Eqs. (62) and (63) above was a little too simplistic. In reality we are considering the isospin breaking combinations

$$\frac{\langle B_d|Q^d - Q^u|B_d\rangle}{M_{B_d}} = f_B^2 B_1 M_{B_d}, \quad \frac{\langle B_d|Q_S^d - Q_S^u|B_d\rangle}{M_{B_d}} = f_B^2 B_2 M_{B_d}, \quad (106)$$

$$\frac{\langle B_d|T^d - T^u|B_d\rangle}{M_{B_d}} = f_B^2 \epsilon_1 M_{B_d}, \quad \frac{\langle B_d|T_S^d - T_S^u|B_d\rangle}{M_{B_d}} = f_B^2 \epsilon_2 M_{B_d}. \quad (107)$$

This definition leads to the cancellation of unwanted penguin contractions, see Fig. 4 and enables thus in principle very precise calculations. For the Λ_b-lifetime our knowledge of the matrix elements is even worse. There is only an exploratory lattice study from Di Pierro, Sachrajda and Michael available, dating back to 1999.[147] Here also any update would be extremely welcome. In that case two matrix elements are arising that are parameterised by L_1 and L_2

$$\frac{\langle \Lambda_b|Q^q|\Lambda_b\rangle}{M_{\Lambda_b}} = f_B^2 M_B L_1, \quad \frac{\langle \Lambda_b|T^q|\Lambda_b\rangle}{M_{\Lambda_b}} = f_B^2 M_B L_2, \quad (108)$$

where the operators were defined in Eqs. (60) and (61). The numerical values obtained in Ref. 147 are shown in Eq. (113). In elder works the colour re-arranged operator was investigated instead of the colour octett operator T^q. There the following definition was used

$$\frac{\langle \Lambda_b | \bar{b}_\alpha \gamma^\mu (1 - \gamma_5) q_\alpha \cdot \bar{q}_\beta \gamma_\mu (1 - \gamma_5) b_\beta \rangle}{M_{\Lambda_b}} = -\frac{f_B^2 M_B}{6} r, \tag{109}$$

$$\frac{\langle \Lambda_b | \bar{b}_\alpha \gamma^\mu (1 - \gamma_5) q_\beta \cdot \bar{q}_\beta \gamma_\mu (1 - \gamma_5) b_\alpha \rangle}{M_{\Lambda_b}} = \tilde{B} \frac{f_B^2 M_B}{6} r. \tag{110}$$

The two parameter sets are related by[p]

$$r = -6L_1, \quad \tilde{B} = -\frac{1}{3} - 2\frac{L_2}{L_1}. \tag{111}$$

Because of the long standing discrepancy between experiment and theory for the Λ_b-lifetime people of course tried different methods to determine the missing matrix elements: Rosner related in 1996 the four-quark matrix elements to results from spectroscopy[149] and found[q]

$$r = \frac{4}{3} \frac{M_{\Sigma_b^*} - M_{\Sigma_b}}{M_{B^*} - M_B}. \tag{112}$$

At that time the values of the masses of the baryons were almost unknown, which resulted in quite rough and large estimates, yielding $r \approx 1.6$. This situation changed completely now and we will use the method of Rosner with new experimental numbers for the baryon masses[r] to update the Λ_b lifetime below. Colangelo and de Fazio applied in 1996[150] the method of QCD sum rules and obtained relatively small numbers for r. Huang, Liu and Zhu managed in 1999,[151] however, to obtain with the same method much larger numbers, that also lead to a lifetime of the Λ_b-baryon, that was compatible with the measurements at that time. Even earlier (1979) estimates within the bag model and the non-relativistic quark model for charmed hadrons from Guberina, Nussinov, Peccei and Rückl[81] pointed towards smaller values of r. All in all currently the following numerical values are available:

$L_1 = -\frac{r}{6}$	L_2 or $\frac{r}{9}$	r	$\tilde{B}$	
$-0.103(10)$	$0.069(7)$	$0.62(6)$	1	2014 Spectroscopy update
$-0.22(4)$	$0.17(2)$	$1.32(24)$	$1.21(34)$	1999 Exploratory Lattice
$-0.22(5)$	$0.14(3)$	$1.3(3)$	1	1999 QCD-SR v1
$-0.60(15)$	$0.40(10)$	$3.6(9)$	1	1999 QCD-SR v2
$-0.033(17)$	$0.022(11)$	$0.2(1)$	1	1996 QCD-SR
≈ -0.03	≈ 0.02	≈ 0.2	1	1979 Bag model
≈ -0.08	≈ 0.06	≈ 0.5	1	1979 NRQM

$$\tag{113}$$

[p]In Ref. 147 the parameters r and $\tilde{B}$ were interchanged in Eq. (111), while the correct relation was given in Ref. 148.

[q]Neubert and Sachrajda[106] quoted in 1996 this formula as $r = 4/3(M_{\Sigma_b^*}^2 - M_{\Sigma_b}^2)/(M_{B^*}^2 - M_B^2)$, which gives values that are about 10% larger than Eq. (112). Cheng[112] quoted in 1997 the same formula as given in Eq. (112).

[r]PDG[1] gives: $M_{\Sigma_b^*} - M_{\Sigma_b} = 21.2(2.0)$ MeV, $M_{\Sigma_b^*} = 5832.1(2.0)$ MeV, $M_{\Sigma_b} = 5811.3(2.0)$ MeV and $M_{B^*} = 5325.20(0.40)$ MeV.

In Ref. 147 the two parameters L_1 and L_2 were calculated, else only r was determined. In the latter case we assumed $\tilde{B} = 1$ (valence quark approximation) in order to determine L_2. Comparing all these numbers we find that two studies obtain values of r larger than one. One is the exploratory lattice calculation. This method could in principle give a reliable value, if an up-to-date study would be made. The second one is the QCD sum rule estimate from Huang, Liu and Zhu in 1999.[151] In principle this is a reliable method, if it is applied properly. The calculation in Ref. 151 seems to be, however, in contradiction with the one from Colangelo and de Fazio.[150] In 1996 also the method from Rosner[149] gave values for r larger than one. This changed with new precise measurements of the $\Sigma_b^{(*)}$-masses. Now Rosner's methods gives a small value in accordance with the QCD sum rule estimate from Colangelo and de Fazio[150] and with the early estimates from Ref. 81. We will vary r between 0.2 (Colangelo and de Fazio) and 1.32 (Di Pierro, Sachrajda and Michael) with a central value of 0.62 (Rosner). The unclear situation with the matrix elements resulted in a broad range of different theory predictions and as long as the experimental values for the Λ_b-lifetime were low — the HFAG average from 2003 was

$$\frac{\tau(\Lambda_b)}{\tau(B_d)} \overset{\text{HFAG 2003}}{=} 0.798 \pm 0.034 \tag{114}$$

— there was a tendency to use preferably the larger values for r in order to see, how far one can "stretch" the HQE. Estimates from that time[152,130,118,153,154,147,151,150] read

Year	Author	$\tau(\Lambda_b)/\tau(B_d)$
2007	Tarantino	0.88 ± 0.05
2004	Petrov *et al.*	0.86 ± 0.05
2002	Rome	0.90 ± 0.05
2000	Körner, Melic	$0.81 \ldots 0.92$
1999	Guberina, Melic, Stefanic	0.90
1999	Di Pierro, Sachrajda, Michael	0.92 ± 0.02
1999	Huang, Liu, Zhu	0.83 ± 0.04
1996	Colangelo, deFazio	> 0.94

(115)

Nowadays it is clear that the low Λ_b-lifetime was a purely experimental issue. On the other hand the precise HQE prediction is still unknown, because we have no reliable calculation of the hadronic matrix elements at hand.

Finally we need matrix elements of dimension six and dimension seven operators that are arising in mixing quantities. The status of the dimension six operators for mixing is considerably more advanced than for the lifetime case; it is discussed in detail in the FLAG review.[54] For the numerically important dimension seven contributions vacuum insertion approximation is used and first studies with QCD sum rules have been performed by Mannel, Pecjak and Pivovarov.[57,58]

4. Status Quo of Lifetimes and the HQE

In this final section we update several of the lifetime predictions and compare them with the most recent data, obtained many times at the LHC experiments.

4.1. *B-meson lifetimes*

The most recent theory expressions for $\tau(B^+)/\tau(B_s)$ and $\tau(B_s)/\tau(B_d)$ are given in Ref. 155 (based on the calculations in Refs. 20, 118, 130, 56). For the charged B-meson we get the updated relation (including α_s-corrections and $1/m_b$-corrections)

$$\frac{\tau(B^+)}{\tau(B_d)}^{\text{HQE 2014}} = 1 + 0.03 \left(\frac{f_{B_d}}{190.5 \text{ MeV}} \right)^2 [(1.0 \pm 0.2)B_1 + (0.1 \pm 0.1)B_2$$

$$- (17.8 \pm 0.9)\epsilon_1 + (3.9 \pm 0.2)\epsilon_2 - 0.26]$$

$$= 1.04^{+0.05}_{-0.01} \pm 0.02 \pm 0.01 . \tag{116}$$

Here we have used the lattice values for the bag parameters from Ref. 56. Using all the values for the bag parameters quoted in Eq. (105), the central value of our prediction for $\tau(B^+)/\tau(B_d)$ varies between 1.03 and 1.09. This is indicated by the first asymmetric error and clearly shows the urgent need for more profound calculations of these non-perturbative parameters. The second error in Eq. (116) stems from varying the matrix elements of Ref. 56 in their allowed range and the third error comes from the renormalisation scale dependence as well as the dependence on m_b.

Next we update also the prediction for the B_s-lifetime given in Ref. 155, by including also $1/m_b^2$-corrections discussed in Eq. (51).

$$\frac{\tau(B_s)}{\tau(B_d)}^{\text{HQE 2014}} = 1.003 + 0.001 \left(\frac{f_{B_s}}{231 \text{ MeV}} \right)^2 [(0.77 \pm 0.10)B_1$$

$$+ (1.0 \pm 0.13)B_2 + (36 \pm 5)\epsilon_1 + (51 \pm 7)\epsilon_2]$$

$$= 1.001 \pm 0.002 . \tag{117}$$

The values in Eqs. (116) and (117) differ slightly from the ones in Ref. 155, because we have used updated lattice values for the decay constants[s] and we included the SU(3)-breaking of the $1/m_b^2$-correction — see Eq. (51) — for the B_s-lifetime, which was previously neglected. Comparing these predictions with the measurements given in Eq. (1), we find a perfect agreement for the B_s-lifetime, leaving thus only a little space for, e.g., hidden new B_s-decay channels, following, e.g., Refs. 68, 69. There is a slight tension in $\tau(B^+)/\tau(B_d)$, which, however, could solely be due to the unknown values of the hadronic matrix elements. A value of, e.g., $\epsilon_1 = -0.092$ — and leaving everything else at the values given in Eq. (78) — would perfectly match the current experimental average from Eq. (1). Such a value of ϵ_2 is within the range of the

[s]We have used $f_{B_s} = 227.7$ MeV.[54]

QCD sum rule predictions[145,146] shown in Eq. (105). Thus, for further investigations updated lattice values for the bag parameters B_1, B_2, ϵ_1 and ϵ_2 are indispensable.

The most recent experimental numbers for these lifetime ratios have been updated by the LHCb Collaboration in 2014.[156]

4.2. b-baryon lifetimes

We discussed already the early stages of the long standing puzzle related to the lifetime of Λ_b-baryon. After 2003 one started to find contradicting experimental values[157–162] — some of them still similarly low as the previous ones and others pointed more to a lifetime comparable to the one of the B_d-meson

Year	Exp	Decay	$\tau(\Lambda_b)$ [ps]	$\tau(\Lambda_b)/\tau(B_d)$
2010	CDF	$J/\psi\Lambda$	1.537 ± 0.047	1.020 ± 0.031
2009	CDF	$\Lambda_c + \pi^-$	1.401 ± 0.058	0.922 ± 0.038
2007	D0	$\Lambda_c \mu\nu X$	1.290 ± 0.150	0.849 ± 0.099
2007	D0	$J/\psi\Lambda$	1.218 ± 0.137	0.802 ± 0.090
2006	CDF	$J/\psi\Lambda$	1.593 ± 0.089	1.049 ± 0.059
2004	D0	$J/\psi\Lambda$	1.22 ± 0.22	0.87 ± 0.17

$$(118)$$

The current HFAG average given in Eq. (2) clearly rules out now the small values of the Λ_b-lifetime. Updating the NLO-calculation from the Rome group[121] and including $1/m_b$-corrections from Ref. 130 we get for the current HQE prediction

$$\frac{\tau(\Lambda_b)}{\tau(B_d)}^{\text{HQE 2014}} = 1 - (0.8 \pm 0.5)\%_{\frac{1}{m_b^2}} - (4.2 \pm 3.3)\%_{\frac{1}{m_b^3}}^{\Lambda_b}$$

$$- (0.0 \pm 0.5)\%_{\frac{1}{m_b^3}}^{B_d} - (1.6 \pm 1.2)\%_{\frac{1}{m_b^4}}$$

$$= 0.935 \pm 0.054, \qquad (119)$$

where we have split up the corrections coming from the $1/m_b^2$-corrections discussed in Eq. (51), the $1/m_b^3$-corrections coming from the Λ_b-matrix elements, the $1/m_b^3$-corrections coming from the B_d-matrix elements and finally $1/m_b^4$-corrections studied in Ref. 130. The number in Eq. (119) is smaller than some of the previous theory predictions because of several reasons: we have used updated, smaller lattice values for the decay constants, which gives a shift of about $+0.01$ in the lifetime ratio. Following our discussion of the dimension six matrix elements, we use three different determinations. Instead of using only the exploratory lattice one,[147] we also take into account the QCD sum rule estimate of Colangelo and de Fazio[150] and the spectroscopy result of Rosner.[149] In 1996 Rosner's method gave a large value of the matrix element. New, precise measurements of the $\Sigma_b^{(*)}$-mass show, however, that the matrix element is much smaller than originally thought. This gives a third enhancement factor. To obtain the final number we also scaled the numerical value of the $1/m_b^4$-correction with the size of r. The current range of the theory prediction in Eq. (119) goes from 0.88 to 0.99. To reduce this large uncertainty, new lattice calculations are necessary. In these calculations also the penguin contractions from Fig. 4 have to be taken into account.

More recent experimental studies of the Λ_b-lifetime further strengthen the case for a value of the lifetime ratio close to one. The most recent and most precise measurement from LHCb gives[163]

$$\frac{\tau(\Lambda_b)}{\tau(B_d)}^{\text{LHCb}} = 0.974 \pm 0.006 \pm 0.004 \,. \tag{120}$$

This results supersedes a previous LHCb measurement.[164] Combined with the world average for the B_d-lifetime one gets

$$\tau(\Lambda_b)^{\text{LHCb}} = 1.479 \pm 0.009 \pm 0.010 \text{ ps} \,. \tag{121}$$

Comparing the accuracy of these new measurements with the HFAG average given in Eq. (2) shows the dramatic experimental progress. LHCb has a further recent investigation of the Λ_b-lifetime[156] — based on different experimental techniques — and there is also a very new TeVatron (CDF) number available[165]

$$\tau(\Lambda_b)^{\text{CDF}} = 1.565 \pm 0.035 \pm 0.020 \text{ ps} \,. \tag{122}$$

All in all, now the new measurements of the Λ_b-lifetime are in nice agreement with the HQE result. This is now a very strong confirmation of the validity of the HQE and this makes also the motivation of many of the studies trying to explain the Λ_b-lifetime puzzle, e.g., Refs. 111–113, invalid.

In Ref. 20 it was shown that the lifetime ratio of the Ξ_b-baryons can be in principle be determined quite precisely, because here the above mentioned problems with penguin contractions do not arise, the diagrams from Fig. 4 cancel. Unfortunately there exists no non-perturbative determination of the matrix elements for Ξ_b-baryons. Cheng[112] suggested to use the relation

$$r_{\Xi_b} = \frac{4}{3} \frac{M_{\Xi_b^*} - M_{\Xi_b}}{M_{B^*} - M_B} \,, \tag{123}$$

but there are no data available yet for the Ξ_b^*-mass. So, we are left with the possibility of assuming that the matrix elements for Ξ_b are equal to the ones of Λ_b. In that case we can give a rough estimate for the expected lifetime ratio — we update here a numerical estimate from 2008.[166] In order to get rid of unwanted $s \to u$-transitions we define (following Ref. 20)

$$\frac{1}{\bar{\tau}(\Xi_b)} = \bar{\Gamma}(\Xi_b) = \Gamma(\Xi_b) - \Gamma(\Xi_b \to \Lambda_b + X) \,. \tag{124}$$

For a numerical estimate we scan over the the results for the Λ_b-matrix elements obtained on the lattice by the study of Di Pierro, Michael and Sachrajda,[147] the

QCD sum rule estimate of Colangelo and de Fazio[150] and the update of the spectroscopy method of Rosner.[149] Using also recent values for the remaining input parameters we obtain

$$\frac{\bar{\tau}(\Xi_b^0)}{\bar{\tau}(\Xi_b^+)}^{\text{HQE 2014}} = 0.95 \pm 0.04 \pm 0.01 \pm ??? \,, \tag{125}$$

where the first error comes from the range of the values used for r, the second denotes the remaining parametric uncertainty and ??? stands for some unknown systematic errors, which comes from the approximation of the Ξ_b-matrix elements by the Λ_b-matrix elements. We expect the size of these unknown systematic uncertainties not to exceed the error stemming from r, thus leading to an estimated overall error of about ± 0.06. As soon as Ξ_b-matrix elements are available the ratio in Eq. (125) can be determine more precisely than $\tau(\Lambda_b)/\tau(B_d)$.

If we further approximate $\bar{\tau}(\Xi_b^0) = \tau(\Lambda_b)$ — here similar cancellations are expected to arise as in τ_{B_s}/τ_{B_d} —, then we arrive at the following prediction

$$\frac{\tau(\Lambda_b)}{\bar{\tau}(\Xi_b^+)}^{\text{HQE 2014}} = 0.95 \pm 0.06 \,. \tag{126}$$

From the new measurements of the LHCb Collaboration[3,4] (see also the CDF update[165]), we deduce

$$\frac{\tau(\Xi_b^0)}{\tau(\Xi_b^+)}^{\text{LHCb 2014}} = 0.92 \pm 0.03 \,, \tag{127}$$

$$\frac{\tau(\Xi_b^0)}{\tau(\Lambda_b)}^{\text{LHCb 2014}} = 1.006 \pm 0.021 \,, \tag{128}$$

$$\frac{\tau(\Lambda_b)}{\tau(\Xi_b^+)}^{\text{LHCb 2014}} = 0.918 \pm 0.028 \,, \tag{129}$$

which is in perfect agreement with the predictions above in Eqs. (125) and (126), within the current uncertainties.

4.3. *D-meson lifetimes*

In Ref. 9 the NLO-QCD corrections for the D-meson lifetimes were completed. Including $1/m_c$-corrections as well as some assumptions about the hadronic matrix elements one obtains

$$\frac{\tau(D^+)}{\tau(D^0)}^{\text{HQE 2013}} = 2.2 \pm 0.4^{(\text{hadronic})}{}^{+0.03(\text{scale})}_{-0.07} \,, \tag{130}$$

$$\frac{\tau(D_s^+)}{\tau(D^0)}^{\text{HQE 2013}} = 1.19 \pm 0.12^{(\text{hadronic})}{}^{+0.04(\text{scale})}_{-0.04} \,, \tag{131}$$

being very close to the experimental values shown in Eq. (3). Therefore this result seems to indicate that one might apply the HQE also to lifetimes of D-mesons, but definite conclusions cannot not be drawn without a reliable non-perturbative determination of the hadronic matrix elements, which is currently missing.

4.4. *Mixing quantities*

The current status of mixing quantities, both in the B- and the D-system, was very recently reviewed in Ref. 167. The arising set of observables allows for model-independent searches for new physics effects in mixing, see e.g. Refs. 168, 169. We discuss here only the decay rate differences $\Delta\Gamma_s$, because this provided one of the strongest proofs of the HQE. The HQE prediction — based on the NLO-QCD corrections[97,120,121,170] and sub-leading HQE corrections[127,128] gave in 2011[155]

$$\Delta\Gamma_s^{\text{HQE 2011}} = (0.087 \pm 0.021)\ \text{ps}^{-1}. \tag{132}$$

$\Delta\Gamma_s$ was measured for the first time in 2012 by the LHCb Collaboration.[171] The current average from HFAG[2] reads

$$\Delta\Gamma_s^{\text{Exp.}} = (0.091 \pm 0.09)\ \text{ps}^{-1}, \tag{133}$$

it includes the measurements from LHCb,[172,173] ATLAS,[174,175] CMS,[176] CDF[177] and D0.[178] Experiment and theory agree perfectly for $\Delta\Gamma_s$, excluding thus huge violations of quark hadron duality. The new question is now: how precisely does the HQE work? The experimental uncertainty will be reduced in future, while the larger theory uncertainty is dominated from unknown matrix elements of dimension seven operators, see Refs. 170, 155. Here a first lattice investigation or a continuation of the QCD sum rule study in Refs. 57, 58 would be very welcome.

5. Conclusion

We have started this review by giving a very basic introduction into lifetimes of weakly decaying particles, followed by a detailed discussion of the individual terms appearing in the HQE. Next we focused on the historical development of the theory, which we summarise briefly as: early investigations of the HQE are based on the work by Voloshin and Shifman[86,87] in the early 1980s. A real systematic expansion was only possible after some conceptual issues have been solved in 1992 by Bigi and Uraltsev,[93] which was proven in 1998 by Beneke *et al.*[97] in an explicit calculation. The HQE in its present form was developed in 1992 by Bigi, Uraltsev and Vainshtein[98] and about the same time by Blok and Shifman.[92,22] For semi-leptonic decays the absence of $1/m_q$-corrections was already shown in 1990 by Chay, Georgi and Grinstein[94] and by Luke.[95]

Since 1992 several discrepancies were arising, that shed some doubt on the validity of the HQE: inclusive non-leptonic decays (in particular predictions for the semi-leptonic branching ratio and the missing charm puzzle) and the Λ_b-lifetime were two prominent examples. We have discussed in detail, how all these issues were resolved. For the semi-leptonic branching ratio NLO-QCD corrections including finite charm-quark mass effect were crucial. The remaining small difference, see Eq. (25) vs. Eq. (26) is probably due to unknown NNLO-QCD effects. The problem of the Λ_b-lifetime was experimentally solved in the last months. One of the most convincing tests of the HQE was, however, the measurement of $\Delta\Gamma_s$ from 2012 onwards — see Eq. (133) — in perfect agreement with the prediction stemming from early 2011 — see Eq. (132).

Thus, the theory in whose development Kolya played such a crucial role, has just now passed numerous non-trivial tests and its validity holds beyond any doubt. This makes also the motivation for looking for some modification of the HQE, see e.g. Refs. 101, 111–113 invalid. The new question is now: how precise is the HQE? This question is not only of academic interest, but it has practical consequences in searches for new physics. The quantification of a statistical significance of a possible discrepancy depends strongly on the intrinsic uncertainty of the HQE. Hence further studies in that direction are crucial. As a starting point for such an endeavour we have updated several theory predictions for lifetime ratios. In order to see, how these predictions could be further improved, we compare for different observables what components of the theory prediction are currently known.

	$\dfrac{\tau(B+)}{\tau(B_d)}$	Γ_{12}	$\dfrac{\tau(\Lambda_b)}{\tau(B_d)}$	$\dfrac{\tau(D^+)}{\tau(D_0)}$	$\dfrac{\bar{\tau}(\Xi_b^0)}{\bar{\tau}(\Xi_b^+)}$
$\Gamma_3^{(0)}$	+	+	+	+	+
$\Gamma_3^{(1)}$	+	+	0	+	+
$\Gamma_3^{(2)}$	−	−	−	−	−
$\langle\Gamma_3\rangle$	0	+	0	−	−
$\Gamma_4^{(0)}$	+	+	+	+	+
$\Gamma_4^{(1)}$	−	−	−	−	−
$\langle\Gamma_4\rangle$	−	0	−	−	−
$\Gamma_5^{(0)}$	−	+	+	−	+
$\langle\Gamma_5\rangle$	−	−	−	−	−

For all these observables the LO-QCD term $\Gamma_3^{(0)}$ and also the NLO-QCD corrections $\Gamma_3^{(1)}$ are known. For the Λ_b-baryon, however, a part of the NLO-QCD calculation is still missing. NNLO-QCD corrections — denoted by $\Gamma_3^{(2)}$ — have not been calculated for any of these observables; a first step for Γ_{12} has been done in Ref. 123. The biggest problem are currently the non-perturbative matrix elements. Concerning the dimension 6 term $\langle\Gamma_3\rangle$ we have only for Γ_{12} several independent lattice calculations. For $\tau(B^+)/\tau(B_d)$ the latest lattice number stems from 2001,[56] for $\tau(\Lambda_b)/\tau(B_d)$ we have only an exploratory lattice study from 1999[147] and for the D-meson lifetimes

A. Lenz

we have no lattice investigations at all. For the b-hadrons also several QCD sum rule determinations of these matrix elements are available.[145,146,150,151]

Concerning the power suppressed $1/m_b$ corrections, we see that the LO-QCD term $\Gamma_4^{(0)}$ is known for all observables and $\Gamma_5^{(0)}$ is also known for some of the observables. The matrix elements of the dimension seven operators, $\langle \Gamma_4 \rangle$, have been determined by vacuum insertion approximation — a first step of a QCD sum rule calculation for $\Delta\Gamma_s$ has be done in Refs. 58, 57.

For all lifetime ratios the uncertainty due to the unknown matrix elements of the dimension six operators is dominant. For $\Delta\Gamma_s$ these operators have already been determined by several groups and thus the dominant uncertainty stems now from Γ_4. Here a full non-perturbative determination of the matrix elements of the dimension seven operators would be very desirable, as well as calculation of the corresponding NLO-QCD corrections, denoted by $\Gamma_4^{(1)}$. Increasing the precision of the HQE will also help in shrinking the allowed space for new physics effects in tree-level decays,[179] a topic that has also profound implications for other branches of flavour physics.

Kolya left us a very promising but also challenging legacy, which might in the end provide the way to identify new physics in the flavour sector.

Acknowledgments

I would like to thank Martin Beneke, Christoph Bobeth, Markus Bobrowski, Gerhard Buchalla, Christoph Greub, Uli Haisch, Fabian Krinner, Uli Nierste, Ben Pecjak, Thomas Rauh, Johann Riedl, Jürgen Rohrwild and Gilberto Tetlalmatzi-Xolocotzi for collaborating on topics related to the determination of lifetime predictions. Valerie Khoze, Ikaros Bigi, Jonathan Rosner and Mikhail Voloshin for helpful comments on the history of lifetimes. Many thanks to Gilberto Tetlalmatzi-Xolocotzi for producing most of the Feynman diagrams, to Fabian Krinner and Thomas Rauh for numerical updates of the coefficient c_3 and to Paolo Gambino, Valery Khoze, Uli Nierste, Thomas Rauh, Jonathan Rosner and Gilberto Tetlalmatzi-Xolocotzi for proof-reading.

References

1. Particle Data Group Collab. (J. Beringer *et al.*), *Phys. Rev. D* **86**, 010001 (2012), online update at http://pdg.lbl.gov/index.html.
2. Heavy Flavor Averaging Group Collab. (Y. Amhis *et al.*), arXiv:1207.1158 [hep-ex], online update at http://www.slac.stanford.edu/xorg/hfag/.
3. LHCb Collab. (R. Aaij *et al.*), *Phys. Rev. Lett.* **113**, 032001 (2014), arXiv:1405.7223 [hep-ex].
4. LHCb Collab. (R. Aaij *et al.*), *Phys. Rev. Lett.* **113**, 242002 (2014), arXiv:1409.8568 [hep-ex].
5. LHCb Collab. (R. Aaij *et al.*), *Phys. Lett. B* **736**, 154 (2014), arXiv:1405.1543 [hep-ex].
6. S. Stone, arXiv:1406.6497 [hep-ex].

7. A. Lenz, arXiv:hep-ph/0011258.
8. M. Beneke and A. Lenz, *J. Phys. G* **27**, 1219 (2001), arXiv:hep-ph/0012222.
9. A. Lenz and T. Rauh, *Phys. Rev. D* **88**, 034004 (2013), arXiv:1305.3588 [hep-ph].
10. I. I. Y. Bigi, M. A. Shifman, N. G. Uraltsev and A. I. Vainshtein, *Phys. Rev. D* **50**, 2234 (1994), arXiv:hep-ph/9402360.
11. M. Beneke and V. M. Braun, *Nucl. Phys. B* **426**, 301 (1994), arXiv:hep-ph/9402364.
12. M. Shifman, arXiv:1310.1966 [hep-th].
13. W. A. Bardeen, A. J. Buras, D. W. Duke and T. Muta, *Phys. Rev. D* **18**, 3998 (1978).
14. I. I. Y. Bigi, M. A. Shifman, N. G. Uraltsev and A. I. Vainshtein, *Phys. Rev. D* **52**, 196 (1995), arXiv:hep-ph/9405410.
15. I. I. Y. Bigi, M. A. Shifman, N. Uraltsev and A. I. Vainshtein, *Phys. Rev. D* **56**, 4017 (1997), arXiv:hep-ph/9704245.
16. M. Beneke, *Phys. Lett. B* **434**, 115 (1998), arXiv:hep-ph/9804241.
17. A. H. Hoang, Z. Ligeti and A. V. Manohar, *Phys. Rev. Lett.* **82**, 277 (1999), arXiv:hep-ph/9809423.
18. A. H. Hoang, Z. Ligeti and A. V. Manohar, *Phys. Rev. D* **59**, 074017 (1999), arXiv:hep-ph/9811239.
19. F. Krinner, A. Lenz and T. Rauh, *Nucl. Phys. B* **876**, 31 (2013), arXiv:1305.5390 [hep-ph].
20. M. Beneke, G. Buchalla, C. Greub, A. Lenz and U. Nierste, *Nucl. Phys. B* **639**, 389 (2002), arXiv:hep-ph/0202106.
21. M. Gourdin and X.-Y. Pham, *Nucl. Phys. B* **164**, 399 (1980).
22. B. Blok and M. A. Shifman, *Nucl. Phys. B* **399**, 459 (1993), arXiv:hep-ph/9209289.
23. I. I. Y. Bigi, B. Blok, M. A. Shifman and A. I. Vainshtein, *Phys. Lett. B* **323**, 408 (1994), arXiv:hep-ph/9311339.
24. I. I. Y. Bigi, B. Blok, M. A. Shifman, N. Uraltsev and A. I. Vainshtein, *B Decays*, revised 2nd edition, ed. S. Stone, pp. 132-0157, and CERN Geneva — TH.-7132 (94/01, rec. Jan.) 28p. Notre Dame U. – UND-HEP-94-01 (94/01, rec. Jan.) 28p. Minnesota U. Minneapolis – TPI-MINN-94-01-T (94/01, rec. Jan.) 28p. Minnesota U. Minneapolis – UMN-TH-94-1234 (94/01, rec. Jan.) 28p. Haifa Isr. Inst. Technol. – TECHNION-PH-94-01 (94/01, rec.Jan.) 28p. arXiv:hep-ph/9401298.
25. A. F. Falk, Z. Ligeti, M. Neubert and Y. Nir, *Phys. Lett. B* **326**, 145 (1994), arXiv:hep-ph/9401226.
26. CKMfitter Group Collab. (J. Charles *et al.*), *Eur. Phys. J. C* **41**, 1 (2005), arXiv:hep-ph/0406184.
27. M. Ciuchini, G. D'Agostini, E. Franco, V. Lubicz, G. Martinelli, F. Parodi, P. Roudeau and A. Stocchi, *JHEP* **0107**, 013 (2001), arXiv:hep-ph/0012308.
28. G. Buchalla, A. J. Buras and M. E. Lautenbacher, *Rev. Mod. Phys.* **68**, 1125 (1996), arXiv:hep-ph/9512380.
29. A. J. Buras, arXiv:hep-ph/9806471.
30. G. Buchalla, arXiv:hep-ph/0202092.
31. A. Grozin, arXiv:1311.0550 [hep-ph].
32. G. Altarelli and S. Petrarca, *Phys. Lett. B* **261**, 303 (1991).
33. M. B. Voloshin, *Phys. Rev. D* **51**, 3948 (1995), arXiv:hep-ph/9409391.
34. Q. Ho-Kim and X.-Y. Pham, *Ann. Phys.* **155**, 202 (1984).
35. E. Bagan, P. Ball, V. M. Braun and P. Gosdzinsky, *Nucl. Phys. B* **432**, 3 (1994), arXiv:hep-ph/9408306.
36. E. Bagan, P. Ball, B. Fiol and P. Gosdzinsky, *Phys. Lett. B* **351**, 546 (1995), arXiv:hep-ph/9502338.

37. A. Lenz, U. Nierste and G. Ostermaier, *Phys. Rev. D* **56**, 7228 (1997), arXiv:hep-ph/9706501.
38. C. Greub and P. Liniger, *Phys. Rev. D* **63**, 054025 (2001), arXiv:hep-ph/0009144.
39. C. Greub and P. Liniger, *Phys. Lett. B* **494**, 237 (2000), arXiv:hep-ph/0008071.
40. M. Gorbahn and U. Haisch, *Nucl. Phys. B* **713**, 291 (2005), arXiv:hep-ph/0411071.
41. T. van Ritbergen, *Phys. Lett. B* **454**, 353 (1999), arXiv:hep-ph/9903226.
42. K. Melnikov, *Phys. Lett. B* **666**, 336 (2008), arXiv:0803.0951 [hep-ph].
43. A. Pak and A. Czarnecki, *Phys. Rev. Lett.* **100**, 241807 (2008), arXiv:0803.0960 [hep-ph].
44. A. Pak and A. Czarnecki, *Phys. Rev. D* **78**, 114015 (2008), arXiv:0808.3509 [hep-ph].
45. R. Bonciani and A. Ferroglia, *JHEP* **0811**, 065 (2008), arXiv:0809.4687 [hep-ph].
46. S. Biswas and K. Melnikov, *JHEP* **1002**, 089 (2010), arXiv:0911.4142 [hep-ph].
47. A. Czarnecki, M. Slusarczyk and F. V. Tkachov, *Phys. Rev. Lett.* **96**, 171803 (2006), arXiv:hep-ph/0511004.
48. Belle Collab. (C. Oswald *et al.*), *Phys. Rev. D* **87**, 072008 (2013), arXiv:1212.6400 [hep-ex].
49. M. Neubert, *Adv. Ser. Direct High Energy Phys.* **15**, 239 (1998), arXiv:hep-ph/9702375.
50. M. E. Luke and M. J. Savage, *Phys. Lett. B* **321**, 88 (1994), arXiv:hep-ph/9308287.
51. P. Gambino and C. Schwanda, *Phys. Rev. D* **89**, 014022 (2014), arXiv:1307.4551 [hep-ph].
52. N. Uraltsev, *Phys. Lett. B* **545**, 337 (2002), arXiv:hep-ph/0111166.
53. I. I. Bigi, T. Mannel and N. Uraltsev, *JHEP* **1109**, 012 (2011), arXiv:1105.4574 [hep-ph].
54. G. Colangelo *et al.*, *Eur. Phys. J. C* **71**, 1695 (2011), arXiv:1011.4408 [hep-lat], web update: itpwiki.unibe.ch/flag; S. Aoki *et al.*, arXiv:1310.8555 [hep-lat].
55. P. Gelhausen, A. Khodjamirian, A. A. Pivovarov and D. Rosenthal, *Phys. Rev. D* **88**, 014015 (2013), arXiv:1305.5432 [hep-ph].
56. D. Becirevic, *PoS* **HEP2001**, 098 (2001), arXiv:hep-ph/0110124.
57. T. Mannel, B. D. Pecjak and A. A. Pivovarov, arXiv:hep-ph/0703244.
58. T. Mannel, B. D. Pecjak and A. A. Pivovarov, *Eur. Phys. J. C* **71**, 1607 (2011).
59. B. Chibisov, R. D. Dikeman, M. A. Shifman and N. Uraltsev, *Int. J. Mod. Phys. A* **12**, 2075 (1997), arXiv:hep-ph/9605465.
60. I. I. Y. Bigi and N. Uraltsev, *Int. J. Mod. Phys. A* **16**, 5201 (2001), arXiv:hep-ph/0106346.
61. I. I. Y. Bigi, M. A. Shifman, N. Uraltsev and A. I. Vainshtein, *Phys. Rev. D* **59**, 054011 (1999), arXiv:hep-ph/9805241.
62. R. F. Lebed and N. G. Uraltsev, *Phys. Rev. D* **62**, 094011 (2000), arXiv:hep-ph/0006346.
63. B. Grinstein and R. F. Lebed, *Phys. Rev. D* **57**, 1366 (1998), arXiv:hep-ph/9708396.
64. B. Grinstein and R. F. Lebed, *Phys. Rev. D* **59**, 054022 (1999), arXiv:hep-ph/9805404.
65. B. Grinstein, *Phys. Rev. D* **64**, 094004 (2001), arXiv:hep-ph/0106205.
66. B. Grinstein, *Phys. Lett. B* **529**, 99 (2002), arXiv:hep-ph/0112323.
67. A. Lenz, arXiv:1205.1444 [hep-ph].
68. C. Bobeth, U. Haisch, A. Lenz, B. Pecjak and G. Tetlalmatzi-Xolocotzi, arXiv:1404.2531 [hep-ph].
69. C. Bobeth and U. Haisch, *Acta Phys. Polon. B* **44**, 127 (2013), arXiv:1109.1826 [hep-ph].

70. I. I. Y. Bigi, V. A. Khoze, N. G. Uraltsev and A. I. Sanda, *Adv. Ser. Direct High Energy Phys.* **3**, 175 (1989).
71. J. S. Hagelin and M. B. Wise, *Nucl. Phys. B* **189**, 87 (1981).
72. J. S. Hagelin, *Nucl. Phys. B* **193**, 123 (1981).
73. A. J. Buras, W. Slominski and H. Steger, *Nucl. Phys. B* **245**, 369 (1984).
74. E598 Collab. (J. J. Aubert *et al.*), *Phys. Rev. Lett.* **33**, 1404 (1974).
75. SLAC-SP-017 Collab. (J. E. Augustin *et al.*), *Phys. Rev. Lett.* **33**, 1406 (1974).
76. N. N. Nikolaev, *Pisma Zh. Eksp. Teor. Fiz.* **18**, 447 (1973).
77. M. K. Gaillard, B. W. Lee and J. L. Rosner, *Rev. Mod. Phys.* **47**, 277 (1975).
78. R. L. Kingsley, S. B. Treiman, F. Wilczek and A. Zee, *Phys. Rev. D* **11**, 1919 (1975).
79. J. R. Ellis, M. K. Gaillard and D. V. Nanopoulos, *Nucl. Phys. B* **100**, 313 (1975) [Erratum: *ibid.* **104**, 547 (1976)].
80. G. Altarelli, N. Cabibbo and L. Maiani, *Phys. Rev. Lett.* **35**, 635 (1975).
81. B. Guberina, S. Nussinov, R. D. Peccei and R. Ruckl, *Phys. Lett. B* **89**, 111 (1979).
82. T. Kobayashi and N. Yamazaki, *Prog. Theor. Phys.* **65**, 775 (1981).
83. M. Bander, D. Silverman and A. Soni, *Phys. Rev. Lett.* **44**, 7 (1980) [Erratum: *ibid.* **44**, 962 (1980)].
84. H. Fritzsch and P. Minkowski, *Phys. Lett. B* **90**, 455 (1980).
85. W. Bernreuther, O. Nachtmann and B. Stech, *Z. Phys. C* **4**, 257 (1980).
86. V. A. Khoze and M. A. Shifman, *Sov. Phys. Usp.* **26**, 387 (1983).
87. M. A. Shifman and M. B. Voloshin, *Sov. J. Nucl. Phys.* **41**, 120 (1985) [*Yad. Fiz.* **41**, 187 (1985)].
88. N. Bilic, B. Guberina and J. Trampetic, *Nucl. Phys. B* **248**, 261 (1984).
89. M. A. Shifman and M. B. Voloshin, *Sov. Phys. JETP* **64**, 698 (1986) [*Zh. Eksp. Teor. Fiz.* **91**, 1180 (1986)].
90. B. Guberina, R. Ruckl and J. Trampetic, *Z. Phys. C* **33**, 297 (1986).
91. V. A. Khoze, M. A. Shifman, N. G. Uraltsev and M. B. Voloshin, *Sov. J. Nucl. Phys.* **46**, 112 (1987) [*Yad. Fiz.* **46**, 181 (1987)].
92. B. Blok and M. A. Shifman, *Nucl. Phys. B* **399**, 441 (1993), arXiv:hep-ph/9207236.
93. I. I. Y. Bigi and N. G. Uraltsev, *Phys. Lett. B* **280**, 271 (1992).
94. J. Chay, H. Georgi and B. Grinstein, *Phys. Lett. B* **247**, 399 (1990).
95. M. E. Luke, *Phys. Lett. B* **252**, 447 (1990).
96. M. Ademollo and R. Gatto, *Phys. Rev. Lett.* **13**, 264 (1964).
97. M. Beneke, G. Buchalla, C. Greub, A. Lenz and U. Nierste, *Phys. Lett. B* **459**, 631 (1999), arXiv:hep-ph/9808385.
98. I. I. Y. Bigi, N. G. Uraltsev and A. I. Vainshtein, *Phys. Lett. B* **293**, 430 (1992) [Erratum: *ibid.* **297**, 477 (1993)], arXiv:hep-ph/9207214.
99. I. I. Y. Bigi, B. Blok, M. A. Shifman, N. G. Uraltsev and A. I. Vainshtein, arXiv:hep-ph/9212227.
100. W. F. Palmer and B. Stech, *Phys. Rev. D* **48**, 4174 (1993).
101. A. F. Falk, M. B. Wise and I. Dunietz, *Phys. Rev. D* **51**, 1183 (1995), arXiv:hep-ph/9405346.
102. I. I. Y. Bigi and N. G. Uraltsev, *Z. Phys. C* **62**, 623 (1994), arXiv:hep-ph/9311243.
103. A. F. Falk, M. E. Luke and M. J. Savage, *Phys. Rev. D* **53**, 6316 (1996), arXiv:hep-ph/9511454.
104. I. I. Y. Bigi and N. G. Uraltsev, *Nucl. Phys. B* **423**, 33 (1994), arXiv:hep-ph/9310285.
105. N. G. Uraltsev, *Phys. Lett. B* **376**, 303 (1996), arXiv:hep-ph/9602324.
106. M. Neubert and C. T. Sachrajda, *Nucl. Phys. B* **483**, 339 (1997), arXiv:hep-ph/9603202.
107. ALEPH Collab. (D. Buskulic *et al.*), *Phys. Lett. B* **297**, 449 (1992).

108. ALEPH Collab. (D. Buskulic *et al.*), *Phys. Lett. B* **357**, 685 (1995).

109. ALEPH Collab. (R. Barate *et al.*), *Eur. Phys. J. C* **2**, 197 (1998).

110. OPAL Collab. (K. Ackerstaff *et al.*), *Phys. Lett. B* **426**, 161 (1998), arXiv:hep-ex/9802002.

111. G. Altarelli, G. Martinelli, S. Petrarca and F. Rapuano, *Phys. Lett. B* **382**, 409 (1996), arXiv:hep-ph/9604202.

112. H.-Y. Cheng, *Phys. Rev. D* **56**, 2783 (1997), arXiv:hep-ph/9704260.

113. T. Ito, M. Matsuda and Y. Matsui, *Prog. Theor. Phys.* **99**, 271 (1998), arXiv:hep-ph/9705402.

114. M. Beneke and G. Buchalla, *Phys. Rev. D* **53**, 4991 (1996), arXiv:hep-ph/9601249.

115. T. Becher, H. Boos and E. Lunghi, *JHEP* **0712**, 062 (2007), arXiv:0708.0855 [hep-ph].

116. A. Alberti, T. Ewerth, P. Gambino and S. Nandi, *Nucl. Phys. B* **870**, 16 (2013), arXiv:1212.5082.

117. A. Alberti, P. Gambino and S. Nandi, *JHEP (2014)* 1, arXiv:1311.7381 [hep-ph].

118. E. Franco, V. Lubicz, F. Mescia and C. Tarantino, *Nucl. Phys. B* **633**, 212 (2002), arXiv:hep-ph/0203089.

119. Y.-Y. Keum and U. Nierste, *Phys. Rev. D* **57**, 4282 (1998), arXiv:hep-ph/9710512.

120. M. Beneke, G. Buchalla, A. Lenz and U. Nierste, *Phys. Lett. B* **576**, 173 (2003), arXiv:hep-ph/0307344.

121. M. Ciuchini, E. Franco, V. Lubicz, F. Mescia and C. Tarantino, *JHEP* **0308**, 031 (2003), arXiv:hep-ph/0308029.

122. M. Bobrowski, A. Lenz, J. Riedl and J. Rohrwild, *JHEP* **1003**, 009 (2010), arXiv:1002.4794 [hep-ph].

123. H. M. Asatrian, A. Hovhannisyan and A. Yeghiazaryan, *Phys. Rev. D* **86**, 114023 (2012), arXiv:1210.7939 [hep-ph].

124. B. Blok, R. D. Dikeman and M. A. Shifman, *Phys. Rev. D* **51**, 6167 (1995), arXiv:hep-ph/9410293.

125. M. Gremm and A. Kapustin, *Phys. Rev. D* **55**, 6924 (1997), arXiv:hep-ph/9603448.

126. T. Mannel, S. Turczyk and N. Uraltsev, *JHEP* **1011**, 109 (2010), arXiv:1009.4622 [hep-ph].

127. M. Beneke, G. Buchalla and I. Dunietz, *Phys. Rev. D* **54**, 4419 (1996) [Erratum: *ibid.* **83**, 119902 (2011)], arXiv:hep-ph/9605259.

128. A. S. Dighe, T. Hurth, C. S. Kim and T. Yoshikawa, *Nucl. Phys. B* **624**, 377 (2002), arXiv:hep-ph/0109088.

129. F. Gabbiani, A. I. Onishchenko and A. A. Petrov, *Phys. Rev. D* **68**, 114006 (2003), arXiv:hep-ph/0303235.

130. F. Gabbiani, A. I. Onishchenko and A. A. Petrov, *Phys. Rev. D* **70**, 094031 (2004), arXiv:hep-ph/0407004.

131. A. Lenz and U. Nierste, talk at Academia Sinica on 3.10.2003.

132. A. Badin, F. Gabbiani and A. A. Petrov, *Phys. Lett. B* **653**, 230 (2007), arXiv:0707.0294 [hep-ph].

133. H. Georgi, *Phys. Lett. B* **297**, 353 (1992), arXiv:hep-ph/9209291.

134. T. Ohl, G. Ricciardi and E. H. Simmons, *Nucl. Phys. B* **403**, 605 (1993), arXiv:hep-ph/9301212.

135. I. I. Y. Bigi and N. G. Uraltsev, *Nucl. Phys. B* **592**, 92 (2001), arXiv:hep-ph/0005089.

136. I. I. Y. Bigi, M. A. Shifman, N. G. Uraltsev and A. I. Vainshtein, *Int. J. Mod. Phys. A* **9**, 2467 (1994), arXiv:hep-ph/9312359.

137. I. I. Y. Bigi, A. G. Grozin, M. A. Shifman, N. G. Uraltsev and A. I. Vainshtein, *Phys. Lett. B* **339**, 160 (1994), arXiv:hep-ph/9407296.

138. P. Ball and V. M. Braun, *Phys. Rev. D* **49**, 2472 (1994), arXiv:hep-ph/9307291.
139. M. Neubert, *Phys. Lett. B* **389**, 727 (1996), arXiv:hep-ph/9608211.
140. A. S. Kronfeld and J. N. Simone, *Phys. Lett. B* **490**, 228 (2000) [Erratum: *ibid.* **495**, 441 (2000)], arXiv:hep-ph/0006345.
141. M. B. Voloshin, *Surveys High Energy Phys.* **8**, 27 (1995).
142. A. Kapustin, Z. Ligeti, M. B. Wise and B. Grinstein, *Phys. Lett. B* **375**, 327 (1996), arXiv:hep-ph/9602262.
143. N. Uraltsev, *Nucl. Phys. B* **491**, 303 (1997), arXiv:hep-ph/9610425.
144. UKQCD Collab. (M. Di Pierro *et al.*), *Nucl. Phys. B* **534**, 373 (1998), arXiv:hep-lat/9805028.
145. M. S. Baek, J. Lee, C. Liu and H. S. Song, *Phys. Rev. D* **57**, 4091 (1998), arXiv:hep-ph/9709386.
146. H.-Y. Cheng and K.-C. Yang, *Phys. Rev. D* **59**, 014011 (1999), arXiv:hep-ph/9805222.
147. UKQCD Collab. (M. Di Pierro *et al.*), *Phys. Lett. B* **468**, 143 (1999), arXiv:hep-lat/9906031.
148. UKQCD Collab. (M. Di Pierro *et al.*), *Nucl. Phys. B* (*Proc. Suppl.*) **73**, 384 (1999), arXiv:hep-lat/9809083.
149. J. L. Rosner, *Phys. Lett. B* **379**, 267 (1996), arXiv:hep-ph/9602265.
150. P. Colangelo and F. De Fazio, *Phys. Lett. B* **387**, 371 (1996), arXiv:hep-ph/9604425.
151. C.-S. Huang, C. Liu and S.-L. Zhu, *Phys. Rev. D* **61**, 054004 (2000), arXiv:hep-ph/9906300.
152. C. Tarantino, arXiv:hep-ph/0702235.
153. J. G. Korner and B. Melic, *Phys. Rev. D* **62**, 074008 (2000), arXiv:hep-ph/0005141.
154. B. Guberina, B. Melic and H. Stefancic, *Phys. Lett. B* **469**, 253 (1999), arXiv:hep-ph/9907468.
155. A. Lenz and U. Nierste, arXiv:1102.4274 [hep-ph].
156. LHCb Collab. (R. Aaij *et al.*), arXiv:1402.2554 [hep-ex].
157. CDF Collab. (T. Aaltonen *et al.*), *Phys. Rev. Lett.* **106**, 121804 (2011), arXiv:1012.3138 [hep-ex].
158. CDF Collab. (T. Aaltonen *et al.*), *Phys. Rev. Lett.* **104**, 102002 (2010), arXiv:0912.3566 [hep-ex].
159. D0 Collab. (V. M. Abazov *et al.*), *Phys. Rev. Lett.* **99**, 182001 (2007), arXiv:0706.2358 [hep-ex].
160. D0 Collab. (V. M. Abazov *et al.*), *Phys. Rev. Lett.* **99**, 142001 (2007), arXiv:0704.3909 [hep-ex].
161. CDF Collab. (A. Abulencia *et al.*), *Phys. Rev. Lett.* **98**, 122001 (2007), arXiv:hep-ex/0609021.
162. D0 Collab. (V. M. Abazov *et al.*), *Phys. Rev. Lett.* **94**, 102001 (2005), arXiv:hep-ex/0410054.
163. LHCb Collab. (R. Aaij *et al.*), arXiv:1402.6242 [hep-ex].
164. LHCb Collab. (R. Aaij *et al.*), *Phys. Rev. Lett.* **111**, 102003 (2013), arXiv:1307.2476 [hep-ex].
165. CDF Collab. (T. A. Aaltonen *et al.*), arXiv:1403.8126 [hep-ex].
166. A. J. Lenz, *AIP Conf. Proc.* **1026**, 36 (2008), arXiv:0802.0977 [hep-ph].
167. A. Lenz, arXiv:1404.6197 [hep-ph].
168. A. Lenz *et al.*, *Phys. Rev. D* **83**, 036004 (2011), arXiv:1008.1593 [hep-ph].
169. A. Lenz, U. Nierste, J. Charles, S. Descotes-Genon, H. Lacker, S. Monteil, V. Niess and S. T'Jampens, *Phys. Rev. D* **86**, 033008 (2012), arXiv:1203.0238 [hep-ph].
170. A. Lenz and U. Nierste, *JHEP* **0706**, 072 (2007), arXiv:hep-ph/0612167.

171. LHCb Collab. (R. Aaij *et al.*), LHCb-CONF-2012-002.
172. LHCb Collab. (R. Aaij *et al.*), *Phys. Rev. D* **87**, 112010 (2013), arXiv:1304.2600 [hep-ex].
173. LHCb Collab. (R. Aaij *et al.*), arXiv:1411.3104 [hep-ex].
174. ATLAS Collab. (G. Aad *et al.*), *JHEP* **1212**, 072 (2012), arXiv:1208.0572 [hep-ex].
175. ATLAS Collab. (G. Aad *et al.*), *Phys. Rev. D* **90**, 052007 (2014), arXiv:1407.1796 [hep-ex].
176. CMS Collab., CMS-PAS-BPH-13-012.
177. CDF Collab. (T. Aaltonen *et al.*), *Phys. Rev. Lett.* **109**, 171802 (2012), arXiv:1208.2967 [hep-ex].
178. D0 Collab. (V. M. Abazov *et al.*), *Phys. Rev. D* **85**, 032006 (2012), arXiv:1109.3166 [hep-ex].
179. J. Brod, A. Lenz, G. Tetlalmatzi-Xolocotzi and M. Wiebusch, arXiv:1412.1446 [hep-ph].

Fundamental Dynamics:
Past, Present and the Future — Like CP Violation and EDMs

Ikaros I. Bigi

Department of Physics, University of Notre Dame du Lac,
Notre Dame, IN 46556, USA

Working with Kolya Uraltsev was a real 'marvel' for me in general, but in particular about CP and T violation, QCD and its impact on transitions in heavy flavor hadrons and EDMs. The goal was — and still is — to define fundamental parameters for dynamics, how to measure them and compare SM forces with New Dynamics using the best tools including our brains. The correlations of them with accurate data were crucial for Kolya. Here is a review of CP asymmetries in B, D and τ decays, the impact of perturbative and non-perturbative QCD, about EDMs till 2013 — and for the future.

1. About my Collaboration with Kolya

In 1988 Kolya, V. A. Khoze, A. I. Sanda and I wrote the article "The Question of CP Noninvariance — as seen through the Eyes of Neutral Beauty" that was published in the World Scientific book "CP Violation" edited by C. Jarlskog.[1] I am very proud of that article; here are two of the best reasons for me:

- After long discussions about this article with the Nobel prize winner Jack Steinberger at the ARCS2000 in the French alps, he smiled and said: 'Very good work.'
- It discussed CP violation predicted with quarks and how they are affected by 'hard' and 'soft' re-scattering; I will come back to that.

Kolya and I had not met in person in 1988; that happened first in 1990 when he visited me at Notre Dame and we produced our second paper, namely "Induced Multi-Gluon Couplings and The Neutron Electric Dipole Moment."[2] A few months later when Kolya was back in the Leningrad Institute of Nuclear Physics, he sent me a Russian paper with a very similar title about this item and asked me, if I agree with the content.[a] I did — so he put my name as an author and submitted it

---

[a] At that time I could read it in Russian; now you can read that article in English.

to *Zh. Eksp. Teor. Fiz.*, where it was published.[3] I was very impressed by the work
that he had done after the previous paper and I was honored (and still am) that
Kolya put my name on it.

Kolya and I spent long times together both at Notre Dame and CERN, in
particular in the year of 1993/94, when we could work most of the night (and even
in the week from Christmas to New Year, when the CERN rooms were cold).[b]
Vivek Sharma allowed us to use the rooms, computers and printers there.[c] It was
a fruitful year. We had worked with Vainshtein and Shifman by email and phone
about establishing Heavy Quark Symmetry (HQS) and its expansion (HQE) not
only in principle, but also 'practically'. Kolya had produced new theoretical tools
for non-perturbative QCD. He knew that hard theoretical work is needed to produce
trustworthy predictions — and that it often takes a lot of time. He also said that
in the end data are the supreme judges. There is also a basic difference between
pre- and *post-*diction. One of the obvious examples is the history of the ratio of the
lifetimes of beauty baryons vs. mesons. Kolya and his collaborators (and Voloshin
as well) stated many times that HQE gives a ratio of around 0.9–1.0 — and were
steadfast about their prediction when it was in obvious disagreement with the data
during a decade or more. Kolya had thought a long time before arriving at:[4]

$$\frac{\tau(\Lambda_b)}{\tau(B_d)} \simeq 1 - \Delta_b, \quad \Delta_b \simeq 0.03\text{--}0.12. \tag{1}$$

It is important to remember that the limit of HQE is unity here. Therefore our control (or lack) of nonperturbative QCD depends on Δ_b, which shows large theoretical
uncertainties.[d] The most accurate data come from the LHCb collaboration:[6]

$$\frac{\tau(\Lambda_b)}{\tau(B_d)} = 0.974 \pm 0.006 \pm 0.004. \tag{2}$$

Both Kolya (*et al.*)[4] and Voloshin[5] were very happy about it — but not surprised
at all. This was a true *prediction* which was very, very different from data for a
long time. It is not obvious how much work and deep understanding of dynamics
were needed to give these predictions — but it is remarkable.

Before I write about dynamics in some details, I want to say: Kolya was a
wonderful person. I know he was very interested about art and history, as obvious
for a person who was born in Russia, worked in Italy, France, Japan and Siegen (the
painter P. P. Rubens was born in Siegen, not in Holland!). Kolya with his family
and I spent a day in the small city in Arezzo in Italy to see paintings, churches
and architecture there. He liked long discussions about fundamental physics with
passion and honesty. He was a true wonderful friend. He often asked me about my

---

[b]Kolya's wife Lilya knew how to deal with Russian theorists; Kolya could not find a better wife,
and he knew it. Kolya followed Lilya's expeditions to Kola Peninsula in the summers to help
getting food by fishing.
[c]Vivek should get fair credit for that.
[d]Voloshin preferred to say $\Delta_b \simeq 0.0\text{--}0.1$.[5]

family's health situation (including my mother's one) and how he could help, when he was in a bad situation himself. I miss him so much.

Here I give a review mostly of CP symmetry and its asymmetries, impact of QCD, some subtle points about EDMs and write about the future for the next ten years.

2. CP Symmetry and its Violations

In 1964 — fifty years ago – CP violation was found due to the existence of $K_L \to \pi^+\pi^-$. Okun explicitly listed the search for it as a priority for the future in his 1963 textbook[7] — a true prophet, since he was the only one. It has been predicted in 1981[8] that sizable or even large CP violation should be found in $B^0 \to \psi K_S$ transitions when the CKM model of flavor dynamics is a least the leading source of it. The existence of the B mesons had not been established then, never mind top quarks. After 1981 Sanda and I, and in parallel Uraltsev and collaborators, worked out the CP asymmetries in B and D mesons and refined the theoretical tools, without communications between Russia on one side and West Europe/USA on the other side due to the 'iron curtain'. It was predicted that CKM dynamics produce sizable or even large *indirect* CP violation in B_d oscillation, but small one in the B_s^0 mixing.

2.1. *Landscape of CP and T violation between* 1986 *and* 2013

It was known that two neutral B^0 and B_s^0 mesons oscillate, but in different landscapes of $x = \Delta M_B/\Gamma_B$ due to SM dynamics. As pointed out by Azimov, Uraltsev and Khoze,[9] indirect CP violation is small in B_s^0 oscillations. However sizable CP asymmetries can occur in CKM suppressed B_s^0 decays, and therefore one has to look for them. They emphasized the hierarchy of neutral B decays. Their quantitative predictions are based on the experimental claim that top quarks have been found with $M_t = 40 \pm 10$ GeV. It has been found that this claim was wrong, as history shows.[e] Still Azimov, Uraltsev and Khoze had good reasons to be proud of this paper, since the basic idea is correct.

By 1987 the B mesons had been found with $|V_{ub}| \ll |V_{cb}|$ and sizable B_d oscillations;[10] indirectly they gave reasons for the existence of top quarks with 50 GeV $< m_t <$ 200 GeV. The long article[1] by Khoze, Sanda, Uraltsev and me discussed both indirect and direct CP violation in B_d, B_s and D^0. The collaboration of these theorists happened by phones, emails and exchanging files between the US West Coast and Russia.

The 1988 article consisted of five Acts plus Prologue and Epilogue: Act I. The Plots: CP Asymmetries in B Decays; Act II. The Likely Hero: B_d and its Decays; Act III. The Dark Horse: B_s and its Decays; Act IV. The Dark Side — Search Scenarios; Act V. Conclusions and Outlook.

---

[e]When I found the statement from CERN outside our offices in the theoretical HEP group in Aachen, I looked at it and said: "They found it!" Peter Zerwas look at it, read it, thought for a few minutes and said: "It must be wrong, and I give you my reasons." Peter was correct as usual.

It focused on the following crucial points: (i) Three sides of 'the golden' CKM triangle are all of the order of λ^3; it gives CP asymmetries between $\sim 10\%$ and $\sim 80\%$ for B_d and B^+ transitions due to large angles.[f] (ii) Another triangle allows to probe B_s transitions. It gives one small angle, which leads to about 5% indirect CP violation in B_s oscillations due to the λ^2 suppression, but it allows for large direct CP asymmetries. (iii) CP violation in D decays are of the order of λ^4 — i.e. a few $\times 10^{-3}$ in the SM. (iv) There are 'good' and 'bad' signs; I will come back to that later. (v) We have to look for the impact of New Physics (NP) with higher accuracy and/or in rare decays. (vi) To have direct CP violation in two- and three-body final states (FS) one needs final state interactions (FSI), and strong forces provide them. (vii) Penguin diagrams[11] can affect or even produce direct CP asymmetries in $B_{u,d,s}$ decays. However there are subtle points: penguin diagrams are formulated for quark states; to compare those predictions with measured data with hadrons one has to use the concept of 'duality'.

Penguin diagrams were introduced for kaon nonleptonic decays based on their connections with local operators.[12] In B decays they affect inclusive final state with 'hard' re-scattering. For CP asymmetry in exclusive decays one has to deal with 'soft' re-scattering. Based on rough models we predicted $A_{CP}(\bar{B}_d \to K^- \pi^+) \sim 0.1$. It is a decent early prediction about subtle features of B dynamics. Actually this article gave good predictions in general, namely:

- B_d transitions are the 'hero' of true large CP violation in the SM as predicted in Act II. It was stated that in the future the angles ϕ_1 in $B_d \to \psi K_S$ and ϕ_2 in $B_d \to \pi^+\pi^-$ will be measured with sizable or even large values; the latter one will also show sizable direct CP asymmetry.

  CP violation in the B system was established only in 2001 in $B_d \to \psi K_S$; PDG 2013 gives:

$$S_{CP}(B_d \to \psi K_S) = \sin \phi_1 = +0.676 \pm 0.021 \,. \tag{3}$$

Very recent Belle data also show:[13]

$$S_{CP}(B_d \to \pi^+\pi^-) = -0.64 \pm 0.08_{\text{stat.}} \pm 0.03_{\text{syst.}} \,, \tag{4}$$

$$A_{CP}(B_d \to \pi^+\pi^-) = +0.64 \pm 0.33_{\text{stat.}} \pm 0.03_{\text{syst.}} \tag{5}$$

- Indeed B_s decays are 'Dark Horse(s)' as in Act III: (i) The SM gives CP asymmetries in semi-leptonic decays significantly less than 10^{-4}, while NP could produce it 'closer' to 0.01. (ii) Indirect CP violation in $B_s \to \psi\phi$ could be seen around a few percent in the SM, while NP would reach the 10–20% level as Sanda and I had said before 2000. (iii) Direct CP asymmetries in CKM suppressed decays could be large.

---

[f]If a reader is interested, like Kolya, in art and its connection with symmetry, she/he can see 'Musee d'Orsay: Aristide Maillol, The Mediterranean/Thought' about "interlocking triangles" and "return to order."

Very recent LHCb data confirm these 1988 predictions in subtle ways;[14] first:

$$\frac{\Delta\Gamma(B_s)}{\text{ps}} = 0.106 \pm 0.011 \pm 0.007\,, \quad \frac{\Gamma(B_s)}{\text{ps}} = 0.661 \pm 0.004 \pm 0.006\,, \quad y_s \simeq 0.08\,.$$

(6)

Kolya and I were not sure (and I am still) about the small theoretical uncertainty about $\Delta\Gamma(B_s)$; Alex Lenz discusses it in his contribution. I focus on indirect CP violation:

$$\phi_s^{c\bar{c}s}|_{\text{data}} = (0.01 \pm 0.07 \pm 0.01)\ \text{rad} \quad \text{vs.} \quad \phi_s^{c\bar{c}s}|_{\text{SM}} = (-0.0363^{+0.0016}_{-0.0015})\ \text{rad}\,. \quad (7)$$

The data are close to the expected SM values — but also consistent with sizable or even leading NP contributions. Furthermore I disagree with the uncertainty from the SM usually claimed in the literature; below I will explain why.

- Search scenarios for CP violation in neutral heavy mesons were discussed in Act IV about the 'Dark Side' of neutral beauty and charm mesons: (i) It was emphasized that CP violation in inclusive decays is much smaller than in exclusive ones. (ii) CP asymmetries in D^0 decays were discussed.

- In Act V it was pointed out that, as of 1988, the connection between the observables and the underlying electroweak parameters was 'obscured' by the impact of FSI.

- The epilogue gave important comments: if detailed data on B decays could be compared with our rather accurate predictions, and failed them — there would be "no plausible deniability" that the CKM theory could no longer be maintained as the sole or even dominant source of CP violation — therefore NP had to exist.

  Now the landscape has changed: the SM gives at least the leading source of CP violation in B decays; therefore we have to go from 'accuracy' to 'precision'. That is the landscape upon us; it seems to me that 'young' people working in HEP (both on the experimental and theoretical side) would like to produce such a transition (I hope). Furthermore, we need a deeper understanding of charm decays.

We knew that the impact of penguin diagrams on hadronic FS and CP asymmetries is very subtle already in B decays, and even more in D ones. It was discussed at a deeper level in 1990 by Dokshitzer and Uraltsev in Ref. 15, and at the DPF-92 meeting at Fermilab, where Kolya gave a talk on a short paper about FSI phases.[16] One needs only a short time to read it — but a long time to think about the items like inclusive vs. exclusive transitions and 'hard' vs. 'soft' re-scattering.

From my own direct experience during this long period I know that Kolya was a real leader in probing CP violation in beauty and charm decays and in understanding the information given by data. Furthermore Kolya showed us that the first and second rounds are not enough — one has to go further.

I. I. Bigi

2.2. *Duality — The connection of quark and hadronic diagrams*

The issue of 'duality' between quarks and hadronic forces has been used in very different situations. Some are straightforward like for 'jets', while others are subtle: flavor forces depend crucially on the non-perturbative aspects of QCD. Kolya and collaborators have worked on duality as a tool in HQE, mostly for beauty hadrons decays.

It is not enough to give hand-waving statements there — we can defend them with some accuracy to measure CKM angles and to compare inclusive vs. exclusive rates. It is not enough at all to compare sums of measured hadronic FS rates vs. parton model ones with quarks. One of the first papers to deal with this subtle issue dates back to 1986.[17] It gives us insight into the inner structure of strong forces. We had discussed local vs. semi-local duality and how close one has to go to thresholds as discussed in Ref. 18. It will be discussed elsewhere in this Memorial Book; still I list a few important references about Kolya's work about duality.[19] For heavy quarks the ratios of lifetimes of baryons and mesons go to unity in the heavy quark limit like $\sim (\Lambda/m_Q)^2$ in HQE; thus the theoretical uncertainty is 'sizable'. The next steps are to measure $\tau(\Xi_b^0)$ and $\tau(\Xi_b^-)$ with some accuracy. It was predicted:[5] $\tau(\Xi_b^0) \simeq \tau(\Lambda_b) < \tau(B_d) < \tau(\Xi_b^-)$. It shows how much we can control the impact of non-perturbative QCD in a semi-quantitative way on inclusive decays of beauty hadrons. Data told us we can reproduce the lifetimes of charm ones semi-quantitatively — is it just luck?

2.2.1. *Re-scattering and CP violation and CPT constraints*

It is important to think about the connections between *quark diagrams* and measured (or measurable) rates with *hadrons*. There are important, but subtle points:

- In the quark world we use weak dynamics with $b\bar{q} \to q_1\bar{q}_2q_3\bar{q}_4$ (even including 'Weak Annihilation/Scattering'). Using the SM we deal with inclusive FS with $q_i = u, d, s$. Including QCD forces we use $m_u < m_d \ll m_s < \bar{\Lambda}$. The predictions are different due to iso-spin and $SU(3)_{fl}$ violations, however differences are small compared to $\bar{\Lambda}$ for inclusive rates. Measured inclusive FS consist of sums of hadrons; those show little effect of $SU(3)_{fl}$ violation.
- The landscapes for *exclusive* non-leptonic decays are quite different. Two-, three- and four-body, etc. FS can easily show sizable $SU(3)_{fl}$ violation and therefore about CP asymmetries. The impact are due to re-scattering with QCD dynamics, in particular soft re-scattering with non-perturbative forces. It can be calibrated by M_K vs. M_π, the impact of chiral symmetry and its violations.

The connection between the strength of re-scattering, CP asymmetries and CPT invariance has been discussed in Ref. 1; it is explained in Sect. 4.10 in Ref. 20 with much more details, including CPT invariance following the history sketched above:

$$T(P \to a) = e^{i\delta_a} \left[T_a + \sum_{a \neq a_j} T_{a_j} i T^{\text{resc}}_{a_j a} \right],$$ (8)

$$T(\bar{P} \to \bar{a}) = e^{i\delta_a} \left[T_a^* + \sum_{a \neq a_j} T_{a_j}^* i T^{\text{resc}}_{a_j a} \right],$$ (9)

where amplitudes $T^{\text{resc}}_{a_j a}$ describe FSI between a and intermediate states a_j that connect with this FS. Thus one gets for CP asymmetries:

$$\Delta\gamma(a) = |T(\bar{P} \to \bar{a})|^2 - |T(P \to a)|^2 = 4 \sum_{a \neq a_j} T^{\text{resc}}_{a_j a} \operatorname{Im} T_a^* T_{a_j}.$$ (10)

This CP asymmetry has to vanish upon summing over all such states a:

$$\sum_a \Delta\gamma(a) = 4 \sum_a \sum_{a \neq a_j} T^{\text{resc}}_{a_j a} \operatorname{Im} T_a^* T_{a_j} = 0,$$ (11)

since $T^{\text{resc}}_{a_j a}$ and $\operatorname{Im} T_a^* T_{a_j}$ are symmetric and antisymmetric, respectively, in the indices a and a_j. This shows that CPT invariance imposes equality also between *sub*classes of partial widths. There are important points:

- These equations show the non-trivial impact of re-scattering/FSI in general.
- CP asymmetries in two-body FS give us 'only' numbers. Those in three-body FS give us two-dimensional observables, namely measure Dalitz plots.
- In principle (with infinite data) one can probe 'local' CP asymmetries. However we have to be realistic and use tools to reduce the numbers of observables. Therefore we probe 'regional' asymmetries. We can use the definition of 'fractional' asymmetry or 'significance' one or others.[21] Furthermore it depends on the 'landscape' of the FS where we choose it. It needs 'judgement' based on our experience, namely on the impact of resonances and the differences between narrow and broad ones. It helps significantly to use chiral symmetry.

For this work I apply these amplitudes for FS with hadrons and resonances: $P \to h_1[h_2 h_3] + h_2[h_1 h_3] + h_3[h_1 h_2] \Rightarrow h_1 h_2 h_3$.

The above equations apply to amplitudes of hadrons or quarks and also to boundstates of $\bar{q}_i q_j$ (or $q_i q_j q_k$). The crucial point is how to connect 'measurable' hadronic amplitudes with quark and gluon ones. One can show that connection with diagrams; however in quantitative ways it is subtle due to non-perturbative forces. To make it short: in the world of quarks our tools mostly focus on inclusive transitions, unless one can use other theoretical tools; it depends on our 'judgement'. I will discuss it separately in B and D decays.

2.2.2. *Comments about 'hard' vs. 'soft' re-scattering*

There is a large difference between computing penguin diagrams and what they mean for beauty and charm decays (for different reasons) — unlike for kaon transitions. In beauty decays one can calculate *inclusive* CKM suppressed FS with CP asymmetries based on 'hard' re-scattering between FS with local operators. Penguin diagrams can give us the direction about correlations between hadrons in exclusive FS, but not in a truly quantitative way. In charm decays we have penguin diagrams about CP asymmetries, but less control over 'soft' re-scattering as discussed in Ref. 22 in some details.

2.3. *CP violation via Higgs dynamics*

We have known for a long time that non-minimal Higgs models could not contribute sizably to ϵ and/or ϵ' unless 'our' world lives in very tiny corners of Higgs forces. In the 1997 book "Perspectives on Higgs Physics II"[23] there is an article by Sanda, Uraltsev and me about "Addressing the Mysterious with the Obscure — CP Violation via Higgs Dynamics." It focused on EDMs of neutron and electrons and atoms, T *odd* electron–nucleon interaction and $K \to \mu\nu\pi$ and CP violation in B and D and top transitions. One neutral Higgs boson has been found in 2012, but no charged one (yet).

Non-minimal Higgs models are one of leading candidates for NP: they provide us with a road to SUSY (and thus string theory) directly and indirectly. I will discuss it in the next subsection; however I will first comment about the measurable status of the real Higgs.

The amplitude of the SM neutral Higgs state is 100% scalar. ATLAS and CMS have established the existence of a neutral spin 0 boson Φ with a mass of 125.8 ± 0.4(stat.) ± 0.4(syst.) GeV in the FS of 2γ, l^+l^-, $\bar{b}b$ quarks, ZZ^* and WW^*.[24] We know that Φ is at least mostly a scalar boson, and pseudo-scalar contributions are at best subleading. Small pseudo-scalar amplitudes cannot produce sizable rates by themselves, but can sizably contribute to the interference with scalar SM ones — i.e., the $j = 0$ Φ state can mix strongly with the scalar amplitude and weakly with the pseudo-scalar one to produce CP asymmetries.[25] Understanding the Higgs sector is an important project for very high luminosity runs at LHC and also for ILC.

Another comment about production and decay of the established neutral Higgs boson: usually ATLAS and CMS base their analyses on the Higgs width predicted by the SM; however the impact of NP and Dark Matter can hide there. A new idea has appeared, namely to probe $pp \to H + X_1 \to {}^{\prime}ZZ{}^{\prime} + X_2 \to e^+e^-\mu^+\mu^- + X_3$ with $M(e^+e^-\mu^+\mu^-) > 130$ GeV that hardly depends on Γ_H; then one can compare the data from 126 GeV.[26]

2.4. *Probing CP asymmetries in the future*

We know that the SM gives at least the leading source of CP violation in $\Delta B \neq 0$ dynamics. Furthermore the neutral Higgs boson has been found now as expected with $M(H^0) \simeq 125$ GeV. On the other hand the usual reasons for NP exist, namely:

- We need NP to produce 'us', namely huge matter vs. antimatter asymmetry now; forces based on the CKM matrix can*not* do it.
- Theorists tell us extreme fine tuning is necessary to get electro-weak symmetry breaking below 1 TeV, if NP is at a much higher scale, like 10^{10} or 10^{15} or 10^{19} GeV.

The data and our experimental colleagues tell us:

- The three neutrinos have different masses to give oscillations as measured.
- Working to get 'known' matter (and 'us') is a sideshow in 'our' universe, since it produces only around 4% part of our universe.
- Dark matter gives around 23% of it; there are several candidates like SUSY, but none is established (yet).
- Dark energy gives around 73% — but what is it?

Really a lot of work has been done about 'our' universe — but we need even more. CP asymmetries can be based on the interference between SM amplitude and NP one. As usual, indirect searches for the impact of NP can reach much higher scales than direct ones due to inferences of two different amplitudes.

NP can produce at best non-leading source of CP violation in B mesons. Therefore one needs more data with better experimental and theoretical accuracies. First one focus on (quasi-)two-body FS. However I think we have to measure three- and four-body non-leptonic FS and their 'topologies' like the two-dimensional Dalitz plots and the correlations between narrow and broad resonances. We need even more data and more tools, but the data give us much more information about the underlying forces. People who work in Hadro-Dynamics have produced and checked their technologies about $h_1 h_2 \to h_3 h_4$ scattering; now we can apply them to achieve a deeper understanding of fundamental physics. Also we know now that the usual Wolfenstein parameterization is very adequate for leading sources of CP violation, but not for non-leading ones.

The situation is different for charm decays. The SM produces only small asymmetries in singly Cabibbo suppressed (SCS) decays and close to zero in doubly Cabibbo (DCS) ones. The limits from the data tell us that NP can produce only small asymmetries in SCS decays. For DCS decays we have little limits on CP asymmetries. However we need much larger productions of $D_{(s)}$ and Λ_c states. Again we have to probe FS with three- and four-body FS; CPT invariance is usable there.

The SM cannot produce measurable CP asymmetry in τ decays beyond the well measured CP violation in $K^0 - \bar{K}^0$ oscillations. BaBar data show a CP asymmetry in $\tau^- \to \nu K_S \pi^- [+\pi^0]$'s that is opposite to the predicted one — but at only 2.9 sigmas:

$$A_{\text{CP}}(\tau^+ \to \bar{\nu} K_S \pi^+)|_{\text{SM}} = +(0.36 \pm 0.01)\% \qquad \text{(Ref. 27)}, \quad (12)$$

$$A_{\text{CP}}(\tau^+ \to \bar{\nu} K_S \pi^+ [+\pi^0\text{'s}])|_{\text{BaBar2012}} = -(0.36 \pm 0.23 \pm 0.11)\% \quad \text{(Ref. 28)}. \quad (13)$$

Now available data probe only integrated CP asymmetries. It is important to probe regional CP asymmetries in $\tau^- \to \nu[S = -1]$ FS; we have to wait for Belle II (and Super-Tau-Charm Factory if and when it will ever exist). As pointed out last year, it is important to measure the correlations in $D^+ \to K^+\pi^+\pi^-/K^+K^+K^-$, etc.[29]

Finally we have to probe correlations between *known* matter and candidates of *dark* matter in CP asymmetries and rare B and D decays — if we get even more data with precison and understand underlying dynamics with better theoretical tools.

2.4.1. *Better parameterization of the CKM matrix*

With three quark families one constructs six triangles with different shapes, but also the same area. Obviously one can construct them in several ways. One can do it by their three sides (or the ratios); crucial contributions come from $|V_{cb}|$, $|V_{ub}|$, $|V_{td}|$, $|V_{ts}|$, etc. In particular vivid discussions are still happening on the comparison of $|V_{cb}|_{\text{excl}}$ vs. $|V_{cb}|_{\text{incl}}$ and $|V_{ub}|_{\text{excl}}$ vs. $|V_{ub}|_{\text{incl}}$. Kolya was a true leader in predicting these four classes of transitions and understanding the informations the data tell us about the underlying dynamics with accuracy; it is discussed in other articles in this Memorial Book.

PDG and HFAG show the 'exact' CKM matrix with three families of quarks. However experimenters and theorists do not use exact CKM matrix as you can see in their papers and talks.

In Wolfenstein parameterization one gets six triangles that are combined into three classes with four parameters λ, A, $\bar{\eta}$ and $\bar{\rho}$ with $\lambda \simeq 0.223$. Those are probed and measured in K, B, B_s and D transitions: $A \sim 1$, but the two remaining ones are *not* of $\mathcal{O}(1)$: $\bar{\eta} \simeq 0.34$ and $\bar{\rho} \simeq 0.13$. It is assumed — usually without mentioning — that one applies them without expansion of $\bar{\eta}$ and $\bar{\rho}$. Obviously it is a 'smart' parameterization with a clear hierarchy.

Now we need a parameterization of the CKM matrix with more precision for non-leading sources in B decays and very small one for CP asymmetries in D decays with little 'background' from SM. Several 'technologies' were proposed, like the one in Ref. 30 with λ as before, but $f \sim 0.75$, $\bar{h} \sim 1.35$ and $\delta_{\text{QM}} \sim 90°$. Now we get somewhat different six classes, and it is more subtle for CP violation:

$$\begin{pmatrix} 1 - \frac{\lambda^2}{2} - \frac{\lambda^4}{8} - \frac{\lambda^6}{16}, & \lambda, & \bar{h}\lambda^4 e^{-i\delta_{\text{QM}}}, \\[2ex] -\lambda + \frac{\lambda^5}{2}f^2, & \begin{matrix} 1 - \frac{\lambda^2}{2} - \frac{\lambda^4}{8}(1+4f^2) - f\bar{h}\lambda^5 e^{i\delta_{\text{QM}}} \\ + \frac{\lambda^6}{16}(4f^2 - 4\bar{h}^2 - 1), \end{matrix} & \begin{matrix} f\lambda^2 + \bar{h}\lambda^3 e^{-i\delta_{\text{QM}}} \\ -\frac{\lambda^5}{2}\bar{h}e^{-i\delta_{\text{QM}}}, \end{matrix} \\[3ex] f\lambda^3, & \begin{matrix} -f\lambda^2 - \bar{h}\lambda^3 e^{i\delta_{\text{QM}}} \\ + \frac{\lambda^4}{2}f + \frac{\lambda^6}{8}f, \end{matrix} & \begin{matrix} 1 - \frac{\lambda^4}{2}f^2 - f\bar{h}\lambda^5 e^{-i\delta_{\text{QM}}} \\ -\frac{\lambda^6}{2}\bar{h}^2 \end{matrix} \end{pmatrix} + \mathcal{O}(\lambda^7),$$

Class I.1 : $V_{ud}V_{us}^*[\mathcal{O}(\lambda)] + V_{cd}V_{cs}^*[\mathcal{O}(\lambda)] + V_{td}V_{ts}^*[\mathcal{O}(\lambda^{5\&6})] = 0$,   (14)

Class I.2 : $V_{ud}^*V_{cd}[\mathcal{O}(\lambda)] + V_{us}^*V_{cs}[\mathcal{O}(\lambda)] + V_{ub}^*V_{cb}^*[\mathcal{O}(\lambda^{6\&7})] = 0$,   (15)

Class II.1 : $V_{us}V_{ub}^*[\mathcal{O}(\lambda^5)] + V_{cs}V_{cb}^*[\mathcal{O}(\lambda^{2\&3})] + V_{ts}V_{tb}^*[\mathcal{O}(\lambda^2)] = 0$,   (16)

Class II.2 : $V_{cd}^*V_{td}[\mathcal{O}(\lambda^4)] + V_{cs}^*V_{ts}[\mathcal{O}(\lambda^{2\&3})] + V_{cb}^*V_{tb}^*[\mathcal{O}(\lambda^{2\&3})] = 0$,   (17)

Class III.1 : $V_{ud}V_{ub}^*[\mathcal{O}(\lambda^4)] + V_{cd}V_{cb}^*[\mathcal{O}(\lambda^{3\&4})] + V_{td}V_{tb}^*[\mathcal{O}(\lambda^3)] = 0$,   (18)

Class III.2 : $V_{ud}^*V_{td}[\mathcal{O}(\lambda^3)] + V_{us}^*V_{ts}[\mathcal{O}(\lambda^{3\&4})] + V_{ub}^*V_{tb}^*[\mathcal{O}(\lambda^4)] = 0$.   (19)

One finds the same pattern as from Wolfenstein parametrization, namely 'large' CP asymmetries in Class III.1, sizable ones in Class II.1 and 'small' one in Class I.1. However, the pattern is not so obvious, and it is similar only in a semi-quantitive way:

(i) In Class III.1 triangle one usually calls the two angles ϕ_1/β and ϕ_3/γ. They are measured in CP asymmetries in $B_d \to \psi K_S$ and $B^+ \to D_+ K^+$ decays due to interference between two contributions one gets from CKM dynamics. Adapting the refined parametrization one finds that CKM dynamics produce $S(B_d \to \psi K_S) \sim 0.72$ as largest possible value for CP asymmetry with $\delta_{\rm QM} \simeq 100° - 120°$ to compare with the measured

$$S(B_d \to \psi K_S) \sim 0.676 \pm 0.021\,. \qquad (20)$$

When correlations with ϕ_2/α and ϕ_3/γ point to $\phi_1/\beta \simeq 75°\text{--}90°$ one gets $S(B_d \to \psi K_S) = \sin 2\phi_1 \simeq 0.62\text{--}0.68$! Therefore it seems that CKM dynamics give very close to 'maximal' SM CP violation. However the situation is more subtle — as discussed next.

(ii) We are searching for non-leading sources of CP violation in B transitions. NP's impact could 'hide' there in "SM predicted" CP asymmetries. 'Data' given by HFAG, for example, are averaged over values of $|V_{ub}/V_{cb}|$ from inclusive and exclusive semileptonic B decays; actually the 'central' value is closer to $|V_{ub}|_{\rm excl}$ rather than the larger $|V_{ub}|_{\rm incl}$. It is quite possible that the theoretical uncertainties about extracting $|V_{cb}|$, $|V_{ub}|$ and $|V_{ub}/V_{cb}|$ from $B \to l\nu\pi$ vs. $B \to l\nu D^*$ are much larger than claimed; some details are told about it in Ref. 31. A new idea using dispersion relations and chiral symmetry (in a smart way) came up very recently, namely to extract $|V(ub)|$ from data on $B \to l\nu\pi^+\pi^-$;[32] it probes the impact of broad scalar resonances. It gives us more roads to understand the underlying dynamics. One can think also about measuring $B_s \to l^+\nu K_S\pi^-$ and $B \to l\nu K\bar{K}$ and how much you can use chiral symmetry.

(iii) The information on NP from data now and in the future has to be based on accuracy and the correlations among different FS in several B, D and K transitions and rare decays. In particular we can probe Class I.1 with the tiny rates of $K \to \pi\nu\bar{\nu}$ rates. The theoretical uncertainties are under control in that case; the hope is to produce enough data.

(iv) We have to probe correlations with different FS based on CPT invariance. The best fitting of the data do not give us the best information about the underlying dynamics.

2.5. 'Catholic' road to NP — three-body final states

For $D/B \to P_1 P_2 P_3$ or $\tau \to \nu P_1 P_2$ decays there is a single path to 'heaven', namely asymmetries in the Dalitz plots. One can rely on relative rather than absolute CP violation; thus it is much less dependent on production asymmetries. However one needs a lot of statistics — and robust pattern recognition.

2.5.1. CP asymmetries in $B^\pm$ decays

Data of CKM suppressed B^+ decays to charged three-body FS show not surprising rates

$$\text{BR}(B^+ \to K^+\pi^-\pi^+) = (5.10 \pm 0.29) \cdot 10^{-5}, \tag{21}$$

$$\text{BR}(B^+ \to K^+K^-K^+) = (3.37 \pm 0.22) \cdot 10^{-5}. \tag{22}$$

LHCb data show sizable CP asymmetries *averaged* over the FS with correlations:[33]

$$\Delta A_{\text{CP}}(B^\pm \to K^\pm\pi^+\pi^-) = +0.032 \pm 0.008_{\text{stat}} \pm 0.004_{\text{syst}} [\pm 0.007_{\psi K^\pm}],$$
$$\Delta A_{\text{CP}}(B^\pm \to K^\pm K^+K^-) = -0.043 \pm 0.009_{\text{stat}} \pm 0.003_{\text{syst}} [\pm 0.007_{\psi K^\pm}]. \tag{23}$$

It is not surprising that these CP asymmetries come with *opposite* signs, due to the CPT invariance constraint in Eq. (11). Furthermore there are also large *regional* CP asymmetries which refer to a particular region of the phase space:

$$A_{\text{CP}}(B^\pm \to K^\pm\pi^+\pi^-)|_{\text{regional}} = +0.678 \pm 0.078_{\text{stat}} \pm 0.032_{\text{syst}} [\pm 0.007_{\psi K^\pm}],$$
$$A_{\text{CP}}(B^\pm \to K^\pm K^+K^-)|_{\text{regional}} = -0.226 \pm 0.020_{\text{stat}} \pm 0.004_{\text{syst}} [\pm 0.007_{\psi K^\pm}]. \tag{24}$$

The 'regional' CP asymmetries in the LHCb data mean here: (i) positive asymmetry at low $m_{\pi^+\pi^-}$ below m_{ρ^0}; (ii) negative asymmetry both at low and high $m_{K^+K^-}$ values. These statements make very good sense. However I want to emphasize we need some thinking and judgment about the definitions of regional asymmetries and to go beyond analyses of the best fitted data. One has to remember that scalar resonances (like $f_0(500)/\sigma$ and κ) produce broad ones that are not described by Breit–Wigner parametrization; instead they can be described by dispersion relations (or other ways). At the qualitative level one should not be surprised. Probing the topologies of Dalitz plots with accuracy one might find the existence of NP. Most of the data come along the frontiers, while the centers are practically empty. Therefore interferences happen on few places, and regional asymmetries are much larger than averaged ones — but so much? We have to remember that the final goal is to find non-leading sources of CPV in $\Delta B \neq 1, 2$.

One can look at even more CKM suppressed three-body FS:

$$\mathrm{BR}(B^+ \to \pi^+\pi^-\pi^+) = (1.52 \pm 0.14) \cdot 10^{-5}, \tag{25}$$

$$\mathrm{BR}(B^+ \to \pi^+K^-K^+) = (0.52 \pm 0.07) \cdot 10^{-5}. \tag{26}$$

One might guess that penguin diagrams have a smaller impact on $b \to d$ than on $b \to s$ comparing Eq. (26) with Eq. (22). On the other hand one also expects smaller impact on three- or more body FS than in two-body ones due to chiral symmetry.

LHCb has shown these averaged and 'regional' CP asymmetries:[34]

$$A_{\mathrm{CP}}(B^\pm \to \pi^\pm\pi^+\pi^-) = +0.117 \pm 0.021_{\mathrm{stat}} \pm 0.009_{\mathrm{syst}}[\pm 0.007_{\psi K^\pm}],$$
$$A_{\mathrm{CP}}(B^\pm \to \pi^\pm K^+K^-) = -0.141 \pm 0.040_{\mathrm{stat}} \pm 0.018_{\mathrm{syst}}[\pm 0.007_{\psi K^\pm}], \tag{27}$$

$$\Delta A_{\mathrm{CP}}(B^\pm \to \pi^\pm\pi^+\pi^-)|_{\mathrm{reg.}} = +0.584 \pm 0.082_{\mathrm{stat}} \pm 0.027_{\mathrm{syst}}[\pm 0.007_{\psi K^\pm}],$$
$$\Delta A_{\mathrm{CP}}(B^\pm \to \pi^\pm K^+K^-)|_{\mathrm{reg.}} = -0.648 \pm 0.070_{\mathrm{stat}} \pm 0.013_{\mathrm{syst}}[\pm 0.007_{\psi K^\pm}]. \tag{28}$$

Again it is not surprizing that these asymmetries come with opposite signs. However there are two very interesting statements about the data shown:[34]

- $B^\pm \to \pi^\pm\pi^-\pi^+$ decays show CP asymmetries both positive signs with $m^2_{\pi^+\pi^-} > 15$ GeV^2 and $m^2_{\pi^+\pi^-} < 0.4$ GeV^2.
- On the other hand we find negative CP asymmetry in $m^2_{K^+K^-} < 1.5$ GeV^2.

We need more data — they will appear 'soon' —, find other regional asymmetries and work on correlations with other FS. Importantly we need more thinking to understand what the data tell us about the underlying dynamics including the impact of non-perturbative QCD. It seems that the landscape is even more complex than said before and show the impact of really broad resonances.

There will be 'active' discussions about the impact of CPT invariance, namely the duality — averaged and regional transitions — between the worlds of hadrons and quarks. There are several reasons to expect that the impact of penguin diagrams is large. However there is a quantitative question now: How can CP asymmetries in Eq. (27) be larger by a factor of three than in Eq. (23), etc.? Also there are subtle questions, namely the definition of 'regional' transitions: The best-fit result does not give us the best information about underlying dynamics, in particular about non-leading sources; we have to think deeper and use several theoretical tools. At first we need some good judgment to define 'regional' asymmetry with finite data. Later we can test our judgment with even more — but still finite — data and correlations with other data. Still we need thinking — model independent analyses are not always an excellent idea.

We have to think about the impact of penguins diagrams on *exclusive* rates and CP asymmetries. It shows the impact of penguins/re-scattering diagrams, since the FS with $\Delta S \neq 0$ are larger than with $\Delta S = 0$. However one can remember that penguins *operators* show only hard re-scattering and focus on inclusive decays. First one probes averaged CP asymmetries, but later regional ones in the Dalitz plots and probe the correlations with different FS as shown above. Such procedures have been suggested and simulated in the case of three-body FS in $B^\pm$ decays[21] as second step — but this is not the final step in my view. The meaning of the analysis 'being model independent techniques' is very complex with finite data with non-perturbative QCD about non-leading source of CP violation. One needs other theoretical tools like chiral symmetry and/or dispersion relations based on data with low energy collisions and/or correlations with other transitions.

2.5.2. *CP asymmetries in* $D_{(s)}^\pm$ *(and* $\tau^\pm$*) decays*

So far no CP violation has been established in charm hadron decays. CPV can well be probed with two-body FS, but also in three-(and four-)body ones with more data like $D \to K_S\pi\pi$. A visionary paper by Azimov and Iogansen about direct CPV in two-body FS was published in 1981.[35]

Probing three-body SCS and DCS gives more information about fundamental forces. It was pointed in 1989[36] in general. One can disagree on several details, however it is important to think about our tools. $D^\pm$ has two all charged three-body FS on the SCS level — namely $D^\pm \to \pi^\pm\pi^+\pi^-/\pi^\pm K^+K^-$[37] — and also on the DCS one — $D^\pm \to K^\pm\pi^+\pi^-/K^\pm K^+K^-$. $D_s^\pm$ has two ones on the SCS level — $D_s^\pm \to K^\pm\pi^+\pi^-/K^\pm K^+K^-$ — however only one for DCS level — $D_s^\pm \to K^\pm K^\pm\pi^\mp$.

As stated above, for SCS FS the SM gives small 'background' for CPV and very close to zero about DCS ones. We have to probe FS with broad resonances — in particular scalar ones like $f_0(500)/\sigma$ and κ — and their interferences. The 'landscapes' of Dalitz plots in charm decays are different from B decays, because the central regions of phase space are not empty.

There may be a sign — maybe — of NP in τ decays about *averaged* CP asymmetries, see Eqs. (12,13). It is crucial to probe regional ones. Furthermore one has to measure correlations with $D_{(s)}^\pm$ decays.[29]

2.6. *'Protestant' road to NP — four-body final states*

There are several ways to probe CPV in four-body FS and to differentiate the impact of SM vs. NP: the landscapes are even more complex, while our theoretical toolbox is smaller so far. On the hand, when we will have more data on charm and beauty decays, it will enhance — I hope — the interests of young theorists (maybe from hadrodynamics) to produce new tools to probe four-body FS. I focus on charm decays. One can compare T *odd* moments or correlations in D vs. $\bar{D}$. For example one has to measure the angle ϕ between the planes of $\pi^+ - \pi^-$ and $K - \bar{K}$ and describe its dependence:[20,29]

$$\frac{d\Gamma}{d\phi}(D \to K\bar{K}\pi^+\pi^-) = \Gamma_1 \cos^2\phi + \Gamma_2 \sin^2\phi + \Gamma_3 \cos\phi\sin\phi\,, \qquad (29)$$

$$\frac{d\Gamma}{d\phi}(\bar{D} \to K\bar{K}\pi^+\pi^-) = \bar{\Gamma}_1 \cos^2\phi + \bar{\Gamma}_2 \sin^2\phi - \bar{\Gamma}_3 \cos\phi\sin\phi\,. \qquad (30)$$

The partial width for $D[\bar{D}] \to K\bar{K}\pi^+\pi^-$ is given by $\Gamma_{1,2}[\bar{\Gamma}_{1,2}]$: $\Gamma_1 \neq \bar{\Gamma}_1$ and/or $\Gamma_2 \neq \bar{\Gamma}_2$ represents direct CPV in the partial width. Γ_3 and $\bar{\Gamma}_3$ represent T *odd* correlations; by themselves they do not necessarily indicate CPV, since they can be induced by strong FSI; however:[38,20]

$$\Gamma_3 \neq \bar{\Gamma}_3 \to \text{CPV}. \tag{31}$$

Integrated rates give $\Gamma_1 + \Gamma_2$ vs. $\bar{\Gamma}_1 + \bar{\Gamma}_2$; the integrated *forward–backward* asymmetry

$$\langle A \rangle = \frac{\Gamma_3 - \bar{\Gamma}_3}{\pi(\Gamma_1 + \Gamma_2 + \bar{\Gamma}_1 + \bar{\Gamma}_2)} \tag{32}$$

gives full information about CPV. One could disentangle Γ_1 vs. $\bar{\Gamma}_1$ and Γ_2 vs. $\bar{\Gamma}_2$ by tracking the distribution in ϕ. If there is a *production* asymmetry, it gives global $\Gamma_1 = c\bar{\Gamma}_1$, $\Gamma_s = c\bar{\Gamma}_2$ and $\Gamma_3 = -c\bar{\Gamma}_3$ with *global* $c \neq 1$. Furthermore one can applying these observables to $D[\bar{D}] \to 4\pi$ (with CPT invariance) and later for $D^+ \to K^+\pi^-\pi^+\pi^-$ vs. $D^- \to K^-\pi^+\pi^-\pi^+$. There are different definitions of the angle between the planes of two hadrons.

Of course, there are other 'roads' to probe CP asymmetries in four-body FS with one-dimensional observables and compare them using correlations. We have learnt from the history of $K_L \to \pi^+\pi^-\gamma^* \to \pi^+\pi^- e^+ e^-$, where Seghal[40,41] really predicted CPV there around 14% based on $\epsilon_K \simeq 0.002$, where leptons have spin. It helps to discuss that situation in more details with unit vectors:

$$\vec{n}_\pi = \frac{\vec{p}_+ \times \vec{p}_-}{|\vec{p}_+ \times \vec{p}_-|}, \quad \vec{n}_l = \frac{\vec{k}_+ \times \vec{k}_-}{|\vec{k}_+ \times \vec{k}_-|}, \quad \vec{z} = \frac{\vec{p}_+ + \vec{p}_-}{|\vec{p}_+ + \vec{p}_-|}, \tag{33}$$

$$\sin\phi = (\vec{n}_\pi \times \vec{n}_l) \cdot \vec{z} \, [CP = -, T = -],$$

$$\cos\phi = \vec{n}_\pi \cdot \vec{n}_l \, [CP = +, T = +], \tag{34}$$

$$\frac{d\Gamma}{d\phi} \sim 1 - (Z_3 \cos 2\phi + Z_1 \sin 2\phi). \tag{35}$$

Then one measures asymmetry in the moments:

$$\mathcal{A}_\phi = \frac{\left(\int_0^{\pi/2} - \int_{\pi/2}^\pi + \int_\pi^{3\pi/2} - \int_{3\pi/2}^{2\pi} \right) \frac{d\Gamma}{\phi}}{\left(\int_0^{\pi/2} + \int_{\pi/2}^\pi + \int_\pi^{3\pi/2} + \int_{3\pi/2}^{2\pi} \right) \frac{d\Gamma}{\phi}}. \tag{36}$$

There is an obvious reason to probe the angle between the $\pi^+\pi^-$ and e^+e^- planes based on $K_L \to \pi^+\pi^-\gamma^*$ or $K^0 \to \pi^+\pi^-\gamma^*$ vs. $\bar{K}^0 \to \pi^+\pi^-\gamma^*$.

However the situation for non-leptonic D decays is more complex for several reasons; therefore one can use:

$$\frac{d}{d\phi}\Gamma(H_Q \to h_1 h_2 h_3 h_4) = |C_Q|^2 - [B_Q \cos 2\phi + A_Q \sin 2\phi] \tag{37}$$

$$= |C_Q|^2 - [B_Q(2\cos^2\phi - 1) + 2A_Q \sin\phi \cos\phi], \tag{38}$$

$$\frac{d}{d\phi}\Gamma(\bar{H}_Q \to \bar{h}_1\bar{h}_2\bar{h}_3\bar{h}_4) = |\bar{C}_Q|^2 - [\bar{B}_Q \cos 2\phi - \bar{A}_Q \sin 2\phi] \qquad (39)$$

$$= |\bar{C}_Q|^2 - [\bar{B}_Q(2\cos^2\phi - 1) - 2\bar{A}_Q \sin\phi\cos\phi]. \quad (40)$$

Obviously the landscapes are more 'complex'

$$\Gamma(H_Q \to h_1h_2h_3h_4) = |C_Q|^2 \quad \text{vs.} \quad \Gamma(\bar{H}_Q \to \bar{h}_1\bar{h}_2\bar{h}_3\bar{h}_4) = |\bar{C}_Q|^2. \qquad (41)$$

For these moments one gets:

$$\langle A^Q_{\text{CPV}} \rangle = \frac{2(A_Q - \bar{A}_Q)}{|C_Q|^2 + |\bar{C}_Q|^2}; \qquad (42)$$

i.e., no impact from the B_Q and $\bar{B}_Q$ terms. Furthermore one wants to probe semi-regional asymmetries like:

$$A^Q_{\text{CPV}}\Big|_a^b = \frac{\int_a^b d\phi \frac{d\Gamma}{d\phi} - \int_a^b d\phi \frac{d\bar{\Gamma}}{d\phi}}{\int_a^b d\phi \frac{d\Gamma}{d\phi} + \int_a^b d\phi \frac{d\bar{\Gamma}}{d\phi}}, \qquad (43)$$

where B_Q and $\bar{B}_Q$ contribute. Again, the main point is not to choose which gives the best fitting one as discussed about $K_L \to \pi^+\pi^-e^+e^-$.[40,41]

I want to emphasize that many-body FS give us more information about the underlying dynamics. However we have not yet the best tools to get it quantitatively. More data will attract theorists to think about it.

2.7. *Dealing with final states interactions*

Tools about FSI in three-body FS have been produced in the last 15 years, for instance dispersion relations based on low energy collisions with strong forces.[39] Chiral symmetry is a good tool for probing FS for pions. However their impact and the connections of CPT are subtle for $\pi K \Leftrightarrow \pi K$ and $K\bar{K} \Leftrightarrow K\bar{K}$. For four-body FS we need more thinking — but it is very important both on the theoretical and experimental side.

3. Intermezzo: QCD and the strong CP problem

Very shortly I comment about the extraction of V_{cb} and V_{ub} and their correlation with CP asymmetries as discussed above and with EDM below.

Comparing $|V_{cb}|$ and $|V_{ub}|$ from inclusive and exclusive semi-leptonic B decays is a very 'hot' item. It is crucial to constrain the 'golden' CKM triangle (and the 'second triangle' for B_s transitions) with accuracy or even precision. As Kolya stated in his last conference talk in November 2012 (using refined theoretical technologies like BPS and *non*-local correlations), he found no different values for $|V_{cb}|_{\text{incl}}$ and $|V_{cb}|_{\text{excl}}$; however it does not mean that the angles might not show NP in CP asymmetries.

There is no local 'competitor' with QCD to describe strong forces. However, the landscape of QCD forces is more complex and its connection with global symmetries and their violations. Usually we use QCD as a tool to find CP and T violation in B, K, D, top quarks, τ and neutron and leptons decays due to weak or superweak forces. Of course, there are very good reasons to probe the features of the strong forces in details.

It was pointed out in 1976 by 't Hooft[42] that the dimension four operator $G \cdot \tilde{G}$ — with G gluon field strength tensor — can be added to the QCD Lagrangian. If one 'decides' that the coefficent for this operator is zero, then quantum corrections will come back with non-zero value $\bar{\theta}$. Thus strong forces violate both P and T invariance. Similarly, chiral invariance for massless quarks is no longer conserved in quantum field theory. Strong CP[g] and chiral problems are furthermore intertwined by including also weak dynamics. The neutron EDM is described by an operator with dimension five. Thus its dimensionful coefficient d_N can be calculated as a finite quantity, in particular for $d_N \sim \mathcal{O}[(e/M_N)(m_q/M_N)\bar{\theta}] \sim \mathcal{O}(10^{-16}\bar{\theta})$ $e \cdot$ cm.[43] Data give limits about d_N leading to $\bar{\theta} < 10^{-9}$ or less — an 'un-natural' limit as seen by most in high-energy physics. The Peccei–Quinn symmetry can make it 'natural'.[44] No axion has been found, and many members of our community thought that the 'dawn' of the axions go the their 'dusk'. However it seems to me that members of the cosmology community see a much more important role of axions, but not in the old version: Peccei–Quinn symmetry might be broken also by UV dynamics; once axions were produced in the early universe, they constitute (part of) the DM in the present universe.[45] Renaissance for (refined) axions?

4. Subtle Working for EDMs

So far CP and T violations have been established in ΔS and $\Delta B \neq 0$ transitions, but not in flavor diagonal ones (except 'our' existence). There are excellent reasons to probe EDM deeper and deeper in many different states: from elementary leptons (e, μ, τ), to very complex states (heavy atoms and nuclei), with neutron, proton and deuteron in between. It tells us that the *ratio* of NP vs. SM effects can be huge. However one goes after tiny effects in subtle environments. It needs long time commitments of the experimental groups (and the funding agencies). If an EDM has been found and established, it would be a wonderful achievement. Then we have to understand the features of the underlying NP. It would be a golden mine for theorists. We have not found them yet. However theorists might help experimenters to continue their hard work with good ideas to find other systems with non-zero EDMs and later about correlations with other ones.

Again Kolya's broad horizon is apparent: he worked on EDMs and Higgs dynamics in the 1980's, then on CP asymmetries in heavy quark transitions, next on the impact of perturbative and non-perturbative QCD and then on EDMs again.

---

[g]Old problems (like old soldiers) never die — they just fade away!

I know he had thought many times, as you had seen during and afterwards discussions even for talks given by other people; you knew whether Kolya was attending a seminar or not — it was obvious.

4.1. *Early era*

Kolya had worked about neutron EDM with A.A. Anselm in 1984.[46] I had my first meeting with Anselm at SLAC around 1986; he came to my office after a talk I had given about CP violation in B_d decays and just mentioned a paper about neutron EDM due to Higgs exchanges. He told me politely that most people neglected quark interaction with neutral Higgs bosons. It was claimed that these contributions should be proportional to the third power of light *current* quarks. However the nucleon coupling with a neutral Higgs boson depends on nucleon mass at low momenta and does not vanish in the chiral limit. Higgs exchange could not give sizable contribution to CP violation in kaons transitions, but still could produce neutron EDM around 10^{-22} $e \cdot$ cm — i.e., that prediction exceeds the experimental limit by at least around two orders magnitude. There were very subtle statements to understand that; I had never heard that before. Therefore I read that paper[46] right away and was very impressed by it. Later I met Anselm at least twice at Winter Schools close to Moscow and always enjoyed talking and discussing with him. Kolya was a graduate student of Anselm and co-autor of the 1984 paper.

Kolya and I had produced the first paper in 'person' when Kolya was invited to the physics department at Notre Dame in 1990 with the title "Induced Multi-Gluon Couplings and the Neutron EDM."[2] Let us assume that in the future this or another idea will be established 'natural' to make $\bar{\theta} < 10^{-9}$ or less. Then we can deal with a challenge that on the surface is hardly connected with the $G \cdot \tilde{G}$ problem. Non-minimal Higgs models can produce neutron EDM close to the experimental limit while contributing very little to $K_L \to \pi\pi$ due to the emergence of the $G^2\tilde{G}$ operator. In 1990 Kolya generalized and refined these arguments:

(a) The operator $G^2\tilde{G}$ is induced in different classes of models for CP violation. It had also been noted by several authors that typically 'sizable' effects arise there.[47] The CKM dynamics produces a coefficient of the $G^2\tilde{G}$ operator that is utterly tiny.
(b) We had discussed a new method for estimating the relevant matrix element, namely $\langle N | i\bar{q}\sigma_{\mu\nu}qF_{\mu\nu} | N \rangle$ as induced by $G^2\tilde{G}$: it yields a result that is considerably smaller than Weinberg's estimate.
(c) We refined the findings from other authors that QCD radiative corrections suppress rather than enhance of the impact the operator $G^2\tilde{G}$.

We discussed the one-, two- and three-loop situations including QCD radiative corrections. It was non-trival work to get three conclusions: (i) We found strong evidence that contribution from $G^2\tilde{G}$ operator is quite unlikely to be cancelled by additional Peccei–Quinn term. (ii) Finding neutron EDM larger than 10^{-31} $e \cdot$ cm is a clear sign of the existence of NP in T violation — but not of its features. (iii) There are even more (theoretical and experimental) reasons to probe EDM for electrons, atoms, nuclei, μ, τ, etc. coming forward.

An excellent 1991 article, "The electric dipole moment of the electron" by Bernreuther and Suzuki,[48] focused on the electron EDM and how to probe it in atoms and molecules; it also discussed the connection with neutron EDM including Kolya's work.

4.2. *Around* 2000–2013

The prospects of probing EDMs as a direct sign of T violation became even more exciting around 2000 with new ideas and more tools and technologies[49] (and also for sociology reasons, since these experimental collaborations are relatively small). One could see that in conferences and workshops — in particular in the "Flavor in the Era of the LHC" CERN Workshop, November 2005–March 2007; it produced a long and very good proceedings published in *Eur. Phys. J. C* (2008) with three sections.[50] I have enjoyed and learnt from them, in particular EDM and $g - 2$ miniworkshop on Oct. 9–11, 2006.[51] Many discussions happened in 'public' or 'private' — and Kolya liked that also. We need more data, more technologies — and more thinking for leptons, quarks and gluon dynamics.

No EDM has been found yet anywhere in this (relatively) huge landscape in different 'dimensions', namely to probe EDMs in neutron, protons, nucleis, atoms, molecules, charged leptons, etc.[59] The ACME collaboration has given limit on electron EDM: $|d_e| < 8.7 \cdot 10^{-29}$ $e \cdot$ cm.[52] I find it exciting to read how experimental physicists did it. Furthermore it attracts more theorists to think about probing EDMs in different directions (and some of them come back as before including connections about axions).[53,54] We need new ideas as before.

As said before Kolya was a true leader in discussions about the importance of probing EDMs more and more in very different situations. Kolya had produced two papers together with Th. Mannel published in 2012/13. In Ref. 55 they pointed out that the neutron EDM can be generated in the SM already by tree diagrams due to boundstate effects without short-distance penguin diagrams. It produces non-zero chiral limit and does not depend on the difference $m_s - m_d$; they estimated around $d_N \sim 10^{-31}$ $e \cdot$ cm — a value similar to hand-waving arguments given in Ref. 2. In the longer paper[56] they gave more details how they came to two statements: (i) The landscape of neutron EDM is described not by effective CP-odd operators of lowest dimension, but by non-trivial interplay of different amplitudes at low energy scales like 1 GeV for two $\Delta C = 1$ and $\Delta S = 0$ four-quark operators. (ii) Those operators can be probed in D decays.

Kolya and Thomas and many physicists (like me) were excited to find CPV in charm transitions for the first time in $D^0 \to K^+K^-$ vs. $D^0 \to \pi^+\pi^-$; furthermore those data seem for many theorists to be beyond what the SM can generate. However more LHCb data did not confirm this CPV in charm decays. Even so I think in the future CPV will be established in $\Delta C \neq 0$ transitions. My main point is: even non-established data can lead us to think deeper about fundamental forces. These two papers are good examples: *tree* diagrams can produce TV/CPV in boundstates of quarks and come back to an old, but still not mature question, namely the impact of penguin diagrams[11] in general and in particular about nucleon EDMs how they depend on long-distance strong forces.

Again they dealt with subtle points that most people prefer to forget:

(i) We have to 'understand' dynamics with measurable parameters, in particular 'complex' impact on baryons transitions.
(ii) We have to talk about the connection with 'constituent' vs. 'current' quarks as discussed before,[46] but with more tools.
(iii) We have to understand why several chiral suppression of light quarks can be vitiated in composite hadrons like nucleons.
(iv) It is possible that the impact of 'heavy' quarks loops is sizable or even important for nucleon EDMs and for boundstate effects.
(v) It is even more subtle to discuss the connection between neutron EDM and direct CP asymmetries in B, D and μ and τ decays — and their connections with Dark Matter.
(vi) It is important to think about boundstate effects without short-distance dynamics from penguins.

I had worked on that item in the past — and have learnt so much from Kolya about fundamental dynamics.

4.3. *Future era*

As mentioned before here (and many other places like in Ref. 51) we have to find non-zero EDMs in leptons, neutron, deuteron, atoms, molecules, etc. for several reasons:

• The SM produces tiny 'backgrounds' in all these situations.
• EDMs affect many landscapes and with different correlations.
• It is a wonderful challenge for experimenters to apply their tools in a new situation or produce a new technology. It helps 'inventiveness'. Young experimenters will enjoy much more working in small groups than in huge collaborations.

Neither charged Higgs or a second neutral Higgs have been found, and the known one is at least mostly a scalar. Fans of SUSY like me do not give up that this theory exists in our world — but not in the mass region which LHC can probe directly.

Obviously SUSY cannot solve all problems for fundamental forces together. However I do not think that our world prefers the minimal version of SUSY. Non-minimal versions can produce EDMs that can be measured in the future. Probing EDMs can be competitive with the reach of ϵ_K.[57]

In the future experiments at FNAL will measure $(g-2)_\mu$ with more data (and I hope also the muon EDM later) and at the J-PARC Hadron Experimental Facility (Japan) the combined measurements of $(g-2)_\mu$ and d_μ using very new tools. Even if the second experiment fails to reach its goals, the community would learn so much and will come up with new ideas and new technology that can be applied in the 'real' world. That way some of us respond to well known Beckett's skepticism.[h] In one way one can compare $d_e < 105 \times 10^{-29}\ e \cdot$ cm (from PDG) or the recent value $d_e < 8.7 \times 10^{-29}\ e \cdot$ cm[52] with $\delta[F_2(0)/2m_e] \sim 2 \times 10^{-22}\ e \cdot$ cm derived from $\delta[(g-2)/2] \sim 10^{-11}$ or $d_\mu = (-0.1 \pm 0.9) \times 10^{-19}\ e \cdot$ cm.

I have been thinking and working about EDMs and leptonic dynamics as signs of the features of NP and correlations with CP asymmetries in particular in charm hadrons and top quarks. The last two published 2012/2013 papers from Kolya gave me and others new 'directions'. Three very recent papers[57,25,58] point out that partners of SUSY might 'easily' exist at mass scales of ten of TeV and above. The best way to probe those mass scales is to measure EDMs. I am not saying that LHC cannot find NP — but we have to think about it.

5. Kolya's Impact in the Past, Present and for the Future

Kolya has worked for around 35 years about fundamental dynamics in many landscapes and showed his broad horizon: impact of Higgs state(s), CP and T violation, perturbative and non-perturbative QCD and the correlations between landscapes (in particular the subtle ones). He always showed that first and second wins are not enough — one has to get deeper. He always liked the connection with experimental colleagues, explaining why theoretical predictions are good or bad and where more theoretical work is needed. He also showed that real theorists do not act as 'slaves' of the data of the time. Sometimes predictions are correct based on good theoretical tools (and a lot of thinking and working), while data are different; however eventually the data move closer and closer to good real *predictions* — like the ratio of $\tau(\Lambda_b)/\tau(B^0)$ as mentioned above.

Acknowledgments

This work was supported by the NSF under the grant number PHY-1215979.

References

1. I. I. Bigi, V. A. Khoze, N. G. Uraltsev and A. I. Sanda, *CP Violation*, ed. C. Jarlskog (World Scientific, 1988), p. 175–248.
2. I. I. Bigi and N. G. Uraltsev, *Nucl. Phys. B* **353**, 321 (1991).
3. N. G. Uraltsev and I. I. Bigi, *Sov. Phys. JETP* **73**, 198 (1991); I have changed the order of the authors' names, since Kolya had done all the important addition to our previous paper.

---

[h] As said by Samuel Beckett: "Ever tried. Ever failed. No matter. Try again. Failed again. Fail better."

4. N. Uraltsev, *Phys. Lett. B* **376**, 303 (1996); I. I Bigi, M. Shifman and N. Uraltsev, *Annu. Rev. Nucl. Part. Sci.* **47**, 591–661 (1997).

5. M. B. Voloshin, *Phys. Rev. D* **61**, 074026 (2000).

6. LHCb Collab. (R. Aaij *et al.*), arXiv:1402.6242 [hep-ex].

7. L. B. Okun, *Weak Interactions of Elementary Particles* (Pergamon, 1965). The Russian original had appeared in 1963, i.e. clearly *before* the discovery of CP violation.

8. I. I. Bigi and A. I. Sanda, *Nucl. Phys. B* **193**, 85 (1981).

9. Y. I. Azimov, N. G. Uraltsev and V. A. Khoze, *JETP Lett.* **43**, 409–411 (1986).

10. ARGUS Collab. (H. Albert *et al.*), *Phys. Lett. B* **192**, 245 (1987).

11. A. I. Vainshtein, V. I. Zakharov and M. A. Shifman, *JETP Lett.* **22**, 55 (1975); *Nucl. Phys. B* **120**, 316 (1977), it was first used to apply for $\Delta I = \frac{1}{2} \gg \Delta I = \frac{3}{2}$ kaon decays.

12. M. Shifman, ITEP lectures in particle physics, arXiv:hep-ph/9510397, he was especially grateful to B. Ioffe, A. Vainshtein and V. Zakharov.

13. Belle Collab. (J. Dalseno *et al.*), *Phys. Rev. D* **88**, 092003 (2013), arXiv:1302.0551.

14. LHCb WS at CERN, Oct. 14–16, 2013.

15. Yu. L. Dokshitzer and N. G. Uraltsev, *ZhETF Lett.* **52**, 509 (1990).

16. N. G. Uraltsev, *Proceedings of DPF-92*, arXiv:hep-ph/9212233.

17. V. A. Khoze, M. A. Shifman, N. G. Uraltsev and M. B. Voloshin, *Sov. J. Nucl. Phys.* **46**, 112 (1987).

18. I. I. Bigi, M. Shifman, N. Uraltsev and A. Vainshtein, *Phys. Rev. D* **56**, 4017–4030 (1997), arXiv:hep-ph/9704245.

19. I. Bigi, M. Shifman, N. Uraltsev and A. Vainshtein, *Phys. Rev. D* **59**, 054011 (1999), arXiv:hep-ph/9805241; I. I. Bigi and N. Uraltsev, *Nucl. Phys. B* **592**, 92 (2001), arXiv:hep-ph/0005089; R. F. Lebed and N. G. Uraltsev, *Phys. Rev. D* **62**, 094011 (2000), arXiv:hep-ph/0006346; I I. Bigi and N. Uraltsev, *Int. J. Mod. Phys. A* **16**, 5201 (2001), arXiv:hep-ph/0106346.

20. I. I. Bigi and A. I. Sanda, *CP Violation*, 2nd edn., Cambridge Monographs on Particle Physics, Nuclear Physics and Cosmology (Cambridge University Press, 2009).

21. I. Bediaga *et al.*, *Phys. Rev. D* **80**, 096006 (2009), arXiv:0905.4233 [hep-ph]; M. Williams, *Phys. Rev. D* **84**, 054015 (2011); I. Bediaga *et al.*, *Phys. Rev. D* **86**, 036005 (2012), arXiv:1205.3036.

22. I. I. Bigi and N. G. Uraltsev, *Nucl. Phys. B* **592**, 92 (2001).

23. I. I. Bigi, A. I. Sanda and N. G. Uraltsev, *Perspectives on Higgs Physics*, eds. G. L. Kane (World Scientific, 1997), pp. 359–382.

24. ATLAS Collab., *Phys. Lett. B* **716**, 1–29 (2012); CMS Collab., *Phys. Lett. B* **716**, 30–61 (2012).

25. S. Berge, W. Bernreuther, B. Niepelt and H. Spiesberger, *Phys. Rev. D* **84**, 116003 (2011), arXiv:1108.0670; S. Berge, W. Bernreuther and H. Spiesberger, arXiv:1208.1507.

26. J. M. Cambell, R. K. Ellis and C. Williams, arXiv:1311.3589 [hep-ph].

27. I. I. Bigi and A. I. Sanda, *Phys. Lett. B* **625**, 47 (2005); Y. Grossman and Y. Nir, *JHEP* **1204**, 002 (2012).

28. BaBar Collab., *Phys. Rev. D* **85**, 031102 (2012) [Erratum: *ibid.* **85**, 099904 (2012)], arXiv:1109.1527 [hep-ex].

29. I.I. Bigi, TAU 2012 WS, *Nuclear Physics B Proceedings Supplement* 1–4 (2013), arXiv:1210.2968; arXiv:1306.6014v2 [hep-ph].

30. Y. H. Ahn, H.-Y. Cheng and S. Oh, *Phys. Lett. B* **703**, 571 (2011).

31. P. Gambino and Ch. Schwanda, *Phys. Rev. D* **89**, 014022 (2014), arXiv:1307.4551; I.I. Bigi, arXiv:1307.0799.

32. X.-W. Kang, B. Kubis, Ch. Hanhart and U-G. Meissner, arXiv:1312.1193 [hep-ph].
33. LHCb Collab. (R. Aaij *et al.*), *Phys. Rev. Lett.* **111**, 101801 (2013).
34. LHCb Collab. (R. Aaij *et al.*), *Phys. Rev. Lett.* **112**, 011801 (2014).
35. Ya. I. Azimov and A. A. Iogansen, *Sov. J. Nucl. Phys.* **33**(2), 205 (1981).
36. M. Golden and B. Grinstein, *Phys. Lett. B* **222**, 501 (1989).
37. I. Bediaga *et al.*, *Phys. Rev. D* **89**, 074024 (2014), arXiv:1401.3310 [hep-ph].
38. I. I. Bigi, Rio de Janeiro 2000, Heavy quarks at fixed target, arXiv:hep-ph/0012161.
39. S. Gardner and U-G. Meissner, *Phys. Rev. D* **65**, 094004 (2002), arXiv:hep-ph/0112281; M. R. Pennington, *Proc. of MESON2002*, arXiv:hep-ph/0207220; J. R. Pelaez and F. J. Yndurain, *Phys. Rev. D* **71**, 074016 (2005), arXiv:hep-ph/0411334; B. Kubis, The role of final-state interactions in Dalitz plot studies, arXiv:1108.5866; J. R. Pelaez *et al.*, *Proc. of Hadron 2011*, arXiv:1109.2392; Chr. Hanhart, Modelling low-mass resonances in multi-body decays, arXiv:1311.6627.
40. L. M. Seghal and M. Wanninger, *Phys. Rev. D* **46**, 1035 (1992), 5209(E).
41. L. M. Seghal and J. van Leusen, *Phys. Rev. Lett.* **83**, 4933 (1999); L. M. Seghal and J. van Leusen, *Phys. Lett. B* **489**, 300 (2000).
42. G. 't Hooft, *Phys. Rev.* **37**, 8 (1976); *Phys. Rev. Lett. B* **14**, 3432 (1976).
43. V. Baluni, *Phys. Rev. D* **19**, 2227 (1979); R. J. Crewther *et al.*, *Phys. Lett. B* **88**, 123 (1979) [Errata: *ibid.* **91**, 487 (1980)]; R. D. Peccei, *CP Violation*, ed. C. Jarlskog.
44. R. D. Peccei and H. R. Quinn, *Phys. Rev. Lett.* **38**, 1440 (1977); *Phys. Rev. D* **16**, 1791 (1977).
45. K. Choi, Talk given at COSMO 2014, Chicago.
46. A. A. Anselm, V. Bunakov, V. Gudkov and N. G. Uraltsev, *Phys. Lett. B* **152**, 116 (1985).
47. S. Weinberg, *Phys. Rev. Lett.* **63**, 2333 (1989); D. Dicus, *Phys. Rev. D* **41**, 999 (1990); D. Chang, C. S. Li and T. C. Yuan, *Phys. Rev. D* **42**, 867 (1990).
48. W. Bernreuther and M. Suzuki, *Rev. Mod. Phys.* **63**(2), 313 (1991).
49. M. Pospelov and A. Ritz, *Ann. Phys.* **318**, 119 (2005).
50. R. Fleischer, T. Hurth and M. L. Mangano (eds.), *Eur. Phys. J.C* **57**, 1–492 (2008).
51. M. Raidal *et al.*, *Eur. Phys. J. C* **57**, 13–182 (2008).
52. ACME Collab. (J. Baron *et al.*), *Science* **343**, 269 (2014).
53. S. Mantry, M. Pitschmann and M. J. Ramsey-Musolf, arXiv:1401.7339 [hep-ph].
54. W. Dekens *et al.*, arXiv:1404.6082 [hep-ph].
55. Th. Mannel and N. G. Uraltsev, *Phys. Rev. D* **85**, 096002 (2012), arXiv:1202.6270.
56. Th. Mannel and N. G. Uraltsev, *JHEP* **1303**, 064 (2013), arXiv:1205.0233.
57. D. McKeen, M. Pospelov and A. Ritz, *Phys. Rev. D* **87**, 113002 (2013), arXiv:1303.1172 [hep-ph].
58. M. Pospelov and A. Ritz, *Phys. Rev. D* **89**, 056006 (2014), arXiv:1311.5537 [hep-ph].
59. T. Fukuyama, *Int. J. Mod. Phys. A* **27**, 1230015 (2012), arXiv:1201.4252.

Exclusive Decays in the Heavy Quark Expansion: From Models to Real QCD

Thomas Mannel

Theoretical Physics 1 (Particle Physics), Faculty of Science and Technology, University of Siegen, 57072 Siegen, Germany

In this article I discuss a few aspects of exclusive decays of heavy mesons from the specific perspective of Kolya Uraltsev's contributions to this field. The selection of topics as well as their presentation reflects my personal view on Kolya's work and also includes some personal remarks related to my collaboration with him.

1. Instead of an Introduction: Some Personal Remarks

I very well remember the first time I met Kolya, which was in the summer of the year 1994. At this time I was a scientific associate in the CERN Theory Division, working on heavy quark physics applying Heavy Quark Effective Theory (HQET). I had learned this from the collaboration with Ben Grinstein and Howard Georgi during my postdoctoral time at Harvard, which also means that I was part of the "american school" of heavy quark physics. Of course, we all knew the brilliant papers by the "russian school", which were written by Misha Shifman, Arcady Vainshtein, Ikaros Bigi (who is bavarian, not russian; nevertheless I count him for the russian side) and Co-workers, in particular also by Kolya, but the "american school" sticked to their way of formulating the heavy mass expansions.

On one particular day of summer 1994 a person dropped into my office at CERN. He introduced himself as Nikolai Uraltsev and immediately started a loud and fierce discussion on the heavy quark expansion and why the formulation used by the "american school" was inferior compared to the way he and his friends discussed this issue. I was quite surprised about this style, but over the many years of friendship between Kolya and myself I learned that this is his specific way of having a scientific discussion. Not everybody in the scientific community could deal with this way of communication, but once I learned how to deal with this I very much profited from all the discussions with Kolya.

After this first meeting at CERN, we bumped into each other at various occasions, since we were working on very similar projects, however, somehow looking from different perspectives. Our collaboration became very close once Kolya accepted an offer from Siegen university and so he joined our group in Siegen in 2008. In the following years not only a fruitful collaboration emerged, but we also became very good friends. Kolya was a wonderful and in fact also a quite exceptional person, who solved problems often in an unconventional way. His office in the Theory Department looked much more like a laboratory of an experimentalist (he was building and fixing all sorts of electronic equipment), and I remember him flying self-made model planes and helicopters in the aisles of the institute.

Over his last three or four years we had a close collaboration, where I learned to understand the way Kolya was thinking about physics, in particular hadronic matrix elements involving heavy quarks. Shortly before he died we managed to get a large project funded, which had the heavy quark expansion as a subject. Kolya had quite a few ideas concerning not only applications, but also on the fundamental structure of the expansion, including some ideas for a partial resummation of the $1/m$ series, and I was looking forward to a fruitful collaboration.

In the rest of this article I will try to describe the way Kolya used to look at the problem of heavy quarks in QCD, focussing on his contributions to the discussion of exclusive decays. In the next section I will in particular discuss the form factors of the $B \to D$ and $B \to D^*$ transitions.

Finally, in Section 3 I will discuss a few ideas related to the heavy quark expansion, which we kept on discussing, but we never ended up in writing a paper about these ideas, since they were not (yet?) substantiated.

2. Decays of Heavy Hadrons

Heavy quarks Q are defined to be quarks with masses m_Q well above Λ_{QCD}, the confinement scale of QCD. Thus the ratio $\Lambda_{\mathrm{QCD}}/m_Q$ is a small quantity and an expansion of QCD in this quantity is useful. The "american school", mainly Estia Eichten from the lattice QCD perspective,[1] Nathan Isgur, Marc Wise,[2,3] Ben Grinstein and Howard Georgi looked at this from the perspective of spectroscopy and of effective field theories. In particular, the formulation of "Heavy Quark Effective Theory" (HQET) by Ben Grinstein[4] and Howard Georgi[5] started a whole new field which needed to be explored and allowed for many applications. Seen from this point of view, it was quite natural that the obvious first application was to describe exclusive decays, such that the $B \to D$ and $B \to D^*$ transitions are seen as transitions between heavy quarks.

However, the "russian school", mainly Misha Shifman, Arcady Vainshtein, Misha Voloshin, Ikaros Bigi and also Kolya Uraltsev never used the notion of an effective theory to understand the heavy quark limit.[6] The papers by Shifman and Co-workers on heavy quark symmetries, form factor normalizations and related issues were in fact a little earlier than the papers by Isgur and Wise, but I think it is fair to quote all of them as the fathers of heavy quark symmetries. Neverthe-less, the "russian school" certainly had an emphasis in studying inclusive decays of heavy quarks, which was derived from QCD directly, without making use of the idea of an effective field theory,[7,8] although there has also been an early paper by Chay, Georgi and Grinstein using HQET.[9]

The reasoning for inclusive decays is quite simple, and I include a few remarks on this, since it is used for sum rules for form factors of exclusive processes,[10,11] to be discussed later. Making use of the optical theorem, one relates the total decay rate of a heavy hadron H to a forward matrix element of the time ordered product of two effective hamiltonian densities $H_{\text{eff}}(x)$ which mediates the corresponding decay of the heavy quark,

$$\Gamma = \text{Im} \int d^4x \langle H(p_H)|T[H_{\text{eff}}(x)H_{\text{eff}}(0)]|H(p_H)\rangle , \qquad (1)$$

where H_{eff} is the usual four-fermion operator of weak interactions. The remaining matrix element of course depends on the mass of the heavy quark in the heavy hadron H, but this dependence can be controlled perturbatively, since the heavy quark mass m_Q is a perturbative scale, i.e. $\alpha_s(m_Q)$ is a small quantity.

To construct an expansion in inverse powers of the heavy quark mass, one first defines a time-like unit vector which is the velocity of the heavy hadron

$$v = \frac{p_H}{m_H} , \qquad (2)$$

where m_H is the mass of the hadron. The velocity of the heavy quark inside the heavy hadron is not very different from v, so we write

$$p_Q = m_Q v + k , \qquad (3)$$

where k is a residual momentum with components of order Λ_{QCD}. This decompo-sition of the momentum can be implemented for the heavy quark fields $Q(x)$ by defining

$$Q(x) = \exp(-im_Q(vx))Q_v(x) , \qquad (4)$$

$$i\partial_\mu Q(x) = \exp(-im_Q(vx))[m_Q v_\mu + i\partial_\mu]Q_v(x) \qquad (5)$$

which means that the derivative acting on Q_v is actually the residual momentum k. Replacing the heavy quark field in the effective Hamiltonian in (1) one obtains

$$\Gamma = \text{Im} \int d^4x \, e^{-im_Q(vx)} \langle H(p_H)|T[\tilde{H}_{\text{eff}}(x)\tilde{H}_{\text{eff}}(0)]|H(p_H)\rangle , \qquad (6)$$

where the tilde means that the fields Q in H_{eff} are replaced by Q_v.

The heavy mass expansion is constructed from an operator product expansion (OPE), using the fact that m_Q is a large scale and hence one can write

$$\int d^4x \, e^{-im_Q(vx)} T[\tilde{H}_{\text{eff}}(x)\tilde{H}_{\text{eff}}(0)] = \sum_{n,i} \frac{1}{m_Q^n} C_{n,i} O_{n+3,i} , \qquad (7)$$

where $\mathcal{O}_{l,i}$ is a set (labelled by i) of operators of dimension l, and the $C_{n,i}$ are (in terms of $\alpha_s(m_Q)$) perturbatively calculable coefficients.

This particular setup has been formulated by Bigi, Shifman, Vainshtain and co-workers, among which Kolya Uraltsev had significant contributions.[7,8] In fact, the setup of this expansion is very much along the lines of deep inelastic scattering (DIS), however, with one important difference. While in DIS the momentum transfer is a space-like vector q (with $Q^2 = -q^2 \to \infty$), the corresponding quantity here is the momentum $m_Q v$ which is a time-like momentum. I remember very well many discussions with Kolya Uraltsev on the question, if the OPE can be used as well in the time-like region, which at the end boils down to the question of duality, a problem Kolya also was thinking about quite intensively.[12,13]

Some properties of this expansion are in the meantime textbook knowledge. The leading term tuns out not to have any hadronic uncertainty, since it can be related to a single matrix element of the (even in full QCD) conserved current

$$\langle H(p_H)|\bar{Q}\gamma^\mu Q|H(p_H)\rangle = 2p_H^\mu \tag{8}$$

which basically counts the Q-quark flavor inside the hadron H.

All terms of oder $1/m_Q$ can be shown to vanish by heavy quark symmetries and the equations of motion, and the first nontrivial contributions emerge at $1/m_Q^2$. To be specific (and for later use), we shall discuss these matrix elements for the B meson and the bottom quark as the heavy quark. For this case we have

$$2M_B\mu_\pi^2 = -\langle B(p_B)|\bar{b}_v(iD)^2 b_v|B(p_B)\rangle\,, \tag{9}$$

$$2M_B\mu_G^2 = \langle B(p_B)|\bar{b}_v\sigma_{\mu\nu}(iD^\mu)(iD^\nu)b_v|B(p_B)\rangle\,, \tag{10}$$

where we have used the redefinition (4) for the b quark. Note that μ_π corresponds to the residual kinetic energy of the heavy quark inside the heavy meson, while μ_G^2 is the chromomagnetic moment $\vec{\sigma}\cdot\vec{B}$ of the heavy quark inside the heavy meson, where $\vec{B}$ is the chromomagnetic field generated by the light degrees of freedom at the location of the heavy quark.

At order $1/m_b^3$ there are again two matrix elements

$$2M_B\rho_D^3 = -\langle B(p_B)|\bar{b}_v(iD_\mu)(ivD)(iD^\mu)b_v|B(p_B)\rangle\,, \tag{11}$$

$$2M_B\rho_{LS}^3 = \langle B(p_B)|\bar{b}_v\sigma_{\mu\nu}(iD^\mu)(ivD)(iD^\nu)b_v|B(p_B)\rangle \tag{12}$$

which are the Darwin term and the Spin–Orbit term.

Also higher orders have been investigated in this expansion, in particular by Kolya.[14] However, who is not the subject of this article and will be discussed in Sascha Turczyks contribution to this book.

2.1. The Decays $B \to D\ell\bar{\nu}$ and $B \to D^*\ell\bar{\nu}$

Kolyas main interest with regards to heavy quark physics were the inclusive decays, which will be discussed by his friend an colleague Paolo Gambino in a separate article in this book. However, a lot can be inferred from the inclusive calculations also for exclusive decays, mainly by a clever use of sum rules, where Kolya was a real wizard. In particular, the $b \to c$ semileptonic transitions are a nice laboratory, since the two exclusive channels $B \to D\ell\bar\nu$ and $B \to D^*\ell\bar\nu$ exhaust the inclusive rate by about 72%.

In order to understand the impact of the heavy quark expansion on exclusive decays, one has to consider in some more detail the heavy mass limit of QCD. The major result has been formulated in the papers by Shifman and Voloshin[6] and by Isgur and Wise,[2,3] namely the emergence of additional symmetries in the infinite mass limit.

The origin of the heavy quark symmetries can be understood without a lot of formalism. The first symmetry is due to the fact that the interaction between quarks and gluons does not depend on the mass. For light quarks this leads to the usual flavor symmetries such as isospin or the full flavor $SU(3)$ among the u, d and s quarks. However, also for infinite masses this statement applies, since the heavy quark just becomes a static source of a color field, which looks the same for any infinitely heavy, i.e. static quark. Considering bottom and charm to be heavy enough for this limit, an $SU(2)$ symmetry emerges; this heavy flavor symmetry relates bottom and charm quarks moving with the same velocity.

The second symmetry emerges from the fact, that in any gauge theory the spin of a fermion only couples to the magnetic fields via its spin; in the case at hand, the heavy quark couples to the chromomagnetic field through its spin, and the coupling constant is the chromomagnetic moment, which is inversely proportional to the heavy quark mass. Consequently, the interaction of the heavy quark spin vanishes in the infinite mass limit, since its chromomagnetic moment tends to zero. In this way the second heavy quark symmetry, the spin symmetry emerges, which is again an $SU(2)$ symmetry, corresponding to the rotations of the heavy quark spin. In principle one can combine the two $SU(2)$ symmetries into an $SU(4)$ spin–flavor symmetry, but this does not yield any additional information.

The heavy quark symmetries have interesting consequences. First of all, the spectroscopy of heavy hadrons should be discussed in the light of the heavy quark spin symmetry: Since the heavy quark spin decouples, all heavy hadron states should fall into spin symmetry doublets. For the meson ground states this implies that the pseudo scalar and the vector state should form such a doublet. For orbitally excited states this means that for each value of orbital momentum ℓ there will be one doublet for $j_{\text{light}} = \ell - 1/2$ and $j_{\text{light}} = \ell + 1/2$, where j_{light} denotes the total angular momentum of the light degrees of freedom. The lowest spin symmetry doublets for the D mesons thus are

$$\begin{bmatrix} |D(v)\rangle \\ |D^*(v,\epsilon)\rangle \end{bmatrix} \qquad \text{for} \quad j_{\text{light}} = 1/2, \ \ell = 0 \,, \tag{13}$$

$$\begin{bmatrix} |D_{0^+}(v)\rangle \\ |D_{1^+}^*(v,\epsilon)\rangle \end{bmatrix} \text{ and } \begin{bmatrix} |D_{1^+}(v,\epsilon)\rangle \\ |D_{2^+}^*(v,\epsilon)\rangle \end{bmatrix} \quad \text{for} \quad \begin{cases} j_{\text{light}} = 1/2, \ \ell = 1 \\ j_{\text{light}} = 3/2, \ \ell = 1 \end{cases} \tag{14}$$

and likewise for B mesons.

Another important consequence is related to the form factors in transitions among heavy hadron states, where in the following we shall concentrate on a discussion of mesons. A heavy (ground state) meson, consisting of a heavy quark Q and light degrees of freedom q, may be represented by a Dirac matrix

$$|H(v)\rangle \sim M(v) = (Q_\alpha \bar{q}_\beta)_{0^-} = \frac{\sqrt{m_H}}{2}[\gamma_5(\slashed{v} - 1)]_{\alpha\beta} \quad \text{for a } 0^- \text{ meson}, \quad (15)$$

$$|H(v, \epsilon)\rangle \sim M^*(v, \epsilon) = (Q_\alpha \bar{q}_\beta)_{1^-} = \frac{\sqrt{m_H}}{2}[\slashed{\epsilon}(\slashed{v} - 1)]_{\alpha\beta} \quad \text{for a } 1^- \text{ meson}, \quad (16)$$

where ϵ denotes the polarization of the heavy vector meson.

A transition from a meson moving with velocity v to a meson moving with velocity v' involves the heavy quark symmetries for v and v' separately; making use of the representation matrices (15) and (16) one can show that all transitions between heavy ground state mesons $\mathcal{H}$ induced by a current of the form $\bar{Q}_v \Gamma Q'_{v'}$ can be expresses in terms of a single form factor

$$\langle \mathcal{H}(v) | \bar{Q}_v \Gamma Q'_{v'} | \mathcal{H}'(v') \rangle = \xi(vv') \, \text{Tr}[\bar{\mathcal{M}}(v) \Gamma \mathcal{M}(v')] , \quad (17)$$

where $\mathcal{H}'(v)$ is either a pseudo scalar or a vector (ground state) meson, $\mathcal{M}(v)$ the representation matrix for the corresponding meson (moving with velocity v) and Γ is an arbitrary Dirac matrix. Note that this relation can be viewed as a Wigner–Eckart theorem of the heavy quark symmetry. The single form factor $\xi(vv')$ depends only on the scalar product of the meson velocities and is called the Isgur–Wise function.

This has interesting consequences in particular for semileptonic decays. For the transitions $B \to D$ and $B \to D^*$ (in the standard model) we have in general a parametrization of in total six form factors

$$\frac{\langle D(v') | \bar{c}\gamma^\mu b | B(v) \rangle}{\sqrt{m_B m_D}} = h_+(w)(v_B + v_D)^\mu + h_-(w)(v_B - v_D)^\mu , \quad (18)$$

$$\frac{\langle D^*(v', \epsilon) | \bar{c}\gamma^\mu b | B(v) \rangle}{\sqrt{m_B m_{D^*}}} = h_V(w)\varepsilon^{\mu\nu\rho\sigma} v_{B,\nu} v_{D^*,\rho} \epsilon^*_\sigma , \quad (19)$$

$$\frac{\langle D^*(v', \epsilon) | \bar{c}\gamma^\mu \gamma^5 b | B(v) \rangle}{\sqrt{m_B m_{D^*}}} = i h_{A_1}(w)(1 + w)\epsilon^{*\mu}$$

$$- i\left[h_{A_2}(w) v_B^\mu + h_{A_3}(w) v_{D^*}^\mu \right] \epsilon^* \cdot v_B , \quad (20)$$

which we write already as a function of w, the scalar product of the two velocities $vv' = w$.

Making use of the heavy quark symmetries by employing (17) we may relate the six form factors to the Isgur–Wise function

$$h_+(w) = h_{A_1}(w) = h_{A_3}(w) = h_V(w) = \xi(w) , \quad (21)$$

$$h_-(w) = 0 = h_{A_2}(w) . \quad (22)$$

Aside from this drastic reduction of the number of form factors compared to the general case, the Isgur–Wise function has an important property. Due to the heavy quark symmetries, one finds a normalization statement for the Isgur–Wise function at the point $vv' = 1$. While this can be proven formally, it also can be understood by a physical argument. The point $v = v'$ corresponds to the situation, where a heavy meson decays semileptonically at rest into another heavy meson, which is again at rest and the energy and momentum is carried away by the leptons. However in the infinite mass limit for the initial and the final state quark this means that also the heavy quarks remain both at rest. Consequently the light degrees of freedom do not recognize any change and hence their state does not change. Since the Isgur–Wise function corresponds to the overlap of the initial and the final state wave function, we obtain this normalization statement.

This normalization is the basis of the modern extractions of V_{cb} from exclusive decays. However, a competitive precision can only be obtained once corrections to the infinite mass limit are included. In fact, Kolya contributed significantly to this discussion.

Naively one would assume that the corrections are large, since the relevant parameter is $\Lambda_{\rm QCD}/m_c$ which easily can be as large as 30%. However, certain matrix elements are protected by Luke's theorem[15] which is a special case of a more general theorem valid for any continuous symmetry.[16] This theorem states that the matrix elements related to the currents which are the generators of the heavy quark symmetries do not have any first order corrections, which means that the normalization of theses form factors deviates from unity only by terms of order $(\Lambda_{\rm QCD}/m_c)^2$.[a]

The decay rates for the decays into the ground state mesons D is written as

$$\frac{d\Gamma^{B \to D \ell \bar{\nu}}}{dw} = \frac{G_F^2}{48\pi^3}|V_{cb}|^2(m_B + m_D)^2 m_D^3(w^2 - 1)^{3/2}|\eta_{\rm ew}\mathcal{G}(w)|^2 , \tag{23}$$

where the form factor $\mathcal{G}$ is given by

$$\mathcal{G}(w) = h_+(w) - \frac{m_B - m_D}{m_B + m_D}h_-(w) , \tag{24}$$

where $\eta_{\rm ew} = 1.007$ are the electroweak corrections. The differential rate for the decay into D^* reads

$$\frac{d\Gamma^{B \to D^* \ell \bar{\nu}}}{dw} = \frac{G_F^2 m_B^5}{48\pi^3}|V_{cb}|^2(w^2 - 1)^{1/2}(w + 1)^2 r^3(1 - r)^2 P(w)|\eta_{\rm ew}\mathcal{F}(w)|^2 , \tag{25}$$

where $P(w)$ is a phase space factor

$$P(w) = \left(1 + \frac{4w}{w + 1}\frac{1 - 2rw + r^2}{(1 - r)^2}\right) \tag{26}$$

---

[a]In fact, this statement can be found already in Ref. 6, however, without elaborating on it.

and $r = m_{D^*}/m_B$. The form factor $\mathcal{F}$ is a combination of the form factors h_V and h_{A_i} defined in (19), (20) and is given by

$$P(w)|\mathcal{F}(w)|^2 = |h_{A_1}(w)|^2$$

$$\times \left\{ 2\frac{r^2 - 2rw + 1}{(1-r)^2}\left[1 + \frac{w-1}{w+1}R_1^2(w)\right] \right.$$

$$\left. + \left[1 + \frac{w-1}{1-r}(1 - R_2(w))\right]^2 \right\}, \tag{27}$$

where the ratios R_1 and R_2 are given by

$$R_1(w) = \frac{h_V(w)}{h_{A_1}(w)}, \quad R_2(w) = \frac{h_{A_3}(w) + rh_{A_2}(w)}{h_{A_1}(w)}. \tag{28}$$

The ratios R_1 and R_2 are both unity in the heavy quark limit for the b and the c quark, independent of w.

As mentioned above, heavy quark symmetries yield normalization statements for the form factors. While both $\mathcal{G}(w)$ and $\mathcal{F}(w)$ are chosen such that they are both normalized to unity in the heavy mass limit, the power corrections behave differently in both cases. The form factor h_+ is normalized and protected against first order corrections, but for the other form factor we have $h_-(1) = \mathcal{O}(1/m_c)$. Consequently we have

$$\mathcal{G}(1) = \mathcal{O}(1/m_c); \tag{29}$$

however, this contribution is further suppressed by the factor $(m_B - m_D)/(m_B + m_D)$.

The situation is better in case of $\mathcal{F}(w)$, which at $w = 1$ becomes $h_{A_1}(1)$. This form factor is protected by Luke's theorem and hence we have

$$\mathcal{F}(1) = \mathcal{O}(1/m_c^2). \tag{30}$$

A large effort has been made to assess the corrections to the normalizations of $\mathcal{G}(1)$ and $\mathcal{F}(1)$, an Kolya made a few wonderful contributions in the field of non-lattice estimates. In the following sections I will give examples for these contributions.

2.2. Zero recoil sum rules

One way to estimate the corrections to the normalization statements (29) and (30) is to compare this to the inclusive rate. After all, as mentioned above, the inclusive rate is saturated by the two ground-state D mesons by more than 70%, and at the non-recoil point this percentage is even higher. To this end, Kolya suggested to investigate the relation between the inclusive and the exclusive decays close to zero recoil by formulating "zero recoil zum rules."[10]

The zero recoil point is defined as the kinematic point where

$$q = (M_B - M_D)v = (m_b - m_c)v + \mathcal{O}(1/m_c), \tag{31}$$

where the velocity v defines the rest frame of the decaying B meson. The sum rule is set up by making use of the OPE methods described in the introduction to this section, however, it is set up for the hadronic current $\bar{b}\gamma_\mu\gamma_5 c$ instead for the full effective Hamiltonian

$$T_{\mu\nu}(\epsilon) = i \int d^4x \exp(iv x\epsilon)\langle B(p_B)|T[\bar{b}_v(x)\gamma_\mu\gamma_5 c_v(x)\bar{c}_v(0)\gamma_\nu\gamma_5 b_v(0)]|B(p_B)\rangle , \quad (32)$$

where ϵ is the excitation energy

$$\epsilon = vq + m_b - m_c \quad (33)$$

and q is the momentum transfer to the leptons. In the following we restrict ourselves to the spatial directions for μ and ν, where the spatial components of a vector a are defined by $a_\mu^\perp = a_\mu - v_\mu(v \cdot a)$.

For large enough ω one may use the OPE and express the result within a perturbative calculation by the partonic quantities such as quark masses and by the heavy quark expansion parameters (9)–(12). On the other hand, one may also insert a complete set of states between the two currents in (32), in which case one obtains in the rest frame $\vec{v} = \vec{0}$

$$T_{\mu\nu}(\epsilon) = \sum_X (2\pi)^3 \delta^3(\vec{p}_X) \frac{\langle B|\bar{b}_v\gamma_\mu\gamma_5 \, c_v|X\rangle\langle X|\bar{c}_v\gamma_\nu\gamma_5 b_v|B\rangle}{\epsilon - \epsilon_X + i\eta} + \cdots , \quad (34)$$

where the ellipses denote other contributions corresponding to cuts that are not relevant here, and ϵ_X is the excitation energy of the state X.

Clearly the two ground state mesons D and D^* contribute at $\omega = 0$, while the excited (or inelastic) contributions contribute at $\omega > 0$. To this end, we project out the state with excitation energies less than ϵ_M by performing a contour integration in the complex ϵ plane as depicted in Fig. 1. We define

$$I_{\mu\nu}(\epsilon_M) = -\frac{1}{2\pi i} \oint_{|\epsilon|=\epsilon_M} T_{\mu\nu}(\epsilon) . \quad (35)$$

Comparing this to the result expressed in terms of the intermediate states, we get

$$I_{\mu\nu}(\epsilon_M) \sim |\mathcal{F}(1)|^2 + w_{\text{inel}}(\epsilon_M) \quad (36)$$

since only the D^* contributes to the axial vector current.

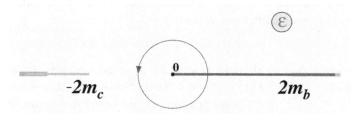

Fig. 1.   Integration contour in the complex ϵ plane. The radius of the circle is ϵ_M.

Making use of the positivity of the inelastic contribution we end up with an upper limit for the form factor; including an estimate for the inelastic contribution one may obtain a value for $\mathcal{F}(1)$. Inserting the OPE expressions, the result reads:[18]

$$|\mathcal{F}(1)|^2 + w_{\text{inel}}(\epsilon_M) = \xi_A^{\text{pert}}(\epsilon_M, \mu) - \Delta_{1/m_Q^2}(\epsilon_M, \mu) - \Delta_{1/m_Q^3}(\epsilon_M, \mu) - \cdots , \quad (37)$$

where ξ_A^{pert} are the perturbative corrections, Δ_{1/m_Q^i} are the nonperturbative contributions at order $1/m_Q^i$, and the ellipses denote terms of higher orders in the $1/m_Q$ expansion. Note that both ξ_A^{pert} and Δ_{1/m_Q^i} depend on the scale μ separating the perturbative from the nonperturbative regime; however the left-hand side does not depend on μ and hence the μ dependences on the right-hand side have to cancel.

The perturbative corrections have been calculated with a Wilsonian cut off and turn out to be[18]

$$\sqrt{\xi_A^{\text{pert}}(\epsilon_M = \mu = 0.75 \text{ GeV})} = 0.985 \pm 0.01 \quad (38)$$

while the nonperturbative corrections are known (at tree level) to order $1/m_Q^5$, the first few read[10,17]

$$\Delta_{1/m_Q^2} = \frac{\mu_G^2}{2m_c^2} + \frac{\mu_\pi^2 - \mu_G^2}{4}\left(\frac{1}{m_c^2} + \frac{2}{3m_c m_b} + \frac{1}{m_b^2}\right), \quad (39)$$

$$\Delta_{1/m_Q^3} = \frac{\rho_D^3 - \frac{1}{3}\rho_{LS}^3}{4m_c^3} + (\rho_D^3 + \rho_{LS}^3)\frac{1}{12m_b}\left(\frac{1}{m_c^2} + \frac{1}{m_c m_b} + \frac{3}{m_b^2}\right). \quad (40)$$

Numerically these two yield

$$\Delta_{1/m_Q^2} + \Delta_{1/m_Q^3} = 0.11 \pm 0.03 \quad (41)$$

while the known higher order terms are indeed estimated to be much smaller

$$\Delta_{1/m_Q^4} + \Delta_{1/m_Q^5} \sim -0.036 . \quad (42)$$

Combining this information one obtains a limit on the form factor[18] $\mathcal{F}(1)$

$$\mathcal{F}(1) \leq 0.93 , \quad (43)$$

where the upper limit is saturated in the case of no inelastic contributions.

Kolya was always keen to give an estimate for the inelastic contributions; however, this needed to set up additional machinery invoking many more matrix elements, including also non-local ones. Without going into all the details, Kolyas best estimate of the inelastic contribution lead to a value of[19]

$$\mathcal{F}(1) = 0.86 \pm 0.04 . \quad (44)$$

It is interesting to note that the central value of the modern lattice determinations is somewhat higher, $\mathcal{F}(1) = 0.906 \pm 0.016;$[20] however, this is only a tension at the level of 1–2σ. Kolya always trusted the non-lattice determinations more, in particular, the lattice value for $\mathcal{F}(1)$ published in 2005 almost saturated the bound (43).

The decay $B \to D^* \ell \bar{\nu}$ is used to extract the absolute value of V_{cb}. Using the value (44) one obtains a value of V_{cb} which is right on top of the central value of the determination based on inclusive decays, however, with a slightly larger uncertainty. On the other hand, as mentioned above, the lattice yields a higher central value, leading to small tension of the extracted V_{cb} compared to the inclusive one.

Kolya also worked on the form factor $\mathcal{G}(1)$, however, here the zero recoil sum rule is not as efficient. We shall discuss in the next section an approach which allowed Kolya to extract also a quite precise number for $\mathcal{G}(1)$.

2.3. *The BPS limit*

Kolya was always interested in getting more information on the matrix elements appearing in higher orders of the heavy quark expansion. Clearly there is a proliferation of new, nonperturbative matrix elements when going to order above $1/m_Q^3$. For this reason Kolya tried to find additional approximation schemes beyond the heavy quark expansion.

One idea that has been suggested by Kolya and which has not really been investigated in detail, is based on the observation, that the relation $\mu_\pi \geq \mu_G$ (valid in the kinetic scheme for both quantities) seems to be almost saturated. To this end, Kolya suggested to start from a limit where $\mu_\pi = \mu_G$ and to treat the difference $\delta = \mu_\pi - \mu_G$ as a small quantity.[21] However, up now no systematic expansion could be formulated, since there is no obvious limit of QCD that could serve as the starting point. If nature is close to this limit, there must be a subtle dynamical reason for this.

In Ref. 21 this limit was extended to other hadronic matrix element by postulating that

$$i \not{D}_\perp b_v | B \rangle = 0 \,, \quad D_\perp^\mu = D_\mu - v_\mu (v \cdot D) \tag{45}$$

for the pseudo-scalar ground state, in this case the B meson. This limit has been called the "BPS limit" (Bogomol'nyi, Prasad, Sommerfield), since there is an analogous relation for the so-called BPS saturated states in string theory.

Obviously (45) implies $\mu_\pi = \mu_G$, since

$$0 = \langle B | \bar{b}_v (i \not{D}_\perp)^2 b_v | B \rangle = 2 M_B (\mu_\pi^2 - \mu_G^2) \,. \tag{46}$$

However, in addition to this we can see that also the higher order matrix elements are constrained, in particular we get

$$0 = \langle B | \bar{b}_v (i \not{D}_\perp)(iv D)(i \not{D}_\perp) b_v | B \rangle = 2 M_B (\rho_D^3 + \rho_{LS}^3) \tag{47}$$

and likewise for all higher-order matrix elements of the heavy quark expansion, including matrix elements involving non-local operators.

This BPS limit has further interesting consequences. The heavy quark expansion implies for the field operators and the Lagrangian[22]

$$b = e^{-im_Q vx}\left[1 + \frac{1}{2m_Q}\not{\!P}_\perp + \left(\frac{1}{2m_Q}\right)^2 (-ivD)\not{\!P}_\perp + \cdots\right]h_v\,, \tag{48}$$

$$\mathcal{L} = \bar{h}_v(ivD)h_v + \frac{1}{2m}\bar{h}_v(i\not{\!P}_\perp)^2 ih_v + \left(\frac{1}{2m}\right)\bar{h}_v(i\not{\!P}_\perp)(-ivD)(i\not{\!P}_\perp)h_v + \cdots. \tag{49}$$

Using (45) we can act with these operators on the state $|B\rangle$ and obtain

$$b(x)|B\rangle = e^{-im_b vx}h_v|B\rangle \quad \text{and thus} \quad (\not{v} - 1)b(x)|B\rangle = 0\,, \tag{50}$$

$$\mathcal{L}|B\rangle = \bar{h}_v(ivD)h_v|B\rangle \tag{51}$$

to all orders in $1/m_b$. Hence we infer that

$$M_B = m_b + \bar{\Lambda}\,, \quad M_D = m_c + \bar{\Lambda}\,, \quad M_B - M_D = m_b - m_c\,, \tag{52}$$

again to all orders in $1/m_Q$.

There are interesting consequences for the form factors as well. Starting from the equation of motion

$$i\partial^\mu(\bar{b}\gamma_\mu c) = (m_b - m_c)(\bar{b}c) \tag{53}$$

we can take a matrix element between the ground states and express this in terms of the form factors

$$i\partial^\mu\langle B|\bar{b}\gamma_\mu c|D\rangle = h_+(w)(M_B - M_D)(w+1) - h_-(w)(M_B + M_D)(w-1). \tag{54}$$

The right-hand side contains the quark-mass difference which becomes in the BPS limit the difference of the 0^- meson masses, see (52). Furthermore, according to (50) in the BPS the lower components vanish and hence we have to all orders in $1/m_Q$:

$$(m_b - m_c)\langle B|\bar{b}c|D\rangle = (M_B - M_D)\langle B|\bar{b}\not{v}c|D\rangle$$
$$= (M_B - M_D)[h_+(w)(w+1) + h_-(w)(w-1)]. \tag{55}$$

Using (53) to equate (54) and (55) we conclude, that in this limit we have

$$h_-(w) = 0 \tag{56}$$

to all orders in $1/m_Q$.

Thus the BPS limit lifts the problem that at subleading order in $1/m_Q$ the form factor $\mathcal{G}(w)$ does not suffer from the potentially large $1/m_Q$ contributions, since h_- vanishes to all orders. In addition, it has been argued in Ref. 21 that the normalization of the form factor also remains unity to all orders in $1/m_Q$. This can be seen from the sum rule

$$|\mathcal{G}(1)|^2 + \text{excited states contributions} = 1 + \frac{\mu_\pi^2 - \mu_G^2}{4}\left(\frac{1}{m_c} - \frac{1}{m_b}\right)^2$$

$$+ \mathcal{O}(\rho_D^3 + \rho_{LS}^3),\qquad(57)$$

where in the BPS limit the right hand side is simply unity. Furthermore, it can be shown that in the BPS limit the contributions of the excited states vanish, in which case we infer

$$h_+(1) = 1\qquad(58)$$

again to all order in $1/m_Q$.

If nature is sufficiently close to the BPS limit, it also opens the road to a precise determination of V_{cb} from the decay $B \to D\ell\bar\nu$. In order to really make use of this, one has to define the small parameter governing the BPS expansion, which is defined to be

$$\beta^2 = \frac{\mu_\pi^2 - \mu_G^2}{\mu_\pi^2}.\qquad(59)$$

It has been argued by Kolya in Ref. 21 that the corrections to the normalization (58) are of order β^2, so it remains to estimate these remaining small corrections. Kolya has performed such an estimate in Ref. 21 with the result

$$\mathcal{G}(1) = \xi_V(\mu) + \beta^2 \frac{\mu_\pi^2}{3\tilde\epsilon}\left(\frac{1}{m_c} - \frac{1}{m_b}\right)\frac{(M_B - M_D)}{(M_B + M_D)} + \mathcal{O}(1/m_Q^2),\qquad(60)$$

where $\xi_V(\mu)$ are the perturbative corrections, calculated in a hard cut-off scheme, and $\tilde\epsilon \sim (500\text{–}700)$ MeV is the average excitation energy of the lowest p wave $j_{\text{light}} = 1/2$ spin symmetry doublet (see (14)).

Including an estimate of the uncertainties, this yields numerically

$$\mathcal{G}(1) = 1.04 \pm 0.02\qquad(61)$$

which turns out to be a competitive possibility to extract V_{cb} from the decays $B \to D\ell\bar\nu$. It is interesting to note that the corresponding lattice results again turn out to have slightly higher central value $\mathcal{G}(1) = 1.074 \pm 0.027$.[23]

3. Still More Ideas on the Heavy Quark Expansion

Kolya had plenty of additional ideas on the heavy quark expansion, on which we did not have the time to work. Aside from ideas to estimate the higher order matrix elements of the heavy quark expansion, there were also a few ideas which deserve to be picked up again and which could give further insight into heavy quark physics.

One issue which frequently came up in our discussions is the question on the structure of the heavy quark expansion. It is an easy exercise to count the number of independent matrix elements (such as μ_π and μ_G) appearing at a given dimension for the operator, since the higher dimensional operators appear at tree level as chains of covariant derivatives, i.e.

$$\langle B|\bar{b}_v(iD_{\mu_1})(iD_{\mu_2})\cdots(iD_{\mu_n})\Gamma b_v|B\rangle \qquad (62)$$

for an operator of dimension $n + 3$. Furthermore, there are spin-singlet operators ($\Gamma = 1$) and spin-triplet operators ($\Gamma = \vec{\sigma}$, the Pauli matrices). It turns out that the number of independent matrix elements grows factorially, to be more precise

$$N_1(n) \approx \frac{1}{2}\sum_{n_g=1}^{[\frac{n}{2}]}(2n_g-1)!!\binom{n-2}{n-2n_g} \qquad \text{(Spin Singlet)}, \qquad (63)$$

$$N_\sigma(n) \approx \frac{1}{2}\sum_{n_g=1}^{[\frac{n}{2}]-1}(2n_g-1)!!\binom{n-2}{n-2n_g-2}\binom{2+2n_g}{2} \qquad \text{(Spin Triplet)}. \qquad (64)$$

If one assumes that the coefficients in front of all these operators are of order unity, this would imply a factorially growing contribution at high orders in the heavy quark expansion, unless there is a conspiracy leading to cancellations. Based on this it has been argued[24] that the heavy quark expansion is in fact an asymptotic expansion in a similar fashion as the perturbative expansion in quantum field theory.

In the perturbative expansion the factorially growing contributions can be identified from "bouble chain" diagrams, which mainly means that in a one-loop calculation the momentum-dependent coupling is inserted. In fact, Kolya also worked on these renormalons in the perturbative expansion, and his contributions to this field are described in the article of Misha Shifman in this volume. Unfortunately, nothing similar is known for the heavy quark expansion, but with a calculational scheme for the higher order matrix elements (as e.g. described in Sascha Turczyk's article in this book) one could get some more detailed idea about the structure of the heavy quark expansion. However, time was too short....

Another related issue are some ideas Kolya developed to achieve a partial resumation of the heavy quark expansion. The heavy quark Lagrangian has a residual symmetry which stems from the original Lorentz-invariance of full QCD. Within the heavy quark expansion, this so called reparametrization invariance relates the coefficients of nonperturbative matrix elements of different orders in the $1/m_Q$ expansion.

The simplest relation implied by reparamtrization invariance has a very simple physical origin, which actually has been pointed out by Kolya. The total decay rate is usually defined in the rest frame of the decaying particle. However, it is the B hadron which is at rest, while the decaying b quark has a residual momentum, leading to a time dilation. In other words, the residual motion results in a factor

$$\Gamma = \frac{m_b}{E_b}\Gamma_0, \qquad (65)$$

where E_b is the energy of the b quark in the meson, including its residual motion, and Γ_0 is the decay rate of the quark in its rest frame, including perturbative corrections. Using the expansion in terms of the residual momentum we get

$$E_b = \sqrt{m_b^2 + \vec{p}^2} \quad \text{thus} \quad \frac{m_b}{E_b} = 1 - \frac{\vec{p}^2}{2m_b^2} + \frac{3(\vec{p}^2)^2}{8m_b^4} + \cdots. \tag{66}$$

This first term involving $\vec{p}^2$ is related to the contribution of the kinetic operator μ_π^2, and hence reparametrization invariance guarantees the relation

$$\Gamma = \Gamma_0 \left(1 - \frac{\mu_\pi^2}{2m_b^2} + \cdots \right). \tag{67}$$

One may continue this argument to higher orders. In fact, the operator $(\vec{p}^2)^2$ translates into the fully symmetrized combination of four covariant derivatives, which corresponds to the single matrix element[14]

$$2M_B m_1 = \langle B | \bar{b}_v (iD_\mu)(iD_\nu)(iD_\alpha)(iD_\beta) b | B \rangle \frac{1}{3} [g_{\mu\nu} g_{\alpha\beta} + g_{\mu\alpha} g_{\nu\beta} + g_{\mu\beta} g_{\alpha\nu}] \tag{68}$$

and thus we find

$$\Gamma = \Gamma_0 \left(1 - \frac{\mu_\pi^2}{2m_b^2} + \frac{3m_1}{8m_b^4} + \cdots \right) \tag{69}$$

which simply can be inferred from reparametrization invariance.

Kolya also suggested to estimate the higher order matrix elements by a ground state saturation approximation, which is described in Sascha Turczyk's article in this book. In this way, one estimates the symmetrized matrix elements as

$$(\vec{p}^2)^n = (\mu_\pi^2)^n \tag{70}$$

which finally allows us to partially resum this part of the the heavy quark expansion into

$$\Gamma = \Gamma_0 \frac{1}{\sqrt{1 + \frac{\mu_\pi^2}{m_b^2}}}. \tag{71}$$

For these two issues (as well as for many other interesting topics), time was too short to elaborate on those together with Kolya. I am sure that some of these issues will be picked up in the future and will give us some more insight into the nature of the heavy mass expansion,

4. Instead of a Conclusion

In this article I covered only very few topics concerning QCD aspects in heavy flavor decays, with some emphasis on the exclusive decays. Many more issues would deserve to be mentioned, which have been covered by Kolyas work, including also additional ideas concerning exclusive decays.

Kolya was thinking intensively about the issue of inclusive decays into orbitally and radially excited D meson states. In particular, he had some ideas on the "1/2-3/2 puzzle," which is a tension between the theoretical prediction and the data for the lowest lying orbitally excited D mesons shown in (14). Related to this there is also the problem of the "missing semileptonic decays," since the known exclusive rates do not sum up to the measured inclusive rate.

Also aside from semileptonic decays and heavy quark physics, Kolya was always interested in all aspects of QCD. With his unconventional views and his very good intuition for nonperturbative effects in QCD, he has also contributed to a broad spectrum of topics, which is reflected in the various contributions to this book. Without Kolyas way of attacking physics problems, we will have a much harder time to make progress in understanding QCD.

Acknowledgments

This work was supported by the Research Unit FOR 1873 "Quark Flavor Physics and Effective Field Theories" funded by "Deutsche Forschungsgemeinschaft" (DFG).

References

1. E. Eichten and B. R. Hill, *Phys. Lett. B* **234**, 511 (1990).
2. N. Isgur and M. B. Wise, *Phys. Lett. B* **232**, 113 (1989).
3. N. Isgur and M. B. Wise, *Phys. Lett. B* **237**, 527 (1990).
4. B. Grinstein, *Nucl. Phys. B* **339**, 253 (1990).
5. H. Georgi, *Phys. Lett. B* **240**, 447 (1990).
6. M. A. Shifman and M. B. Voloshin, *Sov. J. Nucl. Phys.* **47**, 511 (1988) [*Yad. Fiz.* **47**, 801 (1988)].
7. I. I. Y. Bigi, N. G. Uraltsev and A. I. Vainshtein, *Phys. Lett. B* **293**, 430 (1992) [Erratum: *ibid.* **297**, 477 (1993)], arXiv:hep-ph/9207214.
8. I. I. Y. Bigi, M. A. Shifman, N. G. Uraltsev and A. I. Vainshtein, *Phys. Rev. Lett.* **71**, 496 (1993), arXiv:hep-ph/9304225.
9. J. Chay, H. Georgi and B. Grinstein, *Phys. Lett. B* **247**, 399 (1990).
10. I. I. Y. Bigi, M. A. Shifman, N. G. Uraltsev and A. I. Vainshtein, *Phys. Rev. D* **52**, 196 (1995), arXiv:hep-ph/9405410.
11. N. Uraltsev, *Phys. Lett. B* **501**, 86 (2001), arXiv:hep-ph/0011124.
12. R. F. Lebed and N. G. Uraltsev, *Phys. Rev. D* **62**, 094011 (2000), arXiv:hep-ph/0006346.
13. I. I. Y. Bigi and N. Uraltsev, *Int. J. Mod. Phys. A* **16**, 5201 (2001), arXiv:hep-ph/0106346.
14. T. Mannel, S. Turczyk and N. Uraltsev, *JHEP* **1011**, 109 (2010), arXiv:1009.4622 [hep-ph].
15. M. E. Luke, *Phys. Lett. B* **252**, 447 (1990).
16. M. Ademollo and R. Gatto, *Phys. Rev. Lett.* **13**, 264 (1964).
17. A. Kapustin, Z. Ligeti, M. B. Wise and B. Grinstein, *Phys. Lett. B* **375**, 327 (1996), arXiv:hep-ph/9602262.
18. P. Gambino, T. Mannel and N. Uraltsev, *Phys. Rev. D* **81**, 113002 (2010), arXiv:1004.2859 [hep-ph].

19. P. Gambino, T. Mannel and N. Uraltsev, *JHEP* **1210**, 169 (2012), arXiv:1206.2296 [hep-ph].
20. J. A. Bailey *et al.*, arXiv:1403.0635 [hep-lat].
21. N. Uraltsev, *Phys. Lett. B* **585**, 253 (2004), arXiv:hep-ph/0312001.
22. T. Mannel, W. Roberts and Z. Ryzak, *Nucl. Phys. B* **368**, 204 (1992).
23. M. Okamoto *et al.*, *Nucl. Phys. B (Proc. Suppl.)* **140**, 461 (2005), arXiv:hep-lat/0409116.
24. M. A. Shifman, arXiv:hep-ph/9505289.

Isgur–Wise Functions, Bjorken–Uraltsev Sum Rules and Their Lorentz Group Interpretation

L. Oliver* and J.-C. Raynal

Laboratoire de Physique Théorique,[†]
Université de Paris XI, Bâtiment 210, 91405 Orsay Cedex, France
* Luis.Oliver@th.u-psud.fr

In the heavy quark limit of QCD, using the Operator Product Expansion and the non-forward amplitude, as proposed by Nikolai Uraltsev, we formulate sum rules that generalize Bjorken and Uraltsev sum rules. We recover the Uraltsev lower bound for the slope of the Isgur–Wise (IW) function, that we generalize to higher derivatives. We show that these results have a clear interpretation in terms of the Lorentz group, since the IW function is given by an overlap between the initial and final light clouds, related by Lorentz transformations. Both the Lorentz group and the Sum Rules approaches are equivalent. Moreover, we formulate an integral representation of the IW function with a positive measure. Inverting this integral formula, we obtain the measure in terms of the IW function, allowing one to formulate criteria to decide if a given ansatz for the IW function is compatible or not with the sum rule constraints. We compare these theoretical constraints to some forms proposed in the literature.

1. Introduction

Nikolai Uraltsev made very important contributions to Heavy Quark Effective Theory, concerning both the inclusive $\bar{B} \to X_c \ell \nu$ and exclusive processes $\bar{B} \to D(D^*)\ell\nu$. We will be here concerned with his work on exclusive B decays.

In the heavy quark limit,[1] Bjorken formulated a Sum Rule (SR)[2] relating the slope $\rho^2 = -\xi'(1)$ of the elastic IW function $\xi(w)$ to the IW functions at zero recoil for the inelastic transitions changing parity, namely $\tau_{1/2}^{(n)}(w)$, $\tau_{3/2}^{(n)}(w)$, corresponding to transitions of the type $\frac{1}{2}^- \to \frac{1}{2}^+$, $\frac{3}{2}^+$.[3] In the notation of Ref. 3, the SR writes

$$\rho^2 = \frac{1}{4} + \sum_n \left[\left|\tau_{1/2}^{(n)}(1)\right|^2 + 2\left|\tau_{3/2}^{(n)}(1)\right|^2 \right] . \tag{1}$$

This SR implies the famous Bjorken's lower bound on the slope

$$\rho^2 \geq \frac{1}{4} . \tag{2}$$

[†]Unité Mixte de Recherche UMR 8627 – CNRS.

A decade later, came as a great surprise a new SR formulated by N. Uraltsev[4] involving the same inelastic IW functions:

$$\sum_n \left[\left| \tau_{3/2}^{(n)}(1) \right|^2 - \left| \tau_{1/2}^{(n)}(1) \right|^2 \right] = \frac{1}{4}. \tag{3}$$

The combination of both SR (1), (3) implies the much stronger lower bound

$$\rho^2 \geq \frac{3}{4}. \tag{4}$$

The technique used by Uraltsev to obtain the SR (3) was quite original. He considered the T-product of two currents away from the forward direction:

$$i \int d^4x \, e^{-iq \cdot x} \langle \bar{B}(v_f) | T[J(x)J^+(0)] | \bar{B}(v_i) \rangle, \quad J = \bar{c}\Gamma b, \tag{5}$$

where $v_i \neq v_f$, and performed the $1/m_Q$ expansion to obtain the corresponding Operator Product Expansion (OPE) in the direct channel. In other words, he considered non-forward transitions of the type $\bar{B}(v_i) \to D^{(n)}(v') \to \bar{B}(v_f)$, with $v_i \neq v_f$ and v' being the intermediate hadron four-velocity.

By the way, the bound (4) was obtained earlier in a class of quark models[5] that yield covariant current matrix elements in the heavy quark limit.[6] Later on we did realize that this was the case because, in the heavy quark limit, this class of models satisfy IW scaling and also both Bjorken and Uraltsev SR.[7,8]

Other important results in this field were obtained by Uraltsev, in particular the study of the limit in which the elastic slope reaches its lower limit:

$$\rho^2 \to \frac{3}{4}, \tag{6}$$

called by Uraltsev *BPS limit*,[9] on which we will come back below.

Our contribution to this topic began by trying to understand Uraltsev's results and generalize them to higher derivatives of the IW function $\xi(w)$.

2. Non-Forward Direction Sum Rules in the Heavy Quark Limit

Proceeding like Uraltsev, we did consider the non-forward amplitude (5) in a covariant way, and obtained SR that have the general concise form[10]

$$L_{\text{Hadrons}}(w_i, w_f, w_{if}) = R_{\text{OPE}}(w_i, w_f, w_{if}), \tag{7}$$

where

$$w_i = v_i \cdot v', \quad w_f = v_f \cdot v', \quad w_{if} = v_i \cdot v_f, \tag{8}$$

i.e. the consideration of the non-forward amplitude means to extend the *OPE* formula to $w_{if} \neq 1$, while Bjorken's SR was obtained for $w_{if} = 1$.

The l.h.s. of the SR (7) $L_{\text{Hadrons}}(w_i, w_f, w_{if})$ represents the sum over the intermediate $D^{(n)}(v')$ states while the r.h.s. $R_{\text{OPE}}(w_i, w_f, w_{if})$ corresponds to the OPE counterpart.

The domain of the variables (8) is as follows

$$w_i \geq 1, \quad w_f \geq 1,$$

$$w_i w_f - \sqrt{(w_i^2 - 1)(w_f^2 - 1)} \leq w_{if} \leq w_i w_f + \sqrt{(w_i^2 - 1)(w_f^2 - 1)}, \tag{9}$$

and there is the sub-domain:

$$w_i = w_f = w,$$

$$w \geq 1, \quad 1 \leq w_{if} \leq 2w^2 - 1. \tag{10}$$

We make somewhat more explicit the SR (7) considering currents of the form (Γ_i, Γ_f are Dirac matrices)

$$\bar{h}_{v'} \Gamma_i h_{v_i}, \quad \bar{h}_{v_f} \Gamma_f h_{v'}, \tag{11}$$

and one gets:

$$\sum_{D^{(n)}} \langle \bar{B}_f(v_f) | \Gamma_f | D^{(n)}(v') \rangle \langle D^{(n)}(v') | \Gamma_i | \bar{B}_i(v_i) \rangle \xi^{(n)}(w_i) \xi^{(n)}(w_f)$$

$$+ \text{ Other excited states} = -2\xi(w_{if}) \langle \bar{B}_f(v_f) | \Gamma_f P'_+ \Gamma_i | B_i(v_i) \rangle, \tag{12}$$

where the ground state $\frac{1}{2}^-$ and its radial excitations, together with their IW functions $\xi(w)$, $\xi^{(n)}(w)$ ($n \neq 0$) are made explicit and $P'_+ = \frac{1 + \not{v}'}{2}$ is the positive energy projector over the intermediate heavy quark c.

3. Generalized Isgur–Wise Functions and Generalized Sum Rules

We now consider higher excited states with a light cloud angular momentum j and spin J and transitions between the $\bar{B}$, pseudoscalar ground state with $(j^P, J^P) = (\frac{1}{2}^-, 0^-)$ to the whole tower of excited states[11] $D^{(n)}$ with (j^P, J^P), $J = j \pm \frac{1}{2}$, $j = L \pm \frac{1}{2}$, $P = (-1)^{L+1}$. One gets two independent Sum Rules.

Choosing for example

$$\Gamma_i = \not{\psi}_i, \quad \Gamma_f = \not{\psi}_f, \tag{13}$$

gives the *Vector Sum Rule*:

$$(w+1)^2 \sum_{L \geq 0} \frac{L+1}{2L+1} S_L(w, w_{if}) \sum_n \left[\tau_{L+1/2}^{(L)(n)}(w) \right]^2$$

$$+ \sum_{L \geq 1} S_L(w, w_{if}) \sum_n \left[\tau_{L-1/2}^{(L)(n)}(w) \right]^2 = (1 + 2w + w_{if}) \xi(w_{if}), \tag{14}$$

while choosing

$$\Gamma_i = \not{\psi}_i \gamma_5, \quad \Gamma_f = \not{\psi}_f \gamma_5, \tag{15}$$

one obtains the *Axial Sum Rule*:

$$\sum_{L \geq 0} S_{L+1}(w, w_{if}) \sum_n \left[\tau_{L+1/2}^{(L)(n)}(w) \right]^2$$

$$+ (w-1)^2 \sum_{L \geq 1} \frac{L}{2L-1} S_{L-1}(w, w_{if}) \sum_n \left[\tau_{L-1/2}^{(L)(n)}(w) \right]^2$$

$$= -(1 - 2w + w_{if}) \xi(w_{if}), \tag{16}$$

where $\tau_{L\pm1/2}^{(L)(n)}(w)$ are the IW functions corresponding to the transitions $\frac{1}{2}^- \to (L \pm \frac{1}{2})^P$, $P = (-1)^{L+1}$, the function $S_L(w, w_{if})$ is a Legendre polynomial:[10]

$$S_L(w, w_{if}) = \sum_{0 \leq k \leq L/2} C_{L,k} (w^2 - 1)^{2k} (w^2 - w_{if})^{L-2k}, \tag{17}$$

and the coefficient $C_{L,k}$ is given by

$$C_{L,k} = (-1)^k \frac{(L!)^2}{(2L)!} \frac{(2L-2k)!}{k!(L-k)!(L-2k)!}. \tag{18}$$

The function (17) is defined as

$$S_L(w_i, w_f, w_{if}) = v_{f\nu_1} \cdots v_{f\nu_L} T^{v_{f\nu_1} \cdots v_{f\nu_L}, v_{i\mu_1} \cdots v_{i\mu_L}} v_{i\mu_1} \cdots v_{i\mu_L}, \tag{19}$$

for $w_i = w_f = w$. The object $T^{v_{f\nu_1} \cdots v_{f\nu_L}, v_{i\mu_1} \cdots v_{i\mu_L}}$ is the projector on the polarization tensor of integer spin L:

$$T^{v_{f\nu_1} \cdots v_{f\nu_L}, v_{i\mu_1} \cdots v_{i\mu_L}} = \sum_\lambda \epsilon'^{(\lambda)*\nu_1 \cdots \nu_L} \epsilon'^{(\lambda)\mu_1 \cdots \mu_L}, \tag{20}$$

and depends on the intermediate velocity v'. The polarization tensor $\epsilon'^{(\lambda)\mu_1 \cdots \mu_L}$ is symmetric, traceless and transverse, $g_{\mu_i \mu_j} \epsilon'^{(\lambda)\mu_1 \cdots \mu_L} = v'_{\mu_i} \epsilon'^{(\lambda)\mu_1 \cdots \mu_L} = 0$. Although (20) is explicitly very complicated, its contraction (19) can be computed following the method exposed in Ref. 10.

Equations (14) and (16) are two independent SR of the general form (7). Differentiating these SR within the domain (10) and going to the corner of the domain $w_{if} \to 1, w \to 1$

$$\left[\frac{d^{p+q} (L_{\text{Hadrons}} - R_{\text{OPE}})}{dw_{if}^p \, dw^q} \right]_{w_{if}=w=1} = 0, \tag{21}$$

one gets a whole tower of sum rules.

At zero recoil, summing the Vector and the Axial SR one finds

$$\xi^{(L)}(1) = \frac{1}{4}(-1)^L L! \sum_n \left[\frac{L+1}{2L+1} 4 \left[\tau_{L+1/2}^{(L)(n)}(1) \right]^2 + \left[\tau_{L-1/2}^{(L-1)(n)}(1) \right]^2 + \left[\tau_{L-1/2}^{(L)(n)}(1) \right]^2 \right],$$

$$\tag{22}$$

that is a generalization of Bjorken SR (1), that corresponds to $L = 1$.

On the other hand, combining also the Axial SR and (22) one gets

$$\sum_n \left[\frac{L}{2L+1} \left[\tau_{L+1/2}^{(L)(n)}(1) \right]^2 - \frac{1}{4} \left[\tau_{L-1/2}^{(L)(n)}(1) \right]^2 \right] = \sum_n \frac{1}{4} \left[\tau_{L-1/2}^{(L-1)(n)}(1) \right]^2 , \qquad (23)$$

the generalization of Uraltsev SR (3), that is obtained for $L = 1$.

From these SR one finds strong constraints on the derivatives of $\xi(w)$. In particular, one finds

$$(-1)^L \xi^{(L)}(1) \geqslant \frac{(2L+1)!!}{2^{2L}} , \qquad (24)$$

that reduces to the bound (4) for $L = 1$, and generalizes it for all L.

From careful examination of the several equations obtained from (14) and (16) one obtains the improved bound on the curvature in terms of the slope:[12]

$$\xi''(1) \geq \frac{1}{5} \left[4\rho^2 + 3(\rho^2)^2 \right] . \qquad (25)$$

The QCD radiative corrections to this relation have been carefully studied by M. Dorsten.[13]

4. The BPS Limit of HQET

Uraltsev[9] made a very interesting observation by the consideration of the matrix elements of dimension 5 operators in HQET, the kinetic operator and the chromomagnetic operator:

$$\mu_\pi^2 = -\frac{1}{2m_B} \langle \bar{B} | \bar{h}_v (iD)^2 h_v | \bar{B} \rangle , \qquad (26)$$

$$\mu_G^2 = \frac{1}{2m_B} \langle \bar{B} | \frac{g_s}{2} \bar{h}_v \sigma_{\alpha\beta} G^{\alpha\beta} h_v | \bar{B} \rangle . \qquad (27)$$

These matrix elements are given in terms of $\frac{1}{2}^- \to \frac{1}{2}^+, \frac{3}{2}^+$ IW functions $\tau_j^{(n)}$ and level spacings $\Delta E_j^{(n)}$, as obtained from the OPE by I. Bigi, M. Shifman, N. Uraltsev and A. Vainshtein:[14]

$$\mu_\pi^2 = 6 \sum_n \left[\Delta E_{3/2}^{(n)} \right]^2 \left[\tau_{3/2}^{(n)}(1) \right]^2 + 3 \sum_n \left[\Delta E_{1/2}^{(n)} \right]^2 \left[\tau_{1/2}^{(n)}(1) \right]^2 , \qquad (28)$$

$$\mu_G^2 = 6 \sum_n \left[\Delta E_{3/2}^{(n)} \right]^2 \left[\tau_{3/2}^{(n)}(1) \right]^2 - 6 \sum_n \left[\Delta E_{1/2}^{(n)} \right]^2 \left[\tau_{1/2}^{(n)}(1) \right]^2 . \qquad (29)$$

These expressions imply the inequality $\mu_\pi^2 \geqslant \mu_G^2$. From the fit to the inclusive semileptonic decay rate one gets $\mu_\pi^2 \cong 0.40$ GeV^2, while from the meson hyperfine splitting one obtains $\mu_G^2 \cong 0.35$ GeV^2.

From these numerical values, Uraltsev[9] assumed the following limit:

$$\mu_\pi^2 = \mu_G^2 \to \tau_{1/2}^{(n)}(1) = 0 , \qquad (30)$$

that he called *BPS limit* of HQET. Uraltsev made the observation that, in this limit, the slope of the elastic IW function $\xi(w)$ reaches its lower bound (6) $\rho^2 \to \frac{3}{4}$, as can be seen easily from the SR written above and using (30).

Generalizing Uraltsev's arguments we did found the condition on the curvature[15]

$$\tau_{3/2}^{(2)(n)}(1) = 0 \to \xi''(1) = \frac{15}{16}. \tag{31}$$

Moreover, using the whole tower of SR formulated above we did demonstrated by induction that in the BPS limit one obtains[15]

$$\tau_{L-1/2}^{(L)(n)}(1) = 0 \to (-1)^L \xi^{(L)}(1) = \frac{(2L+1)!!}{2^{2L}}. \tag{32}$$

Assuming reasonable continuity regularities this implies the following explicit expression for the elastic IW function in the BPS limit:

$$\xi(w) = \left(\frac{2}{w+1}\right)^{3/2}. \tag{33}$$

As we have demonstrated elsewhere[19] and will expose below, this simple fully explicit form has a transparent group theoretical interpretation in terms of the Lorentz group.

5. The Lorentz Group and the Heavy Quark Limit of QCD

Hadrons with one heavy quark such that $m_Q \gg \Lambda_{\text{QCD}}$ can be thought as a bound state of a light cloud in the color source of the heavy quark. Due to its heavy mass, the latter is unaffected by the interaction with soft gluons.

In this approximation, the decay of a heavy hadron with four-velocity v into another hadron with velocity v', for example the semileptonic decays $\bar{B} \to D^{(*)}\ell\bar{\nu}_\ell$ or $\Lambda_b \to \Lambda_c\ell\bar{\nu}_\ell$, occurs just by free heavy quark decay produced by a current, and rearrangement of the light cloud to follow the heavy quark in the final state and constitute the final heavy hadron.

The dynamics is contained in the complicated light cloud, that concerns long distance QCD and is not calculable from first principles. Therefore, one needs to parametrize this physics through form factors, the IW functions.

The matrix element of a current between heavy hadrons containing heavy quarks Q and Q' can thus be factorized as follows[11,17]

$$\langle H'(v'), J'm'|J^{Q'Q}(q)|H(v), Jm\rangle$$

$$= \left\langle \frac{1}{2}\mu', j'M' \middle| J'm' \right\rangle \left\langle \frac{1}{2}\mu, jM \middle| Jm \right\rangle \left\langle Q'(v'), \frac{1}{2}\mu' \middle| J^{Q'Q}(q) \middle| Q(v), \frac{1}{2}\mu \right\rangle$$

$$\times \langle \text{cloud}, v', j', M'|\text{cloud}, v, j, M\rangle, \tag{34}$$

where v, v' are the initial and final four-velocities, and j, j', M, M' are the angular momenta and corresponding projections of the initial and final light clouds.

The current affects only the heavy quark, and all the soft dynamics is contained in the *overlap* between the initial and final light clouds $\langle v', j', M' | v, j, M \rangle$, that follow the heavy quarks with the same four-velocity. This overlap is independent of the current heavy quark matrix element, and depends on the four-velocities v and v'. The IW functions are given by these light clouds overlaps.

An important hypothesis has been done in writing the previous expression, namely neglecting *hard gluon radiative corrections*.

As we will make explicit below, the light cloud belongs to a Hilbert space, and transforms according to a unitary representation of the Lorentz group. Then, as we have shown,[17] the whole problem of getting rigorous constraints on the IW functions amounts to decompose unitary representations of the Lorentz group into irreducible ones. This allows to obtain, for the IW functions, general integral formulas in which the crucial point is that *the measures are positive*.

In Ref. 17 we did treat the case of a light cloud with angular momentum $j = 0$ in the initial and final states, as happens in the baryon semileptonic decay $\Lambda_b \to \Lambda_c \ell \bar{\nu}_\ell$.

The sum rule method exposed above is completely equivalent to the method based on the Lorentz group, as demonstrated in Ref. 17.

Ignoring spin complications, the IW function writes then simply (e.g. in the special case $j = 0$):

$$\xi(v \cdot v') = \langle U(B_{v'})\phi_0 | U(B_v)\phi_0 \rangle, \tag{35}$$

where B_v and $B_{v'}$ are the corresponding boosts.

One can easily get a physical picture of why the Lorentz group plays an essential role as far as the IW function is concerned. In the limit in which factorization holds, i.e. switching off the hard gluon radiative corrections between the heavy quark and the light cloud, the heavy quark of initial velocity v is strucked by a weak current — a hard process — and the deviated final heavy quark gets a different velocity v'. The factorization means that, *because of confinement* — a soft process — the light cloud *unchanged* state, after the heavy quark is strucked by the current, must be expanded in terms of eigenstates of the Hamiltonian corresponding to the four-velocity v'. This picture corresponds to the so-called "approximation soudaine."[16] Therefore, the overlap of the light cloud before and after the interaction with the current, $\langle \text{cloud}, v' | \text{cloud}, v \rangle$, is constrained only by its quantum numbers and by kinematics.

The crucial point is that *the states of the light component make up a Hilbert space in which acts a unitary representation of the Lorentz group*. In fact, this is more or less implicitly stated, and used in the literature.[11]

To see the point more clearly, let us go into the physical picture which is at the basis of (35). Considering first a heavy hadron *at rest*, with velocity $v_0 = (1, 0, 0, 0)$ its light component is submitted to the interactions between the light particles, light quarks, light antiquarks and gluons, and to the external chromo-electric field generated by the heavy quarks at rest. This chromo-electric field does not depend on the spin μ of the heavy quark nor on its mass. We shall then have a complete orthonormal system of energy eigenstates $|v_0, j, M\rangle$ of the light component, where j and M are the angular momentum quantum numbers, $\langle v_0, j', M', \alpha' | v_0, j, M \rangle = \delta_{j,j'} \delta_{M,M'}$

Now, for a heavy hadron moving with a velocity v, the only thing which changes for the light component is that the external chromo-electric field generated by the heavy quark at rest is replaced by a chromo-electromagnetic field generated by the heavy quark moving with velocity v. Neither the Hilbert space describing the possible states of the light component, nor the interactions between the light particles, are changed. We shall then have *a new complete orthonormal system* of energy eigenstates $|v, j, M\rangle$ in the same Hilbert space. Then, because the color fields generated by a heavy quark for different velocities are related by Lorentz transformations, we may expect that the energy eigenstates of the light component will, for various velocities, be themselves related by Lorentz transformations acting in their Hilbert space.

6. From a Lorentz Representation to Isgur–Wise Functions

For half-integer spin j, as is the case of the ground state mesons $j^P = \frac{1}{2}^-$, the polarization tensor is a Rarita–Schwinger tensor-spinor $\epsilon_\alpha^{\mu_1,\ldots,\mu_{j-1/2}}$ subject to the constraints of symmetry, transversality and tracelessness $v_\mu \epsilon_\alpha^{\mu_1,\ldots,\mu_{j-1/2}} = 0$, $g_{\mu\nu} \epsilon_\alpha^{\mu,\nu\cdots\mu_{j-1/2}} = 0$ and $(\not v)_{\alpha\beta} \epsilon_\beta^{\mu_1,\ldots,\mu_{j-1/2}} = 0$, $(\gamma_{\mu_1})_{\alpha\beta} \epsilon_\beta^{\mu_1,\ldots,\mu_{j-1/2}} = 0$. Then a light cloud scalar product

$$\langle v', j', \epsilon' | v, j, \epsilon \rangle, \tag{36}$$

that gives the IW function, is a covariant function of the vectors v and v' and of the tensors (or tensor-spinors) ϵ'^* and ϵ, bilinear with respect to ϵ'^* and ϵ, and the IW functions, functions of the scalar $v.v'$, are introduced accordingly.

The covariance property of the scalar products is explicitly expressed by the equality

$$\langle \Lambda v', j', \Lambda\epsilon' | \Lambda v, j, \Lambda\epsilon \rangle = \langle v', j', \epsilon' | v, j, \epsilon \rangle, \tag{37}$$

valid for any Lorentz transformation Λ, with the transformation of a tensor-spinor given by

$$(\Lambda\epsilon)_\alpha^{\mu_1,\ldots,\mu_{j-1/2}} = \Lambda_{\nu_1}^{\mu_1} \cdots \Lambda_{\nu_{j-1/2}}^{\mu_{j-1/2}} D(\Lambda)_{\alpha\beta} \epsilon_\beta^{\nu_1,\ldots,\nu_{j-1/2}}. \tag{38}$$

Then, let us *define* the operator $U(\Lambda)$, in the space of the light cloud states, by

$$U(\Lambda)|v, j, \epsilon\rangle = |\Lambda v, j, \Lambda\epsilon\rangle, \tag{39}$$

where here v is a fixed, arbitrarily chosen velocity. Eq. (37) implies that $U(\Lambda)$ is a *unitary operator*, as demonstrated in Ref. 17.

A unitary representation of the Lorentz group emerges thus from the usual treatment of heavy hadrons in the heavy quark theory. For the present purpose, we need to go in the opposite way, namely, to show how, starting from a unitary representation of the Lorentz group, the usual treatment of heavy hadrons and the introduction of the IW functions emerges. What follows is not restricted to the $j = \frac{1}{2}$ case, but concerns any IW function.

So, let us consider some unitary representation $\Lambda \to U(\Lambda)$ of the Lorentz group, or more precisely of the group $SL(2, C)$, in a Hilbert space $\mathcal{H}$, and we have to identify states in $\mathcal{H}$, depending on a velocity v. As said in Ref. 17, we have in $\mathcal{H}$ an additional structure, namely the energy operator of the light component *for a heavy quark at rest*, with $v_0 = (1, 0, 0, 0)$. Since this energy operator is invariant under rotations, we consider the subgroup $SU(2)$ of $SL(2, C)$. By restriction, the representation in $\mathcal{H}$ of $SL(2, C)$ gives a representation $R \to U(R)$ of $SU(2)$, and its decomposition into irreducible representations of $SU(2)$ is needed. We then have the eigenstates $|v_0, j, M\rangle$ of the energy operator, classified by the angular momentum number j of the irreducible representations of $SU(2)$, and *associated with the rest velocity v_0*, since their physical meaning is to describe the energy eigenstates of the light component for a heavy quark at rest.

We need now to express the states $|v, j, \epsilon\rangle$ in terms of the states $|v_0, j, M\rangle$. We begin with $v = v_0$. For fixed j and α, the states $|v_0, j, M\rangle$ are, for $-j \leq M \leq j$, a standard basis of a representation j of $SU(2)$:

$$U(R)|v_0, j, M\rangle = \sum_{M'} D^j_{M',M}(R)|v_0, j, M'\rangle, \qquad (40)$$

where the rotation matrix elements $D^j_{M',M}$ are defined by

$$D^j_{M',M} = \langle j, M'|U_j(R)|j, M\rangle. \qquad (41)$$

On the other hand, the states $|v_0, j, \epsilon\rangle$ constitute, when ϵ goes over all polarization tensors (or tensor-spinors), the whole space of a representation j of $SU(2)$. As emphasized in Ref. 17, the representation of $SU(2)$ in the space of 3-tensors (or 3-tensor-spinors) is not irreducible, but contains the irreducible subspace of spin j, which is precisely the polarization 3-tensor (or 3-tensor-spinor) space selected by the other constraints.

We may then introduce a standard basis $\epsilon^{(M)}$, $-j \leq M \leq j$, for the $SU(2)$ representation of spin j in the space of polarization 3-tensors (or 3-tensor-spinors). As demonstrated in Ref. 17, for any Lorentz transformation Λ we must have

$$|v, j, \epsilon\rangle = \sum_{M}(\Lambda^{-1}\epsilon)_M \, U(\Lambda)|v_0, j, M\rangle, \qquad (42)$$

for Λ such that $\Lambda v_0 = v$, with $v_0 = (1, 0, 0, 0)$.

Equation (42) is *our main result* here, defining, in the Hilbert space $\mathcal{H}$ of a unitary representation of $SL(2, C)$, the states $|v, j, \epsilon\rangle$ whose scalar products define the IW functions, in terms of $|v_0, j, M\rangle$ which occur as $SU(2)$ multiplets in the restriction to $SU(2)$ of the $SL(2, C)$ representation.

7. Irreducible Unitary Representations of the Lorentz Group and their Decomposition Under Rotations

7.1. *Explicit form of the principal series of irreducible unitary representations of the Lorentz group*

Following Naïmark,[18] we have exposed in Ref. 17 an explicit form of the irreducible unitary representations of $SL(2, C)$. Their set X is divided into three sets, the set X_p of representations of the principal series, the set X_s of representations of the supplementary series, and the one-element set X_t made up of the trivial representation. Actually, for the $j = \frac{1}{2}$ case, only the principal series is relevant, and we now consider the principal series, leaving j completely general.

A representation $\chi = (n, \rho)$ in the principal series is labeled by an integer $n \in Z$ and a real number $\rho \in R$. Actually, the representations (n, ρ) and $(-n, -\rho)$ (as given below) turn out to be equivalent so that, in order to have each representation only once, n and ρ will be restricted as follows:

$$n = 0, \quad \rho \geq 0,$$
$$n > 0, \quad \rho \in R. \tag{43}$$

The Hilbert space $\mathcal{H}_{n,\rho}$ is made up of functions of a complex variable z with the standard scalar product

$$\langle \phi' | \phi \rangle = \int \overline{\phi'(z)} \phi(z) d^2 z. \tag{44}$$

with the measure $d^2 z$ in the complex plane being simply $d^2 z = d(\mathrm{Re} z) d(\mathrm{Im} z)$, and therefore $\mathcal{H}_{n,\rho} = L^2(C, d^2 z)$.

The unitary operator $U_{n,\rho}(\Lambda)$ is given by:

$$\left(U_{n,\rho}(\Lambda)\phi \right)(z) = \left(\frac{\alpha - \gamma z}{|\alpha - \gamma z|} \right)^n |\alpha - \gamma z|^{2i\rho - 2} \phi \left(\frac{\delta z - \beta}{\alpha - \gamma z} \right), \tag{45}$$

where α, β, γ, δ are complex matrix elements of $\Lambda \in SL(2, C)$:

$$\Lambda = \begin{pmatrix} \alpha & \beta \\ \gamma & \delta \end{pmatrix}, \quad \alpha\delta - \beta\gamma = 1. \tag{46}$$

7.2. *Decomposition under the rotation group*

Next we need the decomposition of the restriction to the subgroup $SU(2)$ of each irreducible unitary representation of $SL(2, C)$.

Since $SU(2)$ is compact, the decomposition is by a direct sum so that, for each representation $\chi \in X$ we have an *orthonormal basis* $\phi^\chi_{j,M}$ of $\mathcal{H}_\chi$ adapted to $SU(2)$. Having in mind the usual notation for the spin of the light component of a heavy hadron, here we denote by j the spin of an irreducible representation of $SU(2)$. It turns out[17] that each representation j of $SU(2)$ appears in χ with multiplicity 0 or 1, so that $\phi^\chi_{j,M}$ needs no more indices, and the values taken by j are integer and half-integer numbers. For fixed j, the functions $\phi^\chi_{j,M}$, $-j \leq M \leq j$ are choosen as a standard basis of the representation j of $SU(2)$.

It turns out[17] that the functions $\phi^\chi_{j,M}(z)$ are expressed in terms of the rotation matrix elements $D^j_{M',M}$ defined by (41). A matrix $R \in SU(2)$ being of the form

$$R = \begin{pmatrix} a & b \\ -\bar{b} & \bar{a} \end{pmatrix}, \quad |a|^2 + |b|^2 = 1. \tag{47}$$

We shall also consider $D^j_{M',M}$ as a function of a and b, satisfying $|a|^2 + |b|^2 = 1$.

We can now give explicit formulae for the orthonormal basis $\phi^\chi_{j,M}$ of $\mathcal{H}_\chi$. The spins j which appear in a representation $\chi = (n, \rho)$ are:[18,17]

$$\text{all integers} \quad j \geq \frac{n}{2} \quad \text{for } n \text{ even}, \tag{48}$$

$$\text{all half-integers} \quad j \geq \frac{n}{2} \quad \text{for } n \text{ odd}. \tag{49}$$

and such a spin appears with multiplicity 1.

The basis functions $\phi^{n,\rho}_{j,M}(z)$ are given by the expression[17]

$$\phi^{n,\rho}_{j,M}(z) = \frac{\sqrt{2j+1}}{\sqrt{\pi}}(1+|z|^2)^{i\rho-1} D^j_{n/2,M}\left(\frac{1}{\sqrt{1+|z|^2}}, -\frac{z}{\sqrt{1+|z|^2}}\right), \tag{50}$$

or, using the explicit formula for $D^j_{n/2,M}$,

$$\phi^{n,\rho}_{j,M}(z) = \frac{\sqrt{2j+1}}{\sqrt{\pi}}(-1)^{n/2-M}\sqrt{\frac{(j-n/2)!(j+n/2)!}{(j-M)!(j+M)!}}(1+|z|^2)^{i\rho-j-1}$$

$$\times \sum_k (-1)^k \binom{j+M}{k}\binom{j-M}{j-n/2-k} z^{n/2-M+k}\bar{z}^k, \tag{51}$$

where the range for k is limited to $0 \leq k \leq j - \frac{n}{2}$ due to the binomial factors.

8. Irreducible Isgur–Wise Functions for $j = \frac{1}{2}$

We are interested now in the ground state meson case $j = \frac{1}{2}$,[19] for which from (43), (49) one has a fixed value for n

$$j = \frac{1}{2} \Rightarrow n = 1, \quad \rho \in R. \tag{52}$$

Deleting from now on the fixed indices $j = \frac{1}{2}$ and $n = 1$, and particularizing the explicit formula (51) to this case, we have:

$$\phi^\rho_{+\frac{1}{2}}(z) = \sqrt{\frac{2}{\pi}}(1+|z|^2)^{i\rho-\frac{3}{2}}, \quad \phi^\rho_{-\frac{1}{2}}(z) = -\sqrt{\frac{2}{\pi}}z(1+|z|^2)^{i\rho-\frac{3}{2}}. \tag{53}$$

Let us now particularize the $SL(2, C)$ matrix (46) to a boost in the z direction:

$$\Lambda_\tau = \begin{pmatrix} e^{\frac{\tau}{2}} & 0 \\ 0 & e^{-\frac{\tau}{2}} \end{pmatrix}, \quad w = \cosh(\tau) \tag{54}$$

and following the $j = 0$ case studied at length in Ref. 17 let us consider the objects

$$\xi_\rho^{\pm\frac{1}{2},\pm\frac{1}{2}}(w) = \left\langle \phi_{\pm\frac{1}{2}}^\rho \middle| U^\rho(\Lambda_\tau)\phi_{\pm\frac{1}{2}}^\rho \right\rangle . \tag{55}$$

From the transformation law (45) and the explicit form (53), one gets:

$$\left(U^\rho(\Lambda_\tau)\phi_{+\frac{1}{2}}^\rho \right)(z) = \sqrt{\frac{2}{\pi}} e^{(i\rho-1)\tau}(1 + e^{-2\tau}|z|^2)^{i\rho-\frac{3}{2}} ,$$

$$\left(U^\rho(\Lambda_\tau)\phi_{-\frac{1}{2}}^\rho \right)(z) = -\sqrt{\frac{2}{\pi}} e^{(i\rho-1)\tau}e^{-\tau}z(1 + e^{-2\tau}|z|^2)^{i\rho-\frac{3}{2}} \tag{56}$$

and from these expressions one obtains:

$$\xi_\rho^{+\frac{1}{2},+\frac{1}{2}}(w) = \frac{2}{\pi} \int (1 + |z|^2)^{-i\rho-\frac{3}{2}} e^{(i\rho-1)\tau}(1 + e^{-2\tau}|z|^2)^{i\rho-\frac{3}{2}} d^2z ,$$

$$\xi_\rho^{-\frac{1}{2},-\frac{1}{2}}(w) = \frac{2}{\pi} \int e^{-\tau}|z|^2(1 + |z|^2)^{-i\rho-\frac{3}{2}} e^{(i\rho-1)\tau}(1 + e^{-2\tau}|z|^2)^{i\rho-\frac{3}{2}} d^2z . \tag{57}$$

We must now extract the Lorentz invariant IW function $\xi(w)$. To do that, we must decompose into invariants the matrix elements (57) using the spin $\frac{1}{2}$ spinors of the light cloud $u_{\pm\frac{1}{2}}$. Since we have not introduced parity in our formalism we will have the following decomposition:

$$\xi_\rho^{\pm\frac{1}{2},\pm\frac{1}{2}}(w) = \left(\bar{u}_{\pm\frac{1}{2}}(v')u_{\pm\frac{1}{2}}(v) \right)\xi^\rho(w) + \left(\bar{u}_{\pm\frac{1}{2}}(v')\gamma_5 u_{\pm\frac{1}{2}}(v) \right)\tau^\rho(w) , \tag{58}$$

where $\xi^\rho(w)$ is an irreducible $\frac{1}{2}^- \to \frac{1}{2}^-$ elastic IW function, labeled by the index ρ, and $\tau^\rho(w)$ is a function corresponding to the flip of parity $\frac{1}{2}^- \to \frac{1}{2}^+$.

One gets for the spinor bilinear

$$\bar{u}_{+\frac{1}{2}}(v')u_{+\frac{1}{2}}(v) = \bar{u}_{-\frac{1}{2}}(v')u_{-\frac{1}{2}}(v) = \sqrt{\frac{w+1}{2}} , \tag{59}$$

and since $\bar{u}_{+\frac{1}{2}}(v')\gamma_5 u_{+\frac{1}{2}}(v) = -\bar{u}_{-\frac{1}{2}}(v')\gamma_5 u_{-\frac{1}{2}}(v)$,

$$\xi_\rho(w) = \sqrt{\frac{2}{w+1}}\frac{1}{2}\left[\xi_{+\frac{1}{2},+\frac{1}{2}}^\rho(w) + \xi_{-\frac{1}{2},-\frac{1}{2}}^\rho(w) \right] , \tag{60}$$

one obtains finally:

$$\xi_\rho(w) = \frac{1}{1+\cosh(\tau)}\frac{1}{\sinh(\tau)}\frac{4}{4\rho^2+1}\left[\sinh\left(\frac{\tau}{2}\right)\cos(\rho\tau) + 2\rho\cosh\left(\frac{\tau}{2}\right)\sin(\rho\tau) \right] . \tag{61}$$

This is the expression for the elastic $\frac{1}{2}^{-} \rightarrow \frac{1}{2}^{-}$ *irreducible IW functions* we were looking for, parametrized by the real parameter ρ that labels the irreducible representations. The irreducible IW functions satisfy

$$\xi_\rho(1) = 1. \tag{62}$$

Like in the case $j = 0$, analized in great detail in Ref. 17, the elastic $\frac{1}{2}^{-} \rightarrow \frac{1}{2}^{-}$ IW function $\xi(w)$ will be given by the integral over a positive measure $d\nu(\rho)$:

$$\xi(w) = \int_{]-\infty,\infty[} \xi_\rho(w) \, d\nu(\rho), \tag{63}$$

where the measure is normalized acording to

$$\int_{]-\infty,\infty[} d\nu(\rho) = 1. \tag{64}$$

Notice that the range for the parameter ρ that labels the irreducible representations follows from the fact that in the $j = \frac{1}{2}$ case one has $n = 1$ and $\rho \in R$, Eq. (52). Notice also that the IW irreducible function (61) is even in ρ, $\xi_\rho(w) = \xi_{-\rho}(w)$.

The irreducible IW functions (61), parametrized by some fixed value of $\rho = \rho_0$, are legitimate IW functions since the corresponding measure is given by a delta function,

$$d\nu(\rho) = \delta(\rho - \rho_0)d\rho. \tag{65}$$

In the case of the irreducible representation $\rho_0 = 0$ one finds

$$\xi_0(w) = \frac{4\sinh\left(\frac{\tau}{2}\right)}{(1 + \cosh(\tau))\sinh(\tau)} = \left(\frac{2}{1 + w}\right)^{\frac{3}{2}}, \tag{66}$$

that saturates the lower bound for the slope $-\xi'(1) \geq \frac{3}{4}$. This is the so-called BPS limit of the IW function, considered in Section 4 within the Sum Rule approach, where we have seen that in the limit $-\xi'(1) \rightarrow \frac{3}{4}$ one obtains $\xi(w) \rightarrow \xi_0(w)$.

9. Integral Formula for the IW Function $\xi(w)$ and Polynomial Expression for its Derivatives

From the norm and the normalization of the irreducible IW functions one gets the correct value of the IW function at zero recoil $\xi(1) = 1$. The integral formula writes, explicitly,

$$\xi(w) = \frac{1}{1 + \cosh(\tau)} \frac{1}{\sinh(\tau)} \int_{]-\infty,\infty[} \frac{4}{4\rho^2 + 1}$$

$$\times \left[\sinh\left(\frac{\tau}{2}\right)\cos(\rho\tau) + 2\rho\cosh\left(\frac{\tau}{2}\right)\sin(\rho\tau)\right] d\nu(\rho) \tag{67}$$

from which one can find the following polynomial expression for its derivatives:

$$\xi^{(n)}(1) = (-1)^n \frac{1}{2^{2n}(2n+1)!!} \prod_{i=1}^{n} \langle [(2i+1)^2 + 4\rho^2] \rangle \quad (n \geq 1), \quad (68)$$

where the mean value is defined as $\langle f(\rho) \rangle = \int_{]-\infty,\infty[} f(\rho) d\nu(\rho)$. This formula can be demonstrated along the same lines as the corresponding one in the baryon case done in Appendix D of Ref. 17.

10. Lower Bounds on the Derivatives of the IW Function

From (68) one gets immediately the lowest bounds on the derivatives (24) obtained using the SR approach.

To get improved bounds on the derivatives we must, like in Ref. 17, express the derivatives in terms of moments of the *positive* variable ρ^2, that can be read from (68). Calling the moments:

$$\mu_n = \langle \rho^{2n} \rangle \geq 0 \quad (n \geq 0), \quad (69)$$

one gets the successive derivatives in terms of moments:

$$\xi(1) = \mu_0 = 1, \quad \xi'(1) = -\left(\frac{3}{4} + \frac{1}{3}\mu_1\right), \quad \xi''(1) = \frac{15}{16} + \frac{17}{30}\mu_1 + \frac{1}{15}\mu_2, \ldots . \quad (70)$$

The relations (70) can be solved step by step, and the moment μ_n is expressed as a combination of the derivatives $\xi(1), \xi'(1), \ldots, \xi^{(n)}(1)$:

$$\mu_0 = \xi(1) = 1, \quad \mu_1 = -\frac{3}{4}[3 + 4\xi'(1)], \quad \mu_2 = \frac{3}{16}[27 + 136\xi'(1) + 80\xi''(1)], \ldots .$$
$$(71)$$

Since ρ^2 is a positive variable, one can obtain improved bounds on the derivatives from the following set of constraints. For any $n \geq 0$, one has[17]

$$\det\left[(\mu_{i+j})_{0 \leq i,j \leq n}\right] \geq 0, \quad \det\left[(\mu_{i+j+1})_{0 \leq i,j \leq n}\right] \geq 0. \quad (72)$$

Since each moment μ_k is a combination of the derivatives $\xi(1), \xi'(1), \ldots, \xi^{(k)}(1)$, the constraints on the moments translate into constraints on the derivatives. Using (72) one gets positivity conditions of the form

$$\mu_1 \geq 0, \quad \det\begin{pmatrix} 1 & \mu_1 \\ \mu_1 & \mu_2 \end{pmatrix} = \mu_2 - \mu_1^2 \geq 0, \ldots \quad (73)$$

that imply

$$\mu_1 \geq 0, \quad \mu_2 \geq \mu_1^2, \ldots . \quad (74)$$

These constraints imply, in terms of the derivatives:

$$-\xi'(1) \geq \frac{3}{4}, \quad \xi''(1) \geq \frac{1}{5}\left[-4\xi'(1) + 3\xi'(1)^2\right], \ldots . \quad (75)$$

We see that we recover the bounds obtained using the SR method. The method generalizes in a straightforward way to higer derivatives.

11. Inversion of the Integral Representation of the IW Function

Let us now show that the integral formula for the IW function (67) can be inverted, giving the positive measure $d\nu(\rho)$ in terms of the IW function $\xi(w)$. This will allow to formulate criteria to test the validity of a given phenomenological ansatz for $\xi(w)$.

Let us define

$$\hat{\xi}(\tau) = (\cosh(\tau) + 1)\sinh(\tau)\xi(\cosh(\tau)) \tag{76}$$

and similarly for $\hat{\xi}_\rho(\tau)$.

The integral formula then writes

$$\hat{\xi}(\tau) = \int \hat{\xi}_\rho(\tau)\overline{d\nu}(\rho) = \int (\cosh(\tau) + 1)\sinh(\tau)\xi_\rho(\cosh(\tau))d\nu(\rho). \tag{77}$$

One finds, for its derivative, the simple expression:

$$\frac{d}{d\tau}\hat{\xi}_\rho(\tau) = 2\cos(\rho\tau)\cosh\left(\frac{\tau}{2}\right). \tag{78}$$

Defining the function

$$\eta(\tau) = \frac{1}{2\cosh\left(\frac{\tau}{2}\right)}\frac{d}{d\tau}\hat{\xi}(\tau), \tag{79}$$

one sees that the integral formula reads simply

$$\eta(\tau) = \int_{]-\infty,\infty[}\cos(\rho\tau)d\nu(\rho), \tag{80}$$

Computing the Fourier transform

$$\tilde{\eta}(\rho) = \frac{1}{2\pi}\int_{-\infty}^{+\infty}e^{i\tau\rho}\,d\tau\,\eta(\tau) = \int_{]-\infty,\infty[}\frac{1}{2}[\delta(\rho' + \rho) + \delta(\rho' - \rho)]d\nu(\rho'), \tag{81}$$

and defining the function

$$\mu(\rho) = \frac{d\nu(\rho)}{d\rho}, \tag{82}$$

one finds

$$\tilde{\eta}(\rho) = \frac{1}{2}[\mu(\rho) + \mu(-\rho)]. \tag{83}$$

We now assume that the general measure $d\nu(\rho)$ is even, i.e. it has the same parity as the measure $d\rho$, without loss of generality because $\xi_\rho(w)$ is even in ρ.

Then, the function (82) is even $\mu(\rho) = \mu(-\rho)$ and one finally finds for the measure $d\nu(\rho) = \tilde{\eta}(\rho)d\rho$:

$$d\nu(\rho) = \frac{1}{2\pi}\int_{-\infty}^{+\infty}e^{i\tau\rho}\,d\tau\,\frac{1}{2\cosh\left(\frac{\tau}{2}\right)}\frac{d}{d\tau}[(\cosh(\tau) + 1)\sinh(\tau)\xi(\cosh(\tau))]d\rho. \tag{84}$$

This completes the inversion of the integral representation. Equation (84) is the master formula expressing the measure in terms of the IW function.

One can apply this formula to check if a given phenomenological formula for the IW function $\xi(w)$ satisfies the constraint that the corresponding measure $d\nu(\rho)$ must be positive. This provides a powerful consistency test for any proposed ansatz.

12. An Upper Bound on the Isgur–Wise Function

From the integral formula also an upper bound on the whole IW function $\xi(w)$ can be obtained. Defining the function

$$\eta_\rho(\tau) = \frac{1}{2\cosh\left(\frac{\tau}{2}\right)} \frac{d}{d\tau} \hat{\xi}_\rho(\tau),$$ (85)

we have obtained above:

$$\eta_\rho(\tau) = \cos(\rho\tau),$$ (86)

and from it it follows

$$-1 \le \eta_\rho(\tau) \le 1,$$ (87)

and hence

$$-2\cosh\left(\frac{\tau}{2}\right) \le \frac{d}{d\tau}\hat{\xi}_\rho(\tau) \le 2\cosh\left(\frac{\tau}{2}\right).$$ (88)

Integrating this inequality from 0, one gets:

$$-4\sinh\left(\frac{\tau}{2}\right) \le \hat{\xi}_\rho(\tau) \le 4\sinh\left(\frac{\tau}{2}\right),$$ (89)

and since

$$\hat{\xi}_0(\tau) = 4\sinh\left(\frac{\tau}{2}\right),$$ (90)

one finds the inequalities

$$-\hat{\xi}_0(\tau) \le \hat{\xi}_\rho(\tau) \le \hat{\xi}_0(\tau),$$ (91)

that simplify to:

$$-\xi_0(\tau) \le \xi(\tau) \le \xi_0(\tau).$$ (92)

Since $\xi_0(\tau)$ is given by the expression (66), one finally obtains

$$|\xi(w)| \le \left(\frac{2}{1+w}\right)^{\frac{3}{2}}.$$ (93)

This inequality is a strong result because it holds for any value of w.

13. Consistency Tests for any Ansatz of the Isgur–Wise Function: Phenomenological Applications

To illustrate the methods exposed above, we now examine some phenomenological formulas proposed in the literature. In Ref. 19 we have studied a number of other interesting cases.

We will compare these ansätze with the theoretical criteria that we have formulated: the lower bounds on the derivatives at zero recoil, the upper bound obtained for the whole IW function, and the inversion of the integral formula for the IW function in order to check the positivity of the measure.

We must underline that the satisfaction of the bounds on the derivatives and of the upper bound on the whole IW function are *necessary conditions*, while the criterium of the positivity of the measure is a *necessary and sufficient condition* to establish if a given ansatz of the IW function satisfies the Lorentz group criteria.

13.1. *The exponential ansatz*

This form corresponds to the non-relativistic limit for the light quark with the harmonic oscillator potential:[20]

$$\xi(w) = \exp[-c(w-1)].\tag{94}$$

The bound for the slope is satisfied for $c \geq \frac{3}{4}$, the bound for the second derivative is satisfied for $c \geq 2$, while the bound for the third derivative is violated for any value of c. Therefore, this phenomenological ansatz on the IW function is invalid.

The exponential ansatz satisfies nevertheless the upper bound (93).

Let us now examine the criterium based on the positivity of the measure. One needs to compute

$$\eta(\tau) = \frac{1}{c}\left(-\frac{d^2}{d\tau^2} + \frac{1}{4}\right)\cosh\left(\frac{\tau}{2}\right)\exp[-c(\cosh(\tau)-1)].\tag{95}$$

The function $\eta(\tau)$ is bounded for any value of c. The Fourier transform of this function gives

$$d\nu(\rho) = \frac{e^c}{2\pi}\frac{1}{c}\left(\rho^2 + \frac{1}{4}\right)\left[K_{i\rho+\frac{1}{2}}(\rho) + K_{-i\rho+\frac{1}{2}}(\rho)\right]d\rho.\tag{96}$$

Since this function is not positive for any value of c, the exponential ansatz for the IW function violates the consistency criteria exposed above.

13.2. *The "dipole"*

The following shape has been proposed in the literature (see for example Refs. 21 and 22)

$$\xi(w) = \left(\frac{2}{1+w}\right)^{2c}.\tag{97}$$

The bounds for the slope and for the higher derivatives are satisfied for $c \geq \frac{3}{4}$.

Let us now compute the measure (84). One needs first to compute

$$\eta(\tau) = -4(c-1)\left[\cosh\left(\frac{\tau}{2}\right)\right]^{-4c+3} + (4c-3)\left[\cosh\left(\frac{\tau}{2}\right)\right]^{-4c+1}. \tag{98}$$

Since $\eta(\tau)$ has to be bounded, the parameter c must satisfy $c \geq \frac{3}{4}$.

Moreover, we realize that in the particular case

$$c = \frac{3}{4} \rightarrow \eta(\tau) = 1 \rightarrow d\nu(\rho) = \delta(\rho)d\rho. \tag{99}$$

Therefore, one gets in this case a delta-function for the measure. This is a positive measure that corresponds to the explicit formula (66) for the IW function in the BPS limit.

For $c > \frac{3}{4}$ one obtains a function $\eta(\tau)$ that is bounded and integrable. Computing its Fourier transform one gets the measure

$$d\nu(\rho) = \frac{2^{4c-1}}{2\pi}(4c-3)\left(\rho^2 + \frac{1}{4}\right)\frac{\Gamma\left(i\rho + 2c - \frac{3}{2}\right)\Gamma\left(-i\rho + 2c - \frac{3}{2}\right)}{\Gamma(4c-1)}d\rho, \tag{100}$$

that is positive.

In conclusion, the measure $d\nu(\rho)$ for the "dipole" ansatz is positive for $c \geq \frac{3}{4}$ and therefore satisfies all the consistency criteria.

14. Conclusion

We have reviewed a number of important works by Nikolai Uraltsev on Sum Rules in the heavy quark limit of QCD.

We have generalized Bjorken and Uraltsev SR to higher derivatives and we have formulated lower bounds on the successive derivatives of the Isgur–Wise function $\xi(w)$. We have also obtained an explicit form for the IW function in the "BPS limit" considered by Uraltsev.

On the other hand, the Lorentz group acting on the light cloud provides a transparent physical interpretation of the results obtained from the SR. Both methods are completely equivalent.

Within the Lorentz group method we have obtained an integral formula for the IW function in terms of an explicit kernel and *a positive measure*.

From this representation we have reproduced the bounds on the derivatives of the IW function from positivity conditions on moments of a positive variable.

On the other hand, we have inverted the integral formula expressing the positive measure in terms of any given ansatz of the IW function.

As a consence, the "BPS limit" for the IW function obtained from the SR method turns out to have a clear group theoretical interpretation: the positive measure is just a δ-function, and the cloud ground state belongs to a particular irreducible representation of $SL(2, C)$.

Some phenomenological proposals for the shape of the IW function have been compared with the theoretical constraints obtained in this paper.

These different shapes provide illustrations of the method in a rather complete way. The different criteria based on the Lorentz group, i.e. lower limits on the derivatives at zero recoil, positivity of the measure in the inversion formula for the IW function and the upper bound for the whole IW function, have been illustrated.

A main conclusion is that, using a method based on the Lorentz group, completely equivalent to the one of generalized Bjorken–Uraltsev sum rules, one obtains strong constraints on the IW function for the ground state mesons.

Acknowledgments

We are indebted to our colleague Alain Le Yaouanc for discussions on our common work and on the present text.

References

1. N. Isgur and M. Wise, *Phys. Lett. B* **232**, 113 (1989); **237**, 527 (1990).
2. J. D. Bjorken, invited talk at Les Rencontres de la Vallée d'Aoste, La Thuile, SLAC-PUB-5278, 1990.
3. N. Isgur and M. Wise, *Phys. Rev. D* **43**, 819 (1991).
4. N. Uraltsev, *Phys. Lett. B* **501**, 86 (2001); N. Uraltsev, *J. Phys. G* **27**, 1081 (2001).
5. B. Bakamjian and L. H. Thomas, *Phys. Rev.* **92**, 1300 (1953).
6. A. Le Yaouanc, L. Oliver, O. Pène and J.-C. Raynal, *Phys. Lett. B* **365**, 319 (1996).
7. V. Morénas, A. Le Yaouanc, L. Oliver, O. Pène and J.-C. Raynal, *Phys. Lett. B* **408**, 357 (1997).
8. A. Le Yaouanc, L. Oliver, O. Pène, J.-C. Raynal and V. Morénas, *Phys. Lett. B* **520**, 25 (2001).
9. N. Uraltsev, *Phys. Lett. B* **585**, 253 (2004).
10. A. Le Yaouanc, L. Oliver and J.-C. Raynal, *Phys. Rev. D* **67**, 114009 (2003); *Phys. Lett. B* **557**, 207 (2003).
11. A. F. Falk, *Nucl. Phys. B* **378**, 79 (1992).
12. A. Le Yaouanc, L. Oliver and J.-C. Raynal, *Phys. Rev. D* **69**, 094022 (2004).
13. M. P. Dorsten, *Phys. Rev. D* **70**, 096013 (2004).
14. I. Bigi, M. Shifman, N. Uraltsev and A. Vainshtein, *Phys. Rev. D* **52**, 196 (1995).
15. F. Jugeau, A. Le Yaouanc, L. Oliver and J.-C. Raynal, *Phys. Rev. D* **74**, 094012 (2006).
16. A. Messiah, Mécanique quantique, Dunod, Paris (1962).
17. A. Le Yaouanc, L. Oliver and J.-C. Raynal, *Phys. Rev. D* **80**, 054006 (2009).
18. M. A. Naïmark, Les représentations linéaires du groupe de Lorentz, Dunod, Paris (1962).
19. A. Le Yaouanc, L. Oliver and J.-C. Raynal, to appear.
20. F. Jugeau, A. Le Yaouanc, L. Oliver and J.-C. Raynal, *Phys. Rev. D* **70**, 114020 (2004).
21. M. Neubert, V. Rieckert, B. Stech and Q. P. Xu, *Heavy Flavours*, eds. A. J. Buras and H. Lindner (World Scientific, Singapore, 1992).
22. V. Morénas, A. Le Yaouanc, L. Oliver, O. Pène and J.-C. Raynal, *Phys. Rev. D* **56**, 5668 (1997).

Uraltsev's and Other Sum Rules, Theory and Phenomenology of D^{**}'s

Alain Le Yaouanc and Olivier Pène

Laboratoire de Physique Théorique, CNRS et Université Paris-sud XI,
*Bâtiment 210, 91405 Orsay Cedex, France**

We first discuss Uraltsev's and other sum rules constraining the $B \to D^{**}(L = 1)$ weak transitions in the infinite mass limit, and compare them with dynamical approaches in the same limit. After recalling these well established facts, we discuss how to apply infinite mass limit to the physical situation. We provide predictions concerning semileptonic decays and non-leptonic ones, based on quark models. We then present in more detail the dynamical approaches: the relativistic quark model *à la* Bakamjian–Thomas and lattice QCD. We summarise lattice QCD results in the infinite mass limit and compare them to the quark model predictions. We then present preliminary lattice QCD results with finite b and c quark masses. A systematic comparison between theory and experiment is performed. We show that some large discrepancies exist between different experiments. Altogether the predictions at infinite mass are in fair agreement with experiment for non-leptonic decays contrary to what happens for semileptonic decays. We conclude by considering the prospects to clarify both the experimental situation, the theoretical one and the comparison between both.

1. In Remembrance of Kolya

In November 2012 we had in Paris a workshop dedicated to $B \to D^{**}$ decays. Many of the best specialists of the field were there, experimentalists and theorists. We took time to discuss in detail about what we knew, what was still obscure, how to solve the issues. Among the participants was Kolya Uraltsev. He took a prominent part in the discussions. He was so deep, so rigorous, so strong in his statements, and so careful about the yet unknown, about the ambiguities, so positive about what should be done and about the requests to experimentalists.

We have been working for long on the heavy quark limit for the $B \to D^{**}$ decay using a relativistic quark model. In 2000 Kolya derived a very simple and powerful sum rule in the same limit. From there on we started to interact with him. Ikaros Bigi came several times in Orsay and we have published together two

---

*Unité Mixte de Recherche 8627 du Centre Nationale de la Recherche Scientifique.

"memorinos." Bigi was a close collaborator and friend of Kolya. Thanks to him we could communicate more intensively with Kolya. All these discussions were really enlightening.

Kolya did not come to our workshop dinner arguing that he would be tempted by the food and that it was not good for him. We did not realise the case was so serious. In February 2013 we received a mail from Ikaros "Dear Friends, on Wednesday (February 13th 2013) Kolya had passed away in Siegen — he was a wonderful person, theorist and true friend. I cannot speak more. Ikaros." We couldn't say more either, the shock was too violent.

Now the time has come to honour Kolya by speaking about physics, which was so important for him, to which he devoted all his strength and his admirable brain.

2. Historical Elements Concerning the Discussions About D^{**}

The issue of the decays $B \to D^{**}$ has been actively discussed since more than twenty years under its theoretical and experimental aspects and the relation between both, see Refs. 1, 2 for an overlook on these aspects and more recently.[3]

2.1. Well established facts at $m_Q \to \infty$

Let us first recall what are the D^{**} under concern. They are the charmed states with quark model assignment $L = 1$, and most often we are meaning the lowest lying ones. In the infinite mass limit $m_Q \to \infty$, they can be separated into two doublets $j = 1/2, j = 3/2$, then four states, almost degenerate (except for a small spin orbit force), for each level of excitation. While the $j = 1/2$ are broad, the $j = 3/2$ are narrow, having respectively S and D waves in pionic decay. The broad $j = 1/2$ include a 0^+ and a 1^+, while $j = 3/2$ include a 2^+ and a 1^+.

These states are playing a prominent role in the heavy quark sum rules, e.g. the Bjorken sum rule:

$$\rho^2 = 1/4 + \sum_n \left(\left| \tau_{1/2}^{(n)}(w = 1) \right|^2 + 2 \left| \tau_{3/2}^{(n)}(w = 1) \right|^2 \right) \tag{1}$$

n labels successive levels of excitation. ρ^2 is the forward slope of the elastic Isgur–Wise function, while $\tau_{1/2,3/2}$ are similar quantities for the $B \to D^{**}$ transitions.[4] One defines for instance for the $B \to D(0^+)$ transition:

$$\langle 0^+ | A_\lambda | 0^- \rangle = -\frac{1}{\sqrt{v_0 v_0'}} (v_\lambda - v_\lambda') \tau_{1/2}(w), \tag{2}$$

where v, v' are the four-velocity[a] vectors of the mesons. $w = v \cdot v'$, $w = 1$ corresponding to zero recoil. We had observed that this rule is exactly satisfied in a class of quark models.[5]

---

[a]The four-velocity is defined by $v_\mu \equiv p_\mu/m$.

In 2000, Uraltsev[6] discovered a new class of sum rules, similar but for the combination of τ's, which appear now in differences, e.g.:

$$\sum_n \left(\left| \tau_{3/2}^{(n)}(w=1) \right|^2 - \left| \tau_{1/2}^{(n)}(w=1) \right|^2 \right) = 1/4. \tag{3}$$

We had found, also from duality arguments, one similar rule,[7] except for factors of excitation energy, i.e. the sum has now factors of energy minus the ground state energy:

$$\sum_n \left(E_{3/2}^{(n)} - E_0 \right) \left| \tau_{3/2}^{(n)}(w=1) \right|^2 - \left(E_{3/2}^{(n)} - E_0 \right) \left| \tau_{1/2}^{(n)}(w=1) \right|^2 \right) = \cdots . \tag{4}$$

It is retrieved by the systematic approach of Uraltsev in a whole series of sum rules with increasing powers of these energy differences.

At this point, one enters into what will be the main worry in the rest of the Chapter. Both rules were indicating an important trend for the τ's: that one should have $|\tau_{3/2}|$ on the whole larger than $|\tau_{1/2}|$. This is yet a vague statement, which receives a stricter form under the assumption, put forward by Uraltsev, that the lowest level (noted $n=0$) is dominating the sums. Then

$$\left| \tau_{3/2}^{(0)}(w=1) \right| > \left| \tau_{1/2}^{(0)}(w=1) \right|. \tag{5}$$

One must be aware that for the deduction to be valid, the dominance must be rather strict.

The statement Eq. (5) is precisely the conclusion we had obtained, in the heavy quark limit, from our Bakamjian–Thomas (BT) relativistic quark model approach in the heavy quark limit (shortly described in Subsec. 3.1), some years before.

Another, related, point is that, combining the Uraltsev sum rule Eq. (3), with the Bjorken one, one gets a stronger bound on ρ^2:

$$\rho^2 > 3/4. \tag{6}$$

We had also observed previously that precisely this bound is obtained in the "Bakamjian–Thomas" approach. And, in fact, we have also shown that the Uraltsev sum rule is exactly satisfied in the model.[8]

One sees easily that the difference between $\tau_{3/2}(1)$ and $\tau_{1/2}(1)$ is a relativistic effect. Indeed, in an expansion in terms of the light quark internal velocity, v/c, $\tau_{3/2}(1)$ and $\tau_{1/2}(1)$ are of order c/v, they have the form of dipole transitions where the current matrix element is $\propto \vec{r} \propto mc/v$ and does not depend on the spin. Then, in the non-relativistic limit $\tau_{3/2}(1) - \tau_{1/2}(1) = 0$. The difference is subleading in the expansion, and the Uraltsev sum rule displays this: $1/4$ is subleading. But why is the difference positive?

This type of quark model offers a very simple and intuitive explanation of the difference: one ends with the simple expression

$$\tau_{3/2}(1) - \tau_{1/2}(1) = \frac{1}{2} \int_0^\infty dp\, p^2 \phi_1(p) \frac{p}{p_0 + m} \phi_0(p), \tag{7}$$

$\phi_L(p)$ are the radial wave functions. For simplicity, we have taken $\phi_1(p)$ the same for all four states, which amounts to neglect a small spin-orbit force. The two functions $\phi_0(p)$, $\phi_1(p)$ have been chosen positive by a choice of the phase of states, and the τ's are then positive. This choice is possible for the lowest states, but not for the $L = 1$ radially excited states, for which there are zeroes in the radial wave functions.

The $\mathcal{O}(v/c)$ factor $\frac{p}{p_0+m}$ comes from the Wigner rotation of the light (spectator) quark spin which acts differently on $(1/2)^+$ and $(3/2)^+$ states. This Wigner rotation is a typical relativistic effect in bound states.[9]

It is of order $\mathcal{O}(v/c)$, which gives indeed $\mathcal{O}(v/c)^0$ for the difference of squares in Uraltsev sum rule (recall that each term squared is $\mathcal{O}(v/c)^{-2}$). This is compatible with the 1/4 on the r.h.s. of Eq. (3): $\tau_{3/2}(1) - \tau_{1/2}(1)$ is **positive**. Moreover it is large in fact, because the velocity of **light** quarks within mesons is large: relativistic internal quark velocities.

In summary, $\tau_{3/2}(1) - \tau_{1/2}(1)$ is positive and large. Or, as found quantitatively, $\tau_{1/2}(1)$ is small with respect to $\tau_{3/2}(1)$. In fact, using what is the best potential in our opinion, the model of Godfrey and Isgur,[10] in the "Bakamjian–Thomas" formalism, we found:

$$\tau_{3/2}(1) = 0.55\,, \tag{8}$$

$$\tau_{1/2}(1) = 0.225\,. \tag{9}$$

Of course, the quark model is not meant to be an exact approach.[b]

Only a true QCD calculation can allow a safe conclusion, therefore it is very important that the conclusion was indeed confirmed later in lattice QCD, at $N_F = 0$ and then at $N_F = 2$. The latest result is in Eq. (32).

One notes that the agreement of the quark model with lattice is very good for $j = 3/2$ (central lattice value $\simeq 0.53$), but that it is slightly worse for $j = 1/2$: with the lattice central value ($\simeq 0.3$), quark model is sizeably below, and in squares, which are relevant in the above sum rules and for the rates, the ratio between $\tau_{1/2}(1)$ from lattice and the same from the relativistic quark model is not far from 2: 1.7. Still the close agreement between these estimates for $\tau_{3/2}(1)$ and the semiquantitative agreement for $\tau_{1/2}(1)$ is striking and gives us confidence in these figures.

The advantage of lattice QCD is also that one can estimate errors, since one has now a systematic approach to the true result of QCD. By successive improvement of the calculation, one can estimate the errors with respect, for example, to the continuum and chiral limits.

---

[b]We do not quote errors however. This simply means that we cannot calculate errors. Even quoting a range of variation when one varies some parameters of the model could be misleading; one would have to check that the new values of the parameters fit the spectrum and the other reference data as well, which is not done usually. What can be said usefully is that the $\tau_{1/2}$ is more sensitive to the details of the potential model at short distance.

Up to now everything seems consistent and very encouraging. Lattice approach, general heavy quark statements and quark model agree. The lacking protagonist is experiment. And now comes a possible trouble.

2.2. *Phenomenology in the $m_Q \to \infty$ approximation*

The results in the infinite mass limit can be used in two ways:

- Either as purely theoretical tests of the consistency of various approaches, e.g. does one get in a model the necessary Isgur–Wise scaling and the normalisation condition $\xi(1) = 1$? Or, for the lattice approach, where it is not the question, it may be still a practical check of the soundness of the calculation.
- Or, more ambitiously, it could be relevant for phenomenology, as well as model dependent results like the ones from the quark model, or exact dynamical calculations of lattice QCD, taken in the the the $m_Q \to \infty$.

Here, the initial idea is of course that for heavy flavour physics the heavy quarks c, b could be sufficiently heavy for the heavy quark limit to be applied, and that many advantages are obtained in that limit. First, new general statements and simplifying features, then new properties of the quark model and other approaches. Even for lattice QCD, the $m_Q \to \infty$ limit is useful practically because the treatment of the heavy quark line is especially easy (the heavy quark being described by a Wilson line instead of having to solve numerically a Dirac equation).

A striking example of the phenomenological success of the approach is the prediction of broad and narrow $L = 1$ D^{**} states, the $j = 1/2$ having pure S wave pionic decay, while the $j = 3/2$ have it in D wave. One also observes an agreement for $B \to D^{(*)}$ and for $B \to D^{**}(j = 3/2)$ (narrow states). In the last case, this is especially true if one averages over the j multiplet.

Then, one would expect also the above hierarchy $j = 1/2 \ll j = 3/2$ to be observed. In fact, there is at present no clear conclusion, after twenty years of experimental effort.

At this stage however, it must be said more precisely how the $m_Q = \infty$ predictions, which are well defined by themselves, may be used for real physics, at finite masses.

There is admittedly no compelling procedure. The only real safe way would be to calculate at the finite physical masses, but we miss theoretical tools to do so, or to calculate corrections to the infinite mass limit. Precisely, the hope was in the beginning to simplify the problems by a clever use of the $m_Q = \infty$ limit.[c]

Then, one is just proposing recipes, which have succeeded in the above cases $B \to D^{(*)}, D^{**}(j = 3/2)$. To be specific, among the many choices of form factors, what are the ones that we assimilate to their $m_Q = \infty$ limit? Different choices will

---

[c]In addition, at finite mass, one enters new difficulties for the quark model, like loss of covariance or non-conservation of the vector current.

lead to different results for the physical predictions from the same heavy quark theory.

In our papers, we have made a choice based on Eq. (2). This means that one keeps the same expression, but with physical velocities. There is at least some logic in using invariant form factors defined with velocity coefficients. The exact formula at finite masses duely contains two independent form factors, which one defines also with velocity factors:

$$\langle 0^+|A_\lambda|0^-\rangle = \frac{1}{2\sqrt{v_0 v_0'}}((v_\lambda + v_\lambda')g_+(w) + (v_\lambda - v_\lambda')g_-(w)) \tag{10}$$

with $g_+(w) \propto 1/m_Q$ and $g_-(w) = -2\tau_{1/2}(w) + \mathcal{O}(1/m_Q)$. What we do is to cancel plainly the $\mathcal{O}(1/m_Q)$ in $g_{+,-}(w)$, i.e. we set $g_+(w) = 0, g_-(w) = -2\,\tau_{1/2}(w)$.

It is useful to quote the differential rate corresponding to Eq. (10) for later discussion:

$$\frac{d\Gamma}{dw} = |V_{cb}|^2 \frac{G_F^2 m_B^5}{48\pi^3} r^3 (w^2 - 1)^{\frac{3}{2}} [(1+r)g_+ - (1-r)g_-]^2 \tag{11}$$

with $r = m_D^{**}/m_B$.

2.3. *Semileptonic decays; the beginning of controversies about the* *$j = 1/2$ states*

The predictions have been formulated for integrated rates. Indeed, the semileptonic data were not initially sufficiently accurate to make a comparison for differential distributions. Note that, for the safer lattice QCD, at $m_Q = \infty$, one is restricted to $w = 1$ and the semileptonic data are not sufficiently accurate to make a comparison in differential distributions around $w = 1$ even now. Therefore, one is led to use the quark model even presently.

One obtains for the semileptonic rate of $j = 3/2$ in the above mentioned quark model,[9] with the choice of Godfrey and Isgur as the best potential model:

$$BR(B \to D^{(**)}(j = 3/2)2^+ l\nu) = 0.70\% \quad D_2^*, \tag{12}$$

$$BR(B \to D^{(**)}(j = 3/2)1^+ l\nu) = 0.45\% \quad D_1, \tag{13}$$

$$BR(B \to D^{(**)}(j = 3/2)l\nu) = 1.15\% \tag{14}$$

while for $j = 1/2$:

$$BR(B \to D^{(**)}(j = 1/2)1^+ l\nu) = 0.07\% \quad D_1^*, \tag{15}$$

$$BR(B \to D^{(**)}(j = 1/2)0^+ l\nu) = 0.06\% \quad D_0^*, \tag{16}$$

$$BR(B \to D^{(**)}(j = 1/2)l\nu) = 0.13\%. \tag{17}$$

We have indicated on the right the identification of the considered $D^{(**)}$ with the experimental states in the notations of Ref. 11, terms although the notations with $j = 1/2, 3/2$ seem to us much more transparent. Note that the inequality

$\mathrm{BR}(j = 1/2) \ll \mathrm{BR}(j = 3/2)$ is much stronger that the one between τ's, first because one has squares, second because there is an additional kinematical factor in the rates

$$\frac{d\Gamma_{1/2}/dw}{d\Gamma_{3/2}/dw} = \frac{2}{(w+1)^2}\left(\frac{\tau_{1/2}(w)}{\tau_{3/2}(w)}\right)^2, \tag{18}$$

where one has summed within the j multiplets. The kinematical ratio $\frac{2}{(w+1)^2}$ adds a factor $< 1/2$ to the ratio $\left(\frac{\tau_{1/2}(w)}{\tau_{3/2}(w)}\right)^2$.

While $j = 3/2$ states are in rough agreement with experiment from the beginning, especially the sum, $j = 1/2$ ones were strongly disagreeing with the results of Delphi, Table 10 in Ref. 12 at LEP: a large signal was found there for a broad 1^+, an order of magnitude larger than the prediction in Eq. (15): $\mathrm{BR}(B \to D_1^* l\nu) = (1.25 \pm 0.025 \pm 0.027)\%$. The 0^+ was not clearly seen: $\mathrm{BR}(B \to D_0^* l\nu) = (0.42 \pm 0.33 \pm 0.22)\%$.[d]

We have noted that the lattice QCD number for $\tau_{1/2}(1)$ is somewhat higher than the quark model. Therefore, one would expect larger predictions for $\mathrm{BR}(B \to D_1^* l\nu)$ by a factor around two, which would still leave the Delphi data unexplained.

Of course, the discrepancy could originate in the $m_Q \to \infty$ approximation. Therefore, very early, general estimates of $\mathcal{O}(1/m_Q)$ corrections have been given by Leibovich et al.,[13] in a very careful discussion. Concerning $j = 1/2$, as well as the $j = 3/2$, 1^+ they show that the vanishing of the amplitude at zero recoil which happens in the infinite mass limit is no longer valid at finite masses, while for the $j = 3/2$, 2^+, vanishing at zero recoil stays valid at finite masses (it has a general kinematical origin). Then they predict very small enhancement for the $j = 3/2$, 2^+. The enhancement they predict by an estimate of the subleading $1/m_Q$ contributions does not exceed a factor 3 for $j = 3/2$, 1^+ and $j = 1/2$, 0^+ and is numerically very small for $j = 1/2$, 1^+. Their estimate is of course somewhat model dependent.

For the 0^+, one can understand rather simply that a large enhancement is possible from Eq. (11), through a series of effects (following the discussion in Ref. 14). Now $g_+ \neq 0$. The contributions from g_- and g_+ add algebraically and may be interfering constructively as found from HQET for first order correction at zero recoil and also in the quark model. The contribution of g_- is maintained close to its $m_Q \to \infty$ limit according to the quark model, while g_+ becomes sizable (in fact the BT model finds it rather large, but it is doubtful since it outpasses the $\mathcal{O}(1/m_Q)$ magnitude predicted by HQET near zero recoil). The g_+ contribution is enhanced relatively to the latter by the factor $(1 + r)/(1 - r)$. All these effects are squared in the rate. The predicted enhancements deserve consideration, but do not correspond to the conclusions of DELPHI, which point to a very large 1^+, $1/2$. We will return to this issue of finite mass corrections when speaking about lattice results at finite masses.

---

[d]This number differs from the one quoted in the memorinos.[1,2]

Unexpectedly, the new experiments at the two B factories have not clarified the experimental situation. Quite on the contrary: they have added new contradictions: contradictions within experiment for the same semileptonic reaction, and others between the semileptonic and the non-leptonic transitions, newly measured (see below). And the latter were found rather in agreement with the expectations from the $m_Q \to \infty$ approximation.

2.4. Non-leptonic decay. Class I and Class III

Indeed, a new, very interesting piece of data has appeared with the measurement of the $B \to D^{**}$ transitions, first by CLEO. Namely the quasi-two body decays $B \to D^{**}\pi$. And it has been realised first from a paper by Neubert[15] that there was a close connection with the semileptonic decay. Indeed through "factorisation" (i.e. the old SVZ "vacuum insertion"), the $B \to D^{**}\pi$ amplitude can be related to the $B \to D^{**}$ current matrix element

$$B \to D^{**}\pi = f_\pi p_\pi^\mu \langle D^{**}|J_\mu|B\rangle \tag{19}$$

and then the latter to the semileptonic amplitude. In fact the $B \to D^{**}\pi$ rate is found to be:

$$\frac{\Gamma(\bar{B}^0 \to H_c^+ \pi^-)}{d\Gamma(\bar{B}^0 \to H_c^+ \ell^- \bar{\nu})/dq^2\Big|_{q^2=0}} = 6\pi^2 f_\pi^2 |V_{ud}|^2 |a_1|^2 + O\left(\frac{M_\pi^2}{M_B^2}\right), \tag{20}$$

where H_c is any charmed meson, and where a_1 is very close to 1. The statement is approximate because factorisation is only an (intuitive and uncontrolled) approximation. Moreover, the formula Eq. (20) applies properly to the decay $B^0 \to D^{**-}\pi^+$. For $B^+ \to \bar{D}^{**0}\pi^+$, there is another important diagram contributing, obtained by Fierz transformation, with $B \to \pi$ current matrix element and the emission of the $\bar{D}^{**0}$ described by the annihilation constant $f_{D^{**}}$. One distinguishes the two type of decays as being respectively "Class I" and "Class III" decays. Only Class I decay is directly calculable from the $B \to D^{**}$ current matrix element. These decays are discussed in detail in Ref. 16.

Let us then first consider this simplest case of Class I. The predictions of the quark model with Godfrey–Isgur wave functions in the $m_Q \to \infty$ approximation are then:[3]

$$B(\bar{B}^0 \to D_2^{3/2+}\pi^-) = 11 \times 10^{-4},$$

$$B(\bar{B}^0 \to D_1^{3/2+}\pi^-) = 13 \times 10^{-4},$$

$$B(\bar{B}^0 \to D_1^{1/2+}\pi^-) = 1.1 \times 10^{-4}, \tag{21}$$

$$B(\bar{B}^0 \to D_0^{1/2+}\pi^-) = 1.3 \times 10^{-4}.$$

This time, in contrast with the semileptonic case, they were found in rather good agreement with the Belle experiment, the first to give an analysis of the four states. In particular, the striking dissymmetry BR($j = 1/2$) $\ll$ BR($j = 3/2$) was observed. Finite mass corrections have been estimated according to the procedure of Leibovich *et al.* and one has found conclusions similar to the semileptonic case.[16] Namely, only the 0^+ is expected to be affected by a large possible enhancement. Rather recently, the 0^+ was also measured by Babar, with a larger rate but much larger errors, then not modifying the initial conclusion, see Table 2.

Then about Class III. There comes one more statement in the $m_Q \to \infty$ limit: $f_D^{**}(j = 3/2) = 0$ while $f_D^{**}(j = 1/2)$ is not all suppressed. A rough estimate shows that for $j = 1/2$, the contribution with f_D^{**} is large and wins over the Class I, which is suppressed, and that it has the same sign.[16]

A consequence of this is that Class III should be almost the same as Class I for $3/2$, but differ strongly for $1/2$. The suppression of $1/2$ predicted and observed in Class I should not be present in Class III. An estimate from the same paper[16] is:

$$B\left(B^- \to D_2^{3/2\ 0}\pi^-\right) = (8.7 \pm 3.2) \times 10^{-4},$$

$$B\left(B^- \to D_1^{3/2\ 0}\pi^-\right) = (10.2 \pm 2.3) \times 10^{-4},$$

$$B\left(B^- \to D_1^{1/2\ 0}\pi^-\right) = (7.5 \pm 1.7) \times 10^{-4}, \tag{22}$$

$$B\left(B^- \to D_0^{1/2\ 0}\pi^-\right) = (9.1 \pm 2.9) \times 10^{-4}.$$

This agrees with what is observed Table 2.

It is now worth describing and discussing in more details the two dynamical approaches at our disposal, with their properties at physical masses as well as at $m_Q = \infty$, and then the present experimental situation.

3. Quark Model

One big advantage of the quark model in the present discussion is that it provides well defined predictions for all values of w, including the infinite mass limit where it can thus complement the lattice approach. Among the many quark models, we are privileging here a particular approach because of its remarkable properties, and especially those — but not only those — which hold in the $m_Q = \infty$ limit.

3.1. *General framework of Bakamjian–Thomas*

The Bakamjian–Thomas approach may seem unfamiliar especially because the name itself is unfamiliar, but nevertheless it has been used and developed and promoted to a certain extent among nuclear physicists[17] Its null-plane version has been extensively promoted in particle physics, starting as it seems from Terent'ev.[18] We believe that the form with standard quantization on $t = 0$ plane is nevertheless the closest to the intuitions of the quark model. The main idea, starting from the researches of Foldy, is to formulate a relativistic quantum mechanics with a fixed number of particle and an instantaneous interaction. Of course, this is not fully possible, but already what is obtained is significant and it seems to be the best one can do to implement relativity while maintaining the three-dimensional spirit of the quark model.

Indeed, what is obtained is the full set of operators satisfying the Poincaré algebra, which enables to define states in motion from states at rest; the latter being described by wave functions which are eigenvectors of the mass operator.

One can define two sets of dynamical variables: one of global variables, describing the state as whole, $\vec{P}$, $\vec{R}$, $\vec{S}$ ($\vec{R} = -i\partial/\partial\vec{P}$), and one of internal variables, in fact the internal momenta $\vec{k}_i$ (with $\sum_{i=1}^{n} \vec{k}_i = 0$), the conjugate positions, and the spins $\vec{s}_i$, with global variables commuting with internal ones.

The mass operator M must be a rotation invariant function of internal variables only, and then commutes with $\vec{P}$, $\vec{R}$, $\vec{S}$.

The Poincaré generators are constructed from the global operators and the inter-action in a very simple form, i.e. they have the same form as for a free particle, except that the mass (which appears in H and the generator of Lorentz transfor-mations $\vec{K}$) is replaced by the mass operator M for the bound states

$$\vec{J} = \vec{P} \times -i\partial/\partial\vec{P} + \vec{S}, \tag{23}$$

$$H = \sqrt{M^2 + \vec{P}^2}, \tag{24}$$

$$\vec{K} = 1/2[H, -i\partial/\partial\vec{P}]_+ - \frac{\vec{P} \times \vec{S}}{H + M}. \tag{25}$$

3.2. *Matrix elements of currrents*

One looses relativity when one writes matrix elements of some current between multiquark states (see for details the paper[19]).

Multiquark states in motion are described by first writing a wave function in terms of internal variables:

$$\phi_{s_1,\ldots,s_n}(\vec{k}_1,\ldots,\vec{k}_n). \tag{26}$$

These are solutions of the eigenvalue equation $M\phi = m\phi$. Then, the full wave functions in terms of the usual one-particle variables

$$\Psi^{(\mathbf{P})}_{s_1,\ldots,s_n}(\vec{p}_1,\ldots,\vec{p}_n), \quad \sum_{i=1}^{n}\vec{p}_i = \vec{P} \tag{27}$$

are obtained

(1) by including a plane wave $\delta(\sum_i \vec{p}_i - \vec{P})$,
(2) by expressing the internal variables in terms of the one-particle variables, which is obtained by a free-quark boost operation (with quark energy $p_i^0 = \sqrt{|\vec{p}_i|^2 + m_i^2}$), including Wigner rotation of spins $s_1,\ldots,s_n$ and Jacobian fac-tors ensuring unitarity.

Whence the matrix element of a one-body operator.

If one chooses as current operators standard one quark operators like $1_i, \sigma_i, \ldots$, for simplicity and to comply with the basic notion of additivity, then such matrix elements do not respect covariance. In fact, only multibody current operators could implement the covariance. Neither do the one-body currents satisfy the standard Ward identities like vector current conservation.

The remarkable fact is that in the infinite mass limit of one quark, the main difficulties disappear: one finds covariance, current conservation, and also many properties required in this limit: Isgur–Wise scaling, normalisation $\xi(w) = 1$, a certain set of sum rules, namely those implying only the Isgur–Wise functions (like the ones of Bjorken and Uraltsev).[e] Also, one finds that in this limit, there is in fact equivalence with a null plane formalism, with a suitable relation between the wave functions at rest and on the null plane.

It is quite possible to calculate in the BT approach at physical masses. However, for the present, in view of the above many difficulties concerning the current matrix elements, and of the many advantages of the heavy mass limit, it is logical to prefer using this limit, with the procedure described in Subsec. 2.2 to connect it with phenomenology, keeping in mind that non-neglible $\mathcal{O}(1/m_Q)$ corrections are expected, especially for the transition to 0^+ (see Subsec. 2.3).

3.3. Choice of the potential model

To fix entirely the model, one needs of course to specify the mass operator M, which means the potential model or wave equation at rest, i.e., we choose first the general structure:

$$M = \sum_i \sqrt{\vec{k}_i^2 + m_i^2} + V \,. \tag{28}$$

Then, we make specifically the choice of the model of Godfrey et Isgur (GI)[10] for the reason that it has been tested with success on a very large range of states, which is less the case for others. We have checked that the qualitative conclusions are general, by making calculations with other models. We think however that in the present problem, where the wave function matters much, the model of Godfrey et Isgur should be preferred also for the following reason: we have noted a sensitivity to the short-distance behavior of the potential for $\tau_{1/2}(w = 1)$. With the potential of Veseli and Dunietz,[20] which has a strong Coulomb-like potential, one obtains a notably smaller value. The GI model has a duely softened Coulomb-like potential.

In summary, we claim to use a relativistic approach to the quark model. The meaning is twofold: (1) one aspect is the relativistic boost, which allows to treat large velocities of each hadron as a whole; (2) one other is the choice of the internal kinetic energy. It means that the internal velocities may be large.

---

[e]On the other hand, those implying also the energy levels (like the Voloshin sum rule), are not exactly satisfied.

Still other relativistic aspects, which we do not detail, have been included in the potential V by Godfrey and Isgur.

4. Fundamental Methods for QCD: Lattice QCD

There has not been many lattice calculations of $B \to D^{**}$ decay amplitudes. In the infinite mass limit only the calculation at $w = 1$ seems doable. As we shall see the results are in good agreement with the prediction of the Bakamjian–Thomas model.

Only recently the study at finite b and c mass has been undertaken. They have to overcome very noisy signals and only preliminary results can now be presented.

4.1. *Lattice calculation in the infinite mass limit*

A preliminary calculation[21] was performed to test a new method. The infinitely massive quarks are described on the lattice by Wilson lines in the time direction. The heavy quark is at rest. Thus one can only consider the $w = 1$ case in which the initial and final heavy quarks are both at rest. However, the weak interaction matrix element vanishes at zero recoil. We overcome this problem using formulae from Leibovitch *et al.*[13]

$$\langle H_0^*(v)|A_i D_j|H(v)\rangle = ig_{ij}(M_{H_0^*} - M_H)\tau_{\frac{1}{2}}(1) \tag{29}$$

and

$$\langle H_2^*(v)|A_i D_j|H(v)\rangle = -i\sqrt{3}(M_{H_2^*} - M_H)\tau_{\frac{3}{2}}(1)\epsilon_{ij}^* . \tag{30}$$

where the H, H_0^*, H_2^* are the pseudoscalar, scalar and tensor mesons in the infinite mass limit (the mass differences are finite) and ϵ_{ij}^* is the polarisation tensor of $H_2^*(v)$ and v is the four-velocity. A_i is the axial vector and D_j is the covariant derivative which we know how to discretize on the lattice. Let us skip the details of the lattice calculation. The result of this "quenched" lattice calculation[21] was

$$\tau_{1/2}(1) = 0.38(4)(?) \quad \text{and} \quad \tau_{3/2}(1) = 0.53(8)(?), \tag{31}$$

where the question mark represent unknown systematic errors. A similar calculation was performed later[22,23] using gauge configurations of the European twisted mass collaboration with $N_f = 2$ (two light quarks in the sea). The calculation has used several values of the light quark masses and one lattice spacing. The result is

$$\tau_{1/2}(1) = 0.296(26), \qquad\qquad \tau_{3/2}(1) = 0.526(23), \tag{32}$$

$$\frac{\tau_{3/2}(1)}{\tau_{1/2}(1)} = 1.6\cdots1.8, \quad |\tau_{3/2}(1)|^2 - |\tau_{1/2}(1)|^2 \approx 0.17\cdots0.21 . \tag{33}$$

As already mentioned it is really close to the quark model result, Eq. (8). It is also noticeable that the ground state saturates about 80% of Uraltsev's sum rule, Eq. (3).

4.2. *Lattice calculation with finite b and c masses*

The lattice calculation with finite masses is reported in Refs. 24, 25. The calculation is performed using also the gauge configurations of the European twisted mass collaboration with $N_f = 2$. The calculation is performed with two lattice spacings, one value for the charmed meson mass and three values for the "B" meson: 2.5 GeV, 3. GeV and 3.7 GeV.

4.2.1. *The mass spectrum of D** states*

The first issue is to check the mass spectrum predicted by lattice against experiment and to try to relate them with the $j = 1/2$ and $j = 3/2$ states of the infinite mass limit. It is obvious that the D_0^* (D_2^*) corresponds the $j = 1/2$ ($j = 3/2$) in the infinite mass limit. It is not so clear for the $J = 1$ states which mix $j = 1/2$ and $j = 3/2$. This study has been performed carefully by Kalinowski and Wagner.[26] They find the first (second) excitation to be dominated by the $j = 1/2$ ($j = 3/2$) contribution and they are identified to the broad $D_1(2430)$ (narrow $D_1(2420)$). In Ref. 24 an attempt to check the continuum limit for the D_0^* and D_2^* masses is not successful.

4.2.2. *The $B \to D_0^*$ at zero recoil*

Using twisted quarks the D_0^* can mix with the pseudoscalar D. We need to isolate both states using the generalised eigenvalue problem (GEVP). In Subsec. 2.3 it is mentioned that the $B \to D_0^*$ at zero recoil, which vanishes in the infinite mass limit, might give non-vanishing results with finite mass.[13] As discussed in Subsec. 2.3 the size of this zero recoil contribution is obviously an important issue as it could enhance significantly the $B \to D_0^* l\nu$ branching ratio, improving the agreement with experiment for the semileptonic decay but maybe spoiling the non-leptonic one,[f] see Subsec. 2.4. The preliminary study in Refs. 24, 25 has thus concentrated on the zero recoil $B \to D_0^* l\nu$ decay, leaving the full study for later.

The results are shown in Table 1. From these number we get several conclusions

- For the "B meson" masses used and finite lattice spacing the ratios are significant.
- But they decrease with increasing B mass leading to a small signal/noise ratio at the physical B mass
- The extrapolation to the continuum increases the noise leading to more than 100 % uncertainty.

We can thus conclude that there is really a significant zero recoil $B \to D_0^*$ over $B \to D$ amplitude ratio which however, when extrapolated to the physical situation

---

[f]Notice that the non-leptonic decay ($\bar{B}^0 \to D_0^{1/2+}\pi^-$) happens far from zero recoil, indeed at maximum recoil, but it is difficult to believe that a strong zero recoil amplitude becomes very small at maximum recoil.

Table 1.   Amplitude ratios $B \to D_0^*$ over $B \to D$.

Lattice spacing (fm)	Ratio $m_{b(1)}$	Ratio $m_{b(2)}$	Ratio $m_{b(3)}$	Ratio at physical B
0.085(3)	0.33(5)	0.24(4)	0.16(4)	0.06(4)
0.069(2)	0.40(4)	0.31(4)	0.20(4)	0.09(4)
continuum	0.55(16)	0.45(17)	0.29(18)	0.15(20)

We give the amplitude ratios $B \to D_0^*$ over $B \to D$ at zero recoil, averaged over what seems a good plateau: 6–9 (5–11) for lattice spacing $a = 0.085$ fm ($a = 0.069$ fm). The b quarks range from the lightest to heaviest from left to right: $m_B \simeq 2.5, 3, 3.7$ GeV. The right column corresponds to the extrapolation at the physical B mass, 5.2 GeV. The last line corresponds to the extrapolation to the continuum.

is too uncertain. The situation should improve using a third smaller lattice spacing and larger statistics.

4.2.3. *Possible interpretation of D_0^* as a scattering state*

Mohler *et al.*[27] consider the possible interpretation of the D_0^* as a $D\pi$ scattering state. Indeed, what is identified as D_0^* is a broad structure which is certainly strongly coupled to the $D\pi$ channel. Under this hypothesis, the study of the $B \to D_0^* l\nu$ should be totally reconsidered.

4.2.4. *The $B \to D_2^*$ amplitude compared to the infinite mass limit*

The $B \to D_2^*$ decay amplitude vanishes at zero recoil. This is obvious in the continuum since a spin 0 particle cannot decay into a spin 2 one via Axial matrix elements which carry spin 0 or 1. However, when using twisted mass quarks, the parity symmetry is violated. The twisted quarks are gathered into isospin doublets and even the heavy quarks are represented in fake "isospins." There exists an exact symmetry valid at finite lattice spacing consisting in a parity symmetry combined with the flip of the isospin. Thus combining both isospins one gets decay amplitudes which vanish at zero recoil. Of course statistical fluctuations make them not zero. However these fluctuations at zero recoil and non-zero recoil are correlated. Thus we can reduce the fluctuations by subtracting the zero recoil result from the non-zero recoil result.

As a benchmark we use the prediction of the infinite mass limit formula. We show in Fig. 1 one example of such a ratio. A spin 2 meson has 5 possible polarisation. On a lattice these are decomposed into two different discrete symmetry groups: E^+ and T_2^+. E^+ corresponds to appropriate combinations of terms of the type $D_i V_i$ where D_i is the discretised covariant derivative and V_i the vector current, and T_2^+ corresponds to an appropriate combination of terms of the type $D_i V_j$; $j \neq i$, see Ref. 24 for more details.

Figure 1 shows that there is definitely a signal and the agreement between E^+ and T_2^+ is rewarding. However the ratio is very large. Since the infinite mass formula agrees rather well with experiment we would expect a ratio closer to 1. This is not

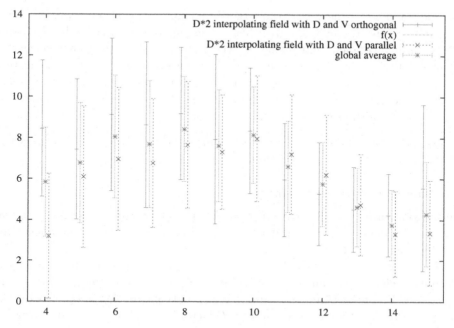

Fig. 1. One example of ratio of the $B \to D_2^*$ amplitude over the value derived from the infinite mass limit, once the zero recoil has been subtracted. The "B meson mass" is 2.5 GeV and the lattice spacing 0.069 fm. The momentum of the B meson corresponds to $w = 1.3$. The blue cross corresponds to the discrete representation E^+ for the D_2^* interpolating field, $D_i V_i$, while the red bar corresponds to the discrete representation T_2^+ i.e. $D_i V_j$, $j \neq i$. The global average is the purple star.

the case, the ratio is one order of magnitude larger than 1 and this does not seem to improve when going towards the physical B mass neither towards continuum. The reason for this problem is not yet understood.

5. Experiment

5.1. *Non-leptonic versus semileptonic*

These respective processes present advantages and drawbacks with respect to each other, so that, on the whole, they should be considered complementary. Indeed:

(α) On the one hand, semileptonic data have have a series of theoretical advantages. They are directly connected with the matrix elements of the currents, which are the true theoretical objects under study. Non-leptonic processes can be related to them only trough the additional assumption of factorisation, which has not a very strong theoretical foundation, although it seems to work well in similar decays like $B \to D^{(*)}\pi$. Moreover, in Class III they are complicated by the additional diagram with D^{**} emission. And in Class I, $B \to D\pi\pi$ contains a non-exotic, strongly resonating $\pi^+\pi^-$ channel which must be extracted off.

(β) One could also believe that the larger branching ratios of semileptonic decays
(10^{-3} versus 10^{-4}) should give them a big superiority in statistics. However,
this is overcompensated by far by the fact that the detection efficiency of
semileptonic decay is much worse. One has four bodies in the final state instead
of three, and a neutrino among them. Finally, the non-leptonic data are much
more accurate. It is only from them that one has been able to determine the
mass and width of the broad states.

Note that one cannot consider the two processes as truly independent, because
of their relation through factorisation.

5.2. *Narrow versus broad states*

The narrow states $j = 3/2$ have always shown quite a good consistency among
experimental measurements. And as already said, the theoretical numbers have
complied with experiment.

On the other hand, for the broad $j = 1/2$, the experimental situation, instead
of becoming clearer, has fallen into a real state of confusion. Indeed, aside from the
attempts by Delphi at LEP, who found a very large contribution of the broad 1^+ in
semileptonic decay, the higher luminosity B factories could be expected to clear up
the question; in fact, contrasting results have been obtained by Belle and Babar,
with different situations for 0^+ versus 1^+ and for non-leptonic versus semileptonic
processes.

The fact that narrow states results are quite consistent, while those for broad
states are not, suggests that it is the broadness which could be responsible for
the difficulties encountered for these states. Indeed, they are very broad: around
300 MeV. It is also why we shall present separately the experimental result for the
two type of states.

5.3. *Results*

We begin, in an unusual way, with the non-leptonic decay because the situation is
clearer, and also to insist on the often underestimated necessity to take them into
account for the discussion of the $j = 1/2$ case. Since there is compatibility between
Belle and Babar where both have measured, we quote only in Table 2 their averaged
results.

One notes the remarkable fact that in Class I, $j = 1/2$ are one order of magnitude
smaller than $j = 3/2$, while all have the same order of magnitude 10^{-3} in Class III.
This is explained clearly by theory, see above, Subsec. 2.4.

For the Class I, the following warnings must be made:

- The large uncertainty (50%) in the 0^+ case, is due to the fact that Babar has
  an appreciably larger value, but with a very large uncertainty. This very large
  uncertainty is itself due to the uncertainty in the extraction of a non-resonant
  contribution.

Table 2. Non-leptonic data $j = 3/2,\ 1/2$.

Decay channel	B_d^0, Class I	B^+, Class III
$\bar{D}_2^* \pi^+$	$(0.49 \pm 0.07) \times 10^{-3}$	$(0.82 \pm 0.11) \times 10^{-3}$
$\bar{D}_1^{3/2} \pi^+$	$(0.82^{+0.25}_{-0.17}) \times 10^{-3}$	$(1.51 \pm 0.34) \times 10^{-3}$
$\bar{D}_1^{1/2} \pi^+$	$< 1 \times 10^{-4}$	$(0.75 \pm 0.17) \times 10^{-3}$
$\bar{D}_0^* \pi^+$	$(1.0 \pm 0.5) \times 10^{-4}$	$(0.96 \pm 0.27) \times 10^{-3}$

Non-leptonic branching ratios $B \to D^{**}\pi$ averaged between Belle and Babar.[28,30,29,31,32] The final strong decay branching fractions have been taken as detailed footnote 6 of Ref. 3.

Table 3. Semileptonic data $j = 3/2$.

Decay channel	Belle	Babar
$\bar{D}_2^*\ l\nu$	$(0.54 \pm 0.12) \times 10^{-2}$	$(0.39 \pm 0.11) \times 10^{-2}$
$\bar{D}_1^{3/2} l\nu$	$(0.93 \pm 0.22) \times 10^{-2}$	$(0.64 \pm 0.15) \times 10^{-2}$

Semileptonic branching ratios for B semileptonic decay into narrow states, from Belle and Babar. See HFAG[33] and references therein. The final strong decay branching fractions have been taken as detailed footnote 6 of Ref. 3.

Table 4. Semileptonic data $j = 1/2$.

Decay channel	Belle	Babar
$\bar{D}_1^{1/2} l\nu$	$< 1.05 \times 10^{-3}$	$(0.40 \pm 0.05) \times 10^{-2}$
$\bar{D}_0^* l\nu$	$(0.36 \pm 0.09) \times 10^{-2}$	$(0.42 \pm 0.09) \times 10^{-2}$

Semileptonic branching ratios for B semileptonic decay into broad states, from Belle and Babar. See HFAG[33] and references therein. The final strong decay branching fractions have been taken as detailed footnote 6 of Ref. 3.

- The 1^+, $j = 1/2,\ 3/2$ ($D^*\pi$ channel) have been measured only by Belle.

For semileptonic decays we quote first the results of $j = 3/2$ (narrow states) in Table 3: Obviously, the two experiments are here quite compatible.

The semileptonic decay into $j = 1/2$ from Babar and Belle is reported in Table 4: There is a strong disagreement for the $1^+(1/2)$ between the two experiments while the results for 0^+ are quite compatible

About 0^+: one could say that in view of the agreement between the two experiments in the semileptonic case, we might trust this large result. But given the rather direct relation between semileptonic and Class I non-leptonic decays, and the strong suppression of 0^+ in non-leptonic (see the above Table 2), this conclusion would be very doubtful.

The general conclusion could be that, not surprisingly, the very broad states raise more difficulties precisely because of their broadness.

6. Conclusion: Discussion on Theory and Experiment, Prospects

6.1. *Discussion on theory and experiment*

One must say that the non-leptonic data, Table 2, seem to support strongly theoretical expectations coming from the $m_Q \to \infty$ approximation:

(1) Suppression of $j = 1/2$ with respect to $j = 3/2$ in Class I, by one order of magnitude. The agreement with Eq. (21) is semiquantitative for both j, and good if summing over the members of the j multiplets.
(2) This suppression is no more present in Class III, due to the diagram with D^{**} which is large in the $j = 1/2$ case, but not for $j = 3/2$, also from a statement of $m_Q \to \infty$ limit. Indeed[16] one observes that the $j = 1/2$ are strongly enhanced in Class III, by one order of magnitude, and the $j = 3/2$ much more slightly (the two diagrams can be shown to add constructively).

The $j = 3/2$ semileptonic data are much larger, as expected from the $m_Q \to \infty$ approximation, and in quantitative agreement when averaging over the members of the multiplet.

The only definite discrepancy with theory in the $m_Q \to \infty$ limit is in the semileptonic case, for the $j = 1/2, 0^+$ state, where the two experiments agree. It could be understood as coming from the broadness of the states and the lower detection efficiency of semileptonic measurements, rendering the identification of the resonance still more difficult (much less observed events).

Of course, part of the discrepancy could be due to the $1/m_Q$ corrections being large. This can be suggested by the finding of Leibovich *et al.*, that precisely this transition can suffer more $1/m_Q$ enhancement. This effect could combined with the fact that lattice QCD finds a somewhat larger $\tau_{1/2}(1)$ than our quark model (see Subsec. 2.1). But then, it would remain to explain the non-leptonic data: why then 0^+ seems so small in Class I. Therefore, still something would remain problematic on one or another side of experiment.

Another problem is the large Babar result for the broad 1^+, $1/2$ (contrasting with the small one from Belle). It would be a real worry if confirmed, because in this case one does not expect any serious enhancement according to the calculations of Leibovich *et al.* (the conclusion is the same in the quark model[14]).

6.2. *Prospects*

• The most urgent problem seems the experimental one:

(1) To solve the discrepancy between the two B factory, see Table 4, concerning the semileptonic decay to $1^+(j = 1/2)$,

(2) To clarify the discrepancy which seems to exist between the semileptonic and the Class I non-leptonic data for the 0^+ ($j = 1/2$) (Tables 2 and 4), if we believe factorisation. There is a strong suppression in the latter with respect to the $j = 3/2$, while 0^+ is of the same order as $j = 3/2$ in semileptonic. It would be already very useful to reduce the uncertainty of Babar measurement of the Class I non-leptonic decay for the 0^+, and to have the Babar measurement of the $1^+(j = 1/2)$ partner.

It would also be very useful to complement the study of the broad states by the one of their strange counterparts, which are narrow, and therefore much easier to identify. This is proposed in the article.[3]

On the other hand, on the theoretical side:

- Since one can "fear" unexpectedly large $1/m_Q$ corrections in the 0^+, lattice QCD at finite mass should be developed as much as possible. For the moment lattice calculations at finite mass make it very likely that a significant zero recoil contribution to $B \to D_0^*$ is there. However the extrapolation to the physical situation leads to more than 100% uncertainty, Table 1. This of course is not the case for $B \to D_2^*$ but there the ratio of the finite mass signal over the infinite mass estimate is larger than expected and than experiment for an unknown reason. The progress will come from using additionally a smaller lattice spacing to make safer the continuum limit, to look in detail into the momentum dependence of all these decays and to chase possible remaining artefacts.

- Since one needs anyway other approaches, mainly the quark model, to understand things at large w (for example $q^2 = 0$ for the pionic weak transition), one should try to improve estimates of $1/m_Q$ corrections.

- Finally, to justify the longstanding efforts dispensed on the problem, both from the part of theorists and of experimentalists, it must be said why it seems so important to clarify the situation for $L = 1$, $j = 1/2$. It is because, on the whole, a serious issue is the **phenomenological relevance of the $m_Q = \infty$ approach in $b \to c$ decay**. This limit has been considered very attractive because allowing several important new theoretical statements, in particular the sum rules, which deeply involve theses states; and at the same time, it has met quite encouraging phenomenological successes in $B \to D^{(*)}$ and $B \to D^{**}$, $L = 1$, $j = 3/2$ transitions, as well as in the strong decays $D^{**} \to D^{(*)}\pi$ with both $j = 1/2$ and $j = 3/2$.

Of course it is expected that the results of $m_Q = \infty$ are a better approximation for the $j = 3/2$, 2+ state than for the other cases. Indeed, in the latter cases, the no recoil amplitude at $m_Q = \infty$ vanishes contrary to general expectation, and a non-zero value should be present, resulting from $1/m_Q$ effects. Indeed, theoretical expectations from certain identities lead to possible large effects at order $\mathcal{O}(1/m_Q)$ in the 1+, 3/2 and 0^+ cases. Unluckily, lattice are not yet able to answer clearly to their size at physical masses, and whether they could fill the large gap with presently observed value for the 0^+ in the semileptonic decay.

Moreover, such a large non-zero value at no recoil which would mean an S-wave between the final D^{**} and the $l\nu$ system, is to be confronted to the success of the $m_Q = \infty$ approximation for the Class I non-leptonic decays. Can this S-wave effect be damped in the maximal recoil kinematics which is that of the $D^{**}\pi$? Models rather point to a soft variation of the corresponding effect.

- This longstanding efforts are also finally justified by the important side effect that the decays considered here have on the estimate of V_{cb}.

References

1. I. I. Bigi, B. Blossier, A. Le Yaouanc, L. Oliver, O. Pene, J.-C. Raynal, A. Oyanguren and P. Roudeau, Memorino on the '1/2 versus 3/2 puzzle' in $\bar{B} \to l\bar{\nu}X(c)$, arXiv:hep-ph/0512270.
2. I. I. Bigi, B. Blossier, A. Le Yaouanc, L. Oliver, O. Pene, J.-C. Raynal, A. Oyanguren and P. Roudeau, Memorino on the '1/2 versus 3/2 puzzle' in $\bar{B} \to l\bar{\nu}X(c)$: A year later and a bit wiser, *Eur. Phys. J. C* **52**, 975–985 (2007), arXiv:0708.1621 [hep-ph].
3. D. Becirevic, A. Le Yaouanc, L. Oliver, J.-C. Raynal, P. Roudeau and J. Serrano, Proposal to study $Bs \to D_{sJ}$ transitions, *Phys. Rev. D* **87**(5), 054007 (2013), arXiv:1206.5869 [hep-ph].
4. N. Isgur and M. B. Wise, *Phys. Rev. D* **43**, 819 (1991).
5. A. Le Yaouanc, L. Oliver, O. Pene and J. C. Raynal, Exact duality and Bjorken sum rule in heavy quark models a la Bakamjian–Thomas, *Phys. Lett. B* **386**, 304–314 (1996), arXiv:hep-ph/9603287.
6. N. Uraltsev, *Phys. Lett. B* **501**, 86 (2001), arXiv:hep-ph/0011124.
7. A. Le Yaouanc, L. Oliver, O. Pene, J. C. Raynal and V. Morenas, *PoS* **HEP2001**, 082 (2001), arXiv:hep-ph/0110372.
8. A. Le Yaouanc, L. Oliver, O. Pene, J. C. Raynal and V. Morenas, Uraltsev sum rule in Bakamjian-Thomas quark models, *Phys. Lett. B* **520**, 25–32 (2001), arXiv:hep-ph/0105247.
9. V. Morenas, A. Le Yaouanc, L. Oliver, O. Pene and J. C. Raynal, *Phys. Rev. D* **56**, 5668 (1997), arXiv:hep-ph/9706265.
10. S. Godfrey and N. Isgur, *Phys. Rev. D* **32**, 189 (1985).
11. Particle Data Group (K. Nakamura *et al.*), *J. Phys. G* **37**, 075021 (2010), and 2011 partial update for the 2012 edition.
12. DELPHI Collab. (J. Abdallah *et al.*), *Eur. Phys. J. C* **45**, 35 (2006), arXiv:hep-ex/0510024.
13. A. K. Leibovich, Z. Ligeti, I. W. Stewart and M. B. Wise, *Phys. Rev. D* **57**, 308 (1998), arXiv:hep-ph/9705467.
14. A. Le Yaouanc, L. Oliver and J. C. Raynal, Finite mass corrections to $B \to D^{(*)}$, D^{**} transitions in the Bakamjian–Thomas approach, to be published.
15. M. Neubert, Theoretical analysis of $\bar{B} \to D^{**}\pi$ decays, *Phys. Lett. B* **418**, 173–180 (1998), arXiv:hep-ph/9709327.
16. F. Jugeau, A. Le Yaouanc, L. Oliver and J.-C. Raynal, The Decays $\bar{B} \to D^{**}\pi$ and the Isgur–Wise functions $\tau(1/2)(w)$, $\tau(3/2)(w)$, *Phys. Rev. D* **72**, 094010 (2005), arXiv:hep-ph/0504206.
17. B. Keister and W. Polyzou, *Adv. Nucl. Phys.* **20**, 225 (1991).
18. M. Terent'ev, *Sov. J. Nucl. Phys.* **24**, 106 (1971).
19. A. Le Yaouanc, L. Oliver, O. Pene and J. C. Raynal, Covariant quark model of form-factors in the heavy mass limit, *Phys. Lett. B* **365**, 319–326 (1996).

20. S. Veseli and I. Dunietz, *Phys. Rev. D* **54**, 6803 (1996).

21. D. Becirevic, B. Blossier, P. Boucaud, G. Herdoiza, J. P. Leroy, A. Le Yaouanc, V. Morenas and O. Pene, *Phys. Lett. B* **609**, 298 (2005), arXiv:hep-lat/0406031.

22. European Twisted Mass Collab. (B. Blossier *et al.*), *JHEP* **0906**, 022 (2009), arXiv:0903.2298 [hep-lat].

23. ETM Collab. (B. Blossier *et al.*), *PoS* **LAT2009**, 253 (2009), arXiv:0909.0858 [hep-lat].

24. M. Atoui, B. Blossier, V. Morenas, O. Pene and K. Petrov, arXiv:1312.2914 [hep-lat].

25. M. Atoui, arXiv:1305.0462 [hep-lat].

26. M. Wagner and M. Kalinowski, arXiv:1310.5513 [hep-lat].

27. D. Mohler, S. Prelovsek and R. M. Woloshyn, *Phys. Rev. D* **87**(3), 034501 (2013), arXiv:1208.4059 [hep-lat].

28. BaBar Collab. (B. Aubert *et al.*), *Phys. Rev. D* **79**, 112004 (2009), arXiv:0901.1291 [hep-ex].

29. BaBar Collab. (P. del Amo Sanchez *et al.*), *PoS* **ICHEP2010**, 250 (2010), arXiv:1007.4464 [hep-ex].

30. Belle Collab. (K. Abe *et al.*), *Phys. Rev. Lett.* **94**, 221805 (2005); Belle Collab. (K. Abe *et al.*), arXiv:hep-ex/0410091.

31. Belle Collab. (K. Abe *et al.*), *Phys. Rev. D* **69**, 112002 (2004), arXiv:hep-ex/0307021.

32. Belle Collab. (A. Kuzmin *et al.*), *Phys. Rev. D* **76**, 012006 (2007), arXiv:hep-ex/0611054.

33. Heavy Flavor Averaging Group Collab. (D. Asner *et al.*), arXiv:1010.1589 [hep-ex].

Reprints of Selected Work

- Nikolai Uraltsev
  New Exact Heavy Quark Sum Rules
  Phys. Lett. B **501** (2001) 86–91.

- I. I. Bigi, M. Shifman N. G. Uraltsev, and A. Vainshtein
  QCD Predictions for Lepton Spectra in Inclusive Heavy Flavor Decays
  Phys. Rev. Lett. **71** (1993) 496–499.

- I. Bigi, M. Shifman N. Uraltsev, and A. Vainshtein
  High Power n of m_b in b-Flavored Widths and $n = 5 \to \infty$ Limit
  Phys. Rev. D **56** (1997) 4017–4030.

ELSEVIER

Physics Letters B 501 (2001) 86–91

PHYSICS LETTERS B

www.elsevier.nl/locate/npe

New exact heavy quark sum rules

Nikolai Uraltsev [a,b,1]

[a] *Department of Physics, University of Notre Dame du Lac, Notre Dame, IN 46556, USA*
[b] *INFN, Sezione di Milano, Milan, Italy*

Received 12 November 2000; received in revised form 8 December 2000; accepted 8 December 2000

Editor: H. Georgi

Abstract

Considering nonforward scattering amplitude off the heavy quark in the Small Velocity limit two exact superconvergent sum rules are derived. The first is the sum rule for spin of the light cloud and leads to the lower bound $\varrho^2 > 3/4$ for the slope of the Isgur–Wise function. It also provides the rationale for the fact that the vector heavy flavor mesons are heavier than the pseudoscalar ones. A spin-nonsinglet analogue $\overline{\Sigma}$ of $\bar{\Lambda} = M_B - m_b$ is introduced. © 2001 Published by Elsevier Science B.V.

Heavy quark symmetry and the heavy quark expansion have played an important role in understanding weak decays of heavy flavors. Recent years witnessed significant success in quantifying strong nonperturbative dynamics in a number of practically important problems via application of Wilson Operator Product Expansion (OPE). It is fair to note, however, that with the analytic solution of QCD still missing, the effect of nonperturbative low-scale domain is more parametrized than computed from the first principles. Therefore, any model-independent constraint on the nonperturbative parameters playing the role in heavy quark decays, is an asset. A number of such relations come from the heavy quark sum rules, in particular for transitions between two sufficiently heavy quarks with no change of velocity (zero-recoil sum rules), or where velocity changes by a small amount (small velocity, or SV sum rules). The unified derivation of the sum rules in the field-theoretic OPE is described in detail in the

dedicated papers [1], with their quantum mechanical interpretation elucidated. A more pedagogical derivation can be found in recent reviews [2,3]. [2]

In this Letter the standard OPE approach is extended to the nonforward SV scattering amplitude of weak currents off the heavy quark. This leads to two new exact heavy quark sum rules related to the spin of the light constituents of the heavy flavor hadron.

We consider the SV scattering amplitude

$$
T(q_0; \vec{v}, \vec{u}) = \frac{1}{2M_{H_Q}} \langle H_Q(\vec{u}) |
$$

$$
\times \int d^3x \, dx_0 \, e^{i\vec{q}\vec{x} - iq_0 x_0}
$$

$$
\times \, iT\{J_0^\dagger(0), \, J_0(x)\} | H_Q(0) \rangle, \quad (1)
$$

where $J_\mu = \overline{Q}\gamma_\mu Q$ and $\vec{q} = m_Q\vec{v}$; likewise $m_Q\vec{u}$ is the momentum of the final heavy flavor hadron. We will also denote $\vec{v}' = \vec{v} - \vec{u}$. Note that J_μ is the nonrelativistic current of heavy quarks, and the

---

E-mail address: uraltsev@undhep.hep.nd.edu (N. Uraltsev).

[1] On leave of absence from St. Petersburg Nuclear Physics Institute, Gatchina, St. Petersburg 188300, Russia.

[2] An interesting introduction to heavy quarks in QCD can be found in lectures [4], with the references therein to the more conventional reviews.

N. Uraltsev / Physics Letters B 501 (2001) 86–91   87

field $Q(x)$ entering it includes only the operator of annihilation of the heavy quark contained in H_Q, but not the creation of $\widetilde{Q}$ which is present in $Q(x)$ from $J_0^\dagger$. Therefore, $J_\mu^\dagger$ is different from J_μ and only the product in the order $J_0^\dagger(0)J_0(x)$ contributes (we adopt the convention where H_Q contains heavy quark and not the antiquark). This can be visualized considering the nondiagonal $b \to c$ transitions with $J_\mu = \bar{c}\gamma_\mu b$, $J_\mu^\dagger = \bar{b}\gamma_\mu c$, however we assume that both m_b, $m_c \to \infty$ and, for simplicity, put $m_b = m_c$.

In the SV limit we retain only terms through second order in $\vec{v}$ and $\vec{u}$ and take the energy variable ϵ according to

$$\epsilon = q_0 - \left(\sqrt{\vec{q}^2 + m_Q^2} - m_Q\right) \simeq q_0 - \frac{m_Q \vec{v}^2}{2}, \qquad (2)$$

the elastic transitions for a free quark would then correspond to $\epsilon = 0$. The amplitude $T(\epsilon; \vec{v}, \vec{v} - \vec{v}')$ can be decomposed into symmetric and antisymmetric in $\vec{v}$, $\vec{v}'$ parts $h_+(\epsilon)$ and $h_-(\epsilon)$; the latter is present if H_Q has nonzero spin of light degrees of freedom j correlated with its total spin, viz. $h_- \sim i\epsilon_{kln}v_k v_l'\langle H_Q|j_n|H_Q\rangle$.

At large (complex) $\epsilon \gg \Lambda_{\text{QCD}}$ the amplitude (1) can be expanded in inverse powers of ϵ. Simultaneously we assume that $\epsilon \ll m_Q$ and can discard all terms suppressed by powers of m_Q. In this limit the OPE simplifies and takes the form

$$-T(\epsilon; \vec{v}, \vec{u})$$
$$= \frac{1}{\epsilon} \frac{1}{2M_{H_Q}} \langle H_Q(\vec{u})|\overline{Q}\left(1 - \frac{\vec{v}^2}{4} + \frac{\vec{v}\vec{u}}{2}\right)Q(0)|H_Q(0)\rangle$$
$$+ \frac{1}{\epsilon^2} \frac{1}{2M_{H_Q}} \langle H_Q(\vec{u})|\overline{Q}(i\vec{D}\vec{v})Q(0)|H_Q(0)\rangle$$
$$+ \sum_{k=0}^{\infty} \frac{(-1)^k}{\epsilon^{k+3}} \frac{1}{2M_{H_Q}}$$
$$\times \langle H_Q(\vec{u})|\overline{Q}\left[(i\vec{D}\vec{v})\pi_0^k(i\vec{D}\vec{v}) - \pi_0^{k+1}(i\vec{D}\vec{v})\right]$$
$$\times Q(0)|H_Q(0)\rangle. \qquad (3)$$

Here π_0 is nonrelativistic energy, $\pi_0 = iD_0 - m_Q$.

On the other hand, the dispersion relation equates the coefficients in the $1/\epsilon$ series to the moments of the absorptive part of $T(\epsilon; \vec{v}, \vec{u})$ (nonforward structure

functions of the hadron H_Q):

$$-T(\epsilon; \vec{v}, \vec{u}) = \sum_{k=0}^{\infty} \frac{1}{\epsilon^{k+1}} \frac{1}{2\pi} \int d\omega\, \omega^k \operatorname{Im} T(\omega; \vec{v}, \vec{u}), \qquad (4)$$

the latter is given by the product of the SV transition amplitudes into the ground and excited stated. T is analytic at negative ϵ and has a cut at positive ϵ. The elastic transitions lead to the pole located at $\epsilon \simeq -\bar{\Lambda}\vec{v}^2/2$. Equating $1/\epsilon^k$ terms in Eqs. (3) and (4) we get the sum rules. We illustrate them on the example of B mesons ($j = 1/2$) taking B^* as H_Q. With $j = 1/2$ the only nontrivial symmetric structure in h_+ is proportional to $(\vec{v}\vec{v}') \cdot \langle H_Q'|H_Q\rangle$. Therefore, it is convenient to introduce

$$-T(\epsilon; \vec{v}, \vec{v} - \vec{v}')$$
$$= \left[\frac{1 - a(\vec{v}^2 + \vec{v}'^2) - b(\vec{v} - \vec{v}')^2}{\epsilon}\right.$$
$$\left. - \frac{c\,\vec{v}^2}{\epsilon^2} + (\vec{v}\,\vec{v}')\,h_+(\epsilon)\right]\frac{\langle H_Q'|H_Q\rangle}{2M_{H_Q}}$$
$$- i\epsilon_{jkl}\, v_j\, v_k'\, h_-(\epsilon)\frac{\langle H_Q'|J_l|H_Q\rangle}{2M_{H_Q}} + \mathcal{O}(\vec{v}^3), \qquad (5)$$

and

$$W_+(\epsilon) = \frac{1}{2\pi} \operatorname{Im} h_+(\epsilon),$$
$$W_-(\epsilon) = \frac{1}{2\pi} \operatorname{Im} h_-(\epsilon), \qquad (6)$$

with $\vec{J}$ denoting the angular momentum of H_Q; constants a, b and c are associated with the elastic transition.

The SV transitions can proceed to $j = 1/2$ (scalar S and axial A) and to $j = 3/2$ (axial D_1 and tensor D_2) "P-wave" states. Following the notations of Ref. [5], the corresponding nonrelativistic amplitudes are given by

$$\langle S(\vec{v}_2)|J_0|B^*(\varepsilon, \vec{v}_1)\rangle = \tau_{1/2}\big(\vec{\varepsilon}(\vec{v}_2 - \vec{v}_1)\big),$$
$$\langle A(e, \vec{v}_2)|J_0|B^*(\varepsilon, \vec{v}_1)\rangle$$
$$= -\tau_{1/2}\, i\epsilon_{jkl}\, e_j^*\, \varepsilon_k\, (v_2 - v_1)_l, \qquad (7)$$

and

$$\langle D_1(e, \vec{v}_2)|J_0|B^*(\varepsilon, \vec{v}_1)\rangle$$
$$= -\frac{1}{\sqrt{2}}\, \tau_{3/2}\, i\epsilon_{jkl}\, e_j^*\, \varepsilon_k\, (v_2 - v_1)_l,$$

$$\langle D_2(e, \vec{v}_2) | J_0 | B^*(\varepsilon, \vec{v}_1) \rangle$$

$$= \sqrt{3} \tau_{3/2} \, e^*_{jk} \, \varepsilon_j (v_2 - v_1)_k. \tag{8}$$

Here ε, e are nonrelativistic 3-vectors of polarization. In the case of D_2, however e is the symmetric rank-2 tensor, with the sum over polarizations

$$\sum_\lambda e^{(\lambda)}_{ij} e^{(\lambda)}_{kl} = -\frac{1}{3} \delta_{ij} \delta_{kl} + \frac{1}{2} (\delta_{ik} \delta_{jl} + \delta_{il} \delta_{jk}). \tag{9}$$

The elastic transition to this order can proceed only to B^*, and the amplitude is given by

$$\langle B^*(\vec{\varepsilon}_2, \vec{v}_2) | J_0 | B^*(\vec{\varepsilon}_1, \vec{v}_1) \rangle$$

$$= 2 M_{B^*} \cdot \xi \big((\vec{v}_2 - \vec{v}_1)^2 \big) \left(1 + \frac{\vec{v}_1^2 + \vec{v}_2^2}{4} \right) (\vec{\varepsilon}_2^* \vec{\varepsilon}_1)$$

$$+ \mathcal{O}(\vec{v}^3), \tag{10}$$

where ξ is the Isgur–Wise function (its slope $\varrho^2 = -2\xi'(0)$ will be used later) and $\vec{\varepsilon}$ are the rest-frame polarizations. The latter are related to the polarization 4-vectors ϵ via $\epsilon_0 = (\vec{v}\vec{\varepsilon})$, $\vec{\epsilon} = \vec{\varepsilon} + \frac{1}{2} \vec{v}(\vec{v}\vec{\varepsilon})$, up to terms cubic in velocity.

The explicit computation yields the following structures for the contributions of the 1/2 and 3/2 multiplets,

$$|\tau_{1/2}|^2 \left\{ (\vec{\varepsilon}'^* \vec{\varepsilon}) (\vec{v}' \, \vec{v}) \right.$$

$$\left. - \big[(\vec{\varepsilon}'^* \vec{v}) (\vec{\varepsilon} \, \vec{v}') - (\vec{\varepsilon}'^* \vec{v}') (\vec{\varepsilon} \, \vec{v}) \big] \right\},$$

$$|\tau_{3/2}|^2 \left\{ 2 (\vec{\varepsilon}'^* \vec{\varepsilon}) (\vec{v}' \, \vec{v}) \right.$$

$$\left. + \big[(\vec{\varepsilon}'^* \vec{v}) (\vec{\varepsilon} \, \vec{v}') - (\vec{\varepsilon}'^* \vec{v}') (\vec{\varepsilon} \, \vec{v}) \big] \right\}, \tag{11}$$

respectively, where $\vec{\varepsilon}'$ is the polarization vector of $B^*(\vec{u})$. We also need the following matrix elements to evaluate the OPE part:

$$\frac{1}{2 M_{B^*}} \langle H_Q(\varepsilon', \vec{u}) | \overline{Q} Q(0) | H_Q(\varepsilon, 0) \rangle$$

$$= \xi(\vec{u}^2) \left(1 + \frac{\vec{u}^2}{4} \right) (\vec{\varepsilon}'^* \vec{\varepsilon}), \tag{12}$$

$$\frac{1}{2 M_{B^*}} \langle H_Q(\varepsilon', \vec{u}) | \overline{Q} \gamma_i Q(0) | H_Q(\varepsilon, 0) \rangle$$

$$= \frac{1}{2} u_i (\vec{\varepsilon}'^* \vec{\varepsilon}) + \frac{1}{2} \big[(\vec{\varepsilon}'^* \vec{u}) \varepsilon_i - \varepsilon_i'^* (\vec{\varepsilon} \, \vec{u}) \big] \tag{13}$$

and

$$\frac{1}{2 M_{B^*}} \langle H_Q(\varepsilon', \vec{u}) | \overline{Q} i D_j Q(0) | H_Q(\varepsilon, 0) \rangle$$

$$= -\frac{\overline{\Lambda}}{2} u_j (\vec{\varepsilon}'^* \vec{\varepsilon}) + \frac{\overline{\Sigma}}{2} \big\{ (\vec{\varepsilon}'^* \vec{u}) \varepsilon_j - \varepsilon_j'^* (\vec{\varepsilon} \, \vec{u}) \big\}$$

$$+ \mathcal{O}(\vec{u}^2). \tag{14}$$

The higher order in $\vec{u}$, $\vec{v}$ terms have been omitted from all the expressions. The last matrix element introduces the new hadronic parameter $\overline{\Sigma}$ which can be viewed as the spin-nonsinglet analogue of $\overline{\Lambda} = \lim_{m_b \to \infty} M_B - m_b$. The part proportional to u_j in Eq. (14) amounts to $-\overline{\Lambda}/2$; this follows from the nonrelativistic equation of motion $(v_\mu i D_\mu) Q(x) = m_Q Q(x)$ which holds in the static limit $m_Q \to \infty$ for the heavy quark moving with velocity v:

$$M_{H_Q} \big(u - u^{(0)} \big)_\mu u_\mu \langle H_Q(\vec{u}) | \overline{Q} Q(0) | H_Q(0) \rangle$$

$$= -(u_\mu i D_\mu) \langle H_Q(\vec{u}) | \overline{Q} Q(x) | H_Q(0) \rangle \big|_{x=0}$$

$$= \langle H_Q(\vec{u}) | m_Q \overline{Q} Q(0) - \overline{Q}(u_\mu i D_\mu) Q(0) | H_Q(0) \rangle,$$

$(u^{(0)} = (1, \vec{0})$ is the restframe four-velocity) which to order $\vec{u}^2$ leads to

$$(M_{H_Q} - m_Q) \frac{\vec{u}^2}{2} \langle H_Q | \overline{Q} Q(0) | H_Q \rangle$$

$$= -u_k \langle H_Q(\vec{u}) | \overline{Q} i D_k Q(0) | H_Q(0) \rangle. \tag{15}$$

Considering the forward scattering amplitude where $\vec{u} = 0$ and $\vec{v}' = -\vec{v}$ selects the symmetric structure function $W_+(\epsilon)$ and yields the Bjorken [6] and Voloshin [7] sum rules for the zeroth and first moments:

$$\varrho^2 - \frac{1}{4} = 2 \sum_m |\tau_{3/2}^{(m)}|^2 + \sum_n |\tau_{1/2}^{(n)}|^2, \tag{16}$$

$$\frac{\overline{\Lambda}}{2} = 2 \sum_m \epsilon_m |\tau_{3/2}^{(m)}|^2 + \sum_n \epsilon_n |\tau_{1/2}^{(n)}|^2, \tag{17}$$

with $\epsilon_k = M^{(k)} - M_B$ denoting the mass gap between the P-wave and the ground state. Setting $\vec{v} = 0$ or $\vec{v}' = 0$ fixes $a = \varrho^2/2$, $b = -1/4$ and $c = \overline{\Lambda}/2$ in Eq. (5), as expected.

The new sum rules emerge for the zeroth and first moments of the antisymmetric structure function W_-:

$$\frac{1}{4} = \sum_m |\tau_{3/2}^{(m)}|^2 - \sum_n |\tau_{1/2}^{(n)}|^2, \tag{18}$$

N. Uraltsev / Physics Letters B 501 (2001) 86–91   89

$$\frac{\overline{\Sigma}}{2} = \sum_m \epsilon_m |\tau_{3/2}^{(m)}|^2 - \sum_n \epsilon_n |\tau_{1/2}^{(n)}|^2. \qquad (19)$$

The higher moments yield the known sum rules [2] for μ_G^2, ρ_{LS}^3, ..., which can be obtained, for instance, considering the usual zero-recoil structure functions of spacelike components of vector currents appearing at order $1/m_Q^2$ [1].

It is often convenient to work in the static limit $m_Q \to \infty$ assuming that Q are spinless; B and B^* in this case are different components of a single spin-1/2 particle, Ψ_0. The derivation of the sum rules in this case proceeds similarly; the antisymmetric structure would be absent from the OPE, but the elastic transition yield it with the opposite sign for the zeroth moment. The $D = 4$ matrix element in this case takes the form

$$\langle \Psi_0(\vec{u}) | \overline{Q} i D_j Q(0) | \Psi_0(0) \rangle$$

$$= -\frac{\bar{\Lambda}}{2} u_j \Psi_0^\dagger \Psi_0 - i \frac{\overline{\Sigma}}{2} \epsilon_{jkl} u_k \Psi_0^\dagger \sigma_l \Psi_0$$

$$+ \mathcal{O}(\vec{u}^2). \qquad (20)$$

The exact magnitude of the nonperturbative hadronic parameter $\overline{\Sigma}$ is not known at the moment, but can be estimated by means of QCD sum rules; or it can be directly measured on the lattices. Comparing the sum rules (18), (19) with the sum rule for the chromomagnetic expectation value μ_G^2 [2]

$$\frac{\mu_G^2}{6} = \sum_m \epsilon_m^2 |\tau_{3/2}^{(m)}|^2 - \sum_n \epsilon_n^2 |\tau_{1/2}^{(n)}|^2,$$

$$\mu_G^2 = \frac{1}{2M_{H_Q}} \langle B | \bar{b} \frac{i}{2} \sigma_{\mu\nu} G^{\mu\nu} b(0) | B \rangle$$

$$\simeq \frac{3}{4}(M_{B^*}^2 - M_B^2) \approx 0.4 \, \text{GeV}^2, \qquad (21)$$

we expect $\overline{\Sigma}$ to be about 0.25 GeV. In the nonrelativistic system $\overline{\Sigma}$ is given by the product of the light mass and the orbital momentum, and would vanish for the ground-state mesons.

The first sum rule (18) which is independent of the strong dynamics at first may look surprising. In the quark models the $1/2$ and $3/2$ states are differentiated only by spin-orbital interaction. The latter naively can be taken arbitrarily small if the light quark in the meson is nonrelativistic. To resolve this apparent paradox we note that in the nonrelativistic

case τ^2 are large scaling like inverse square of the typical velocity of the light quark, $\tau^2 \sim 1/\bar{v}_{\text{sp}}^2$. The relativistic spin-orbital effects must appear at the relative level $\sim \bar{v}_{\text{sp}}^2$ because spin ceases to commute with momentum to this accuracy due to Thomas precession. The latter phenomenon lies behind the sum rule (18) which is the relation for the angular momentum of the light cloud in B meson. These relativistic corrections lead to the terms of order 1 in the first sum rules Eqs. (16), (18). The constant $1/4$ comes in the latter case from the $1/m^2$ LS-term; this is easy to show using the commutation relations between the momentum, coordinate and the nonrelativistic Hamiltonian.

The heavy quark sum rules lead to a number of exact inequalities in the static limit [2,3]. The most familiar one is the Bjorken bound $\varrho^2 > 1/4$. The sum rule (18) leads to the stronger dynamical bound $\varrho^2 > 3/4$. It ensures that at least some of the inelastic amplitudes must be nonzero; therefore, the bound state with nonzero spin of light cloud cannot be structureless (pointlike) regardless of bound-state dynamics. Comparison to the QCD sum rule evaluation $\varrho^2 = 0.7 \pm 0.1$ [8] suggests that this bound can be nearly saturated. We also have the bound $\bar{\Lambda} > 2\overline{\Sigma}$.

The spin sum rule (18) provides the rationale for the experimental fact that vector mesons B^*, D^* are heavier than their hyperfine pseudoscalar partners B, D. Indeed, if the sum rule for μ_G^2 is dominated by the low-lying states then μ_G^2 must be of the same sign as the constant in Eq. (18), which dictates the negative energy of the heavy quark spin interaction in B and positive in B^*.

Let us note that the full matrix element in Eqs. (14), (20) is not well defined in the limit $m_Q \to \infty$ due to ultraviolet divergences in the static theory, and, therefore, the spin-independent part proportional to $\bar{\Lambda}$ depends on regularization. Nevertheless, the antisymmetric part proportional to $\overline{\Sigma}$ is well defined and finite. The new sum rules (18) and (19) are convergent and not renormalized by perturbative corrections; this distinguishes them from all other heavy quark sum rules.

For divergent sum rules it is natural to truncate the sums at some cutoff energy $\epsilon = \mu$; then μ serves as the normalization point of the corresponding operators. In particular, the slope ϱ^2 is μ-dependent [5]. The bound $\varrho^2(\mu) > 3/4$ then applies at any normalization point above the energy scale where the spin sum rule (18) is saturated. This scale is determined by nonperturbative dynamics and is usually assumed to be below or about 1 GeV.

At large $\epsilon \gg \Lambda_{\rm QCD}$ the sums over the excited states are dual to the quark–gluon contribution computed in perturbation theory [3,9]:

$$2 \sum_m \epsilon_m^2 |\tau_{3/2}^{(m)}|^2 + \sum_n \epsilon_n^2 |\tau_{1/2}^{(m)}|^2 \to \frac{8}{9} \frac{\alpha_s(\epsilon)}{\pi} \epsilon \, {\rm d}\epsilon, \tag{22}$$

$$\sum_m \epsilon_m^2 |\tau_{3/2}^{(m)}|^2 - \sum_n \epsilon_n^2 |\tau_{1/2}^{(n)}|^2$$
$$\to -\frac{3\alpha_s(\epsilon)}{2\pi} \frac{{\rm d}\epsilon}{\epsilon}$$
$$\times \left\{ \sum_{\epsilon_m < \epsilon} \epsilon_m^2 |\tau_{3/2}^{(m)}|^2 - \sum_{\epsilon_n < \epsilon} \epsilon_n^2 |\tau_{1/2}^{(n)}|^2 \right\}, \tag{23}$$

where the embraced expression simply amounts to $\frac{1}{6} \mu_G^2(\epsilon)$ if the normalization point is implemented as the cutoff in the energy ϵ. Eq. (22) can be immediately extended to higher orders in α_s, this amounts to using the so-called dipole coupling $\alpha_s^{(d)}(\epsilon)$ introduced in Ref. [10]:

$$\alpha_s^{(d)}(\epsilon) = \alpha_s^{\overline{\rm MS}}\big(e^{-5/3 + \ln 2}\epsilon\big) - \left(\frac{\pi^2}{2} - \frac{13}{4}\right)\frac{\alpha_s^2}{\pi}$$
$$+ \mathcal{O}(\alpha_s^3). \tag{24}$$

It determines μ-dependence of ϱ^2. Using Eq. (23) we can estimate the contribution of the high-energy states in the sum rules (18) and (19):

$$\sum_{\epsilon_m < \mu} |\tau_{3/2}^{(m)}|^2 - \sum_{\epsilon_n < \mu} |\tau_{1/2}^{(m)}|^2$$
$$\simeq \frac{1}{4} + \frac{\alpha_s(\mu)}{8\pi} \frac{\mu_G^2(\mu)}{\mu^2}, \tag{25}$$

$$\sum_{\epsilon_m < \mu} \epsilon_m |\tau_{3/2}^{(m)}|^2 - \sum_{\epsilon_n < \mu} \epsilon_n |\tau_{1/2}^{(m)}|^2$$
$$\simeq \frac{\overline{\Sigma}}{2} + \frac{\alpha_s(\mu)}{4\pi} \frac{\mu_G^2(\mu)}{\mu}, \tag{26}$$

they are power suppressed and presumably small in the perturbative domain.

The transition amplitudes with change of the heavy quark velocity acquire ultraviolet divergences when $m_Q \to \infty$, being suppressed by the nonrelativistic

analogue of the Sudakov form factor

$$S \sim \exp\left[-\frac{4}{9\pi}(\Delta \vec{v})^2 \int \frac{{\rm d}\omega}{\omega} \alpha_s^{(d)}(\omega)\right], \tag{27}$$

where the upper limit of integration is set by m_Q. The nonforward scattering amplitude has the similar universal suppression. This does not affect our consideration since these effects are given only by symmetric combinations of $\vec{v}$ and $\vec{v}'$.

Even the convergent sum rules can, in principle, get a finite perturbative renormalization as it happens, for example, with the Bjorken or Ellis–Jaffe sum rules in DIS where it comes from gluon momenta scaling like $\sqrt{Q^2}$. Since we defined the SV transition amplitudes (τ's) via the flavor-diagonal vector current, such perturbative renormalization is absent from the sum rules.

The heavy quark sum rules can also be considered for other heavy flavor hadrons. In Λ_b the spin of light degrees of freedom vanishes, and no nontrivial relations are obtained with the scattering amplitudes at $\vec{u} \neq 0$. They are informative for the Σ_b-type states with $j = 1$ and can be derived directly applying the quoted equations to the corresponding states. For spin-1 light cloud an additional, spin-dependent amplitude is present in the elastic transitions at order $\vec{v}^2$, yet it does not yield the antisymmetric in $\vec{v}, \vec{v}'$ structure. Since the nonforward amplitude has three tensor structures bilinear in $\vec{v}, \vec{v}'$, there are three sum rules for the contributions of the three types of P-wave states proportional to $|\tau_0|^2$, $|\tau_1|^2$ and $|\tau_3|^2$, respectively, constraining the slope of the Isgur–Wise function and the second formfactor at zero recoil.

The sum rule (18) can be applied in atomic physics; there it is a sum rule for spin-orbital interaction in dipole transitions. There is a difference between the transitions in atom and B meson, though: in the latter case we are studying the amplitudes mediated by the heavy quark currents, whereas in atoms photons are emitted through their interaction with electrons. The two amplitudes are directly related only in the nonrelativistic approximation. In particular, additional relativistic effects emerge due to explicit corrections in the electromagnetic current of electrons.

The various sum rules we discuss for heavy flavor transitions are the operator relations for the equal time commutators of heavy quark currents (the lowest sum rules), or their time derivatives, for higher sum rules.

N. Uraltsev / Physics Letters B 501 (2001) 86–91   91

The OPE allows straightforward calculation of these commutators including possible Schwinger terms.

To summarize, applying the OPE to the nonforward heavy quark scattering amplitude two new superconvergent sum rules are derived intrinsically connected to the spin of light cloud in the heavy flavor hadron. The first sum rule for the spin of light cloud in B leads to the nontrivial bound $\varrho^2 > 3/4$ for the slope of the IW function, and suggests why B^* is heavier than B. The second sum rule bounds the difference $M_B - m_b$ from below. These sum rules are exact in the heavy quark limit and can help to constrain a number of nonperturbative parameters in heavy flavor hadrons. The bound for ϱ^2 applies at any normalization point above the scale where the sum rule (18) is saturated.

Note added

After this Letter was submitted, I knew about paper [11] where it was found that the sum in the r.h.s. of Eq. (19) determines one of the subleading $B \rightarrow D^*$ formfactors. I am grateful to the authors of Ref. [11] for communication.

Acknowledgements

Useful discussion with I. Bigi, M. Eides, M. Shifman and A. Vainshtein are gratefully acknowledged. This work was supported in part by NSF under grant number PHY96-05080 and by RFFI under grant No. 99-02-18355.

References

[1] I.I. Bigi, M. Shifman, N.G. Uraltsev, A. Vainshtein, Phys. Rev. D 52 (1995) 196;
M. Shifman, N.G. Uraltsev, A. Vainshtein, Phys. Rev. D 51 (1995) 2217.

[2] I. Bigi, M. Shifman, N.G. Uraltsev, Ann. Rev. Nucl. Part. Sci. 47 (1997) 591.

[3] N. Uraltsev, hep-ph/0010328, to be published B.I. Festschrift, in: M. Shifman (Ed.), At the Frontier of Particle Physics/Handbook of QCD, World Scientific, Singapore, 2001.

[4] M. Shifman, in: Lectures on Heavy Quarks in Quantum Chromodynamics, ITEP Lectures on Particle Physics and Field Theory, Vol. 1, World Scientific, Singapore, 1999, p. 1, hep-ph/9510377.

[5] N. Isgur, M. Wise, Phys. Rev. D 43 (1991) 819.

[6] J.D. Bjorken, in: M. Greco (Ed.), Proceedings of the 4th Rencontres de Physique de la Vallèe d'Aoste, La Thuille, Italy, 1990, Editions Frontières, Gif-Sur-Yvette, France, 1990, p. 583.

[7] M. Voloshin, Phys. Rev. D 46 (1992) 3062.

[8] B. Blok, M. Shifman, Phys. Rev. D 47 (1993) 2949.

[9] N. Uraltsev, in: I. Bigi, L. Moroni (Eds.), Heavy Flavour Physics: A Probe of Nature's Grand Design, Proceedings of the International School of Physics Enrico Fermi, Course CXXXVII, Varenna, July 7–18, 1997, IOS Press, Amsterdam, 1998, p. 329, hep-ph/9804275.

[10] A. Czarnecki, K. Melnikov, N. Uraltsev, Phys. Rev. Lett. 80 (1998) 3189.

[11] A. Le Yaouanc et al., Phys. Lett. B 480 (2000) 119.

QCD Predictions for Lepton Spectra in Inclusive Heavy Flavor Decays

I. I. Bigi,[1,*] M. Shifman,[2,†] N. G. Uraltsev,[1,3,‡] and A. Vainshtein[2,4,§]

[1]*Department of Physics, University of Notre Dame du Lac, Notre Dame, Indiana 46556*
[2]*Theoretical Physics Institute, University of Minnesota, Minneapolis, Minnesota 55455*
[3]*St. Petersburg Nuclear Physics Institute, Gatchina, St. Petersburg 188350, Russia*[‖]
[4]*Budker Institute of Nuclear Physics, Novosibirsk 630090, Russia*
(Received 5 April 1993)

We derive the lepton spectrum in semileptonic b decays from a nonperturbative treatment of QCD; it is based on an expansion in $1/m_Q$ with m_Q being the heavy flavor quark mass. The leading corrections arising on the $1/m_Q$ level are completely expressed in terms of the difference in the mass of the heavy hadron and the quark. Nontrivial effects appear in $1/m_Q^2$ terms affecting mainly the end-point region; they are different for meson and baryon decays as well as for bottom and charm decays.

PACS numbers: 13.20.Jf, 12.38.Lg

The weak decays of hadrons contain a wealth of information on the fundamental forces of nature. Yet the intervention of the strong interactions has prevented us from extracting this information in a reliable way. Heavy flavor decays promise to be more tractable since the mass of the heavy flavor quark m_Q provides a powerful expansion parameter. Indeed significant successes have been scored by the effective heavy quark theory (EHQT) [1]. Yet that approach has some intrinsic limitations: e.g., it deals with exclusive semileptonic modes only and it requires the presence of heavy quarks both in the initial and in the final state; thus it cannot be applied directly to $b \to u$ transitions. Its model-independent predictions so far include corrections through order $1/m_Q$ only. On the other hand the energy released in heavy flavor decays is much larger than ordinary hadronic energies. Our analysis will make use of this large energy release in treating $Q \to qlv$ transitions with $m_Q, m_Q - m_q \gg \Lambda_{QCD}$.

In previous papers [2] we have shown how nonperturbative contributions to global quantities such as lifetimes and semileptonic branching ratios can be obtained from a model-independent treatment of QCD. The method was based on expanding the weak transition operator into a series of local operators of increasing dimension with coefficients that contain increasing powers of $1/m_Q$. The coefficients depend on the (inclusive) final state; the differences between the decay rates of different heavy flavor hadrons H_Q—charged vs neutral mesons vs baryons—enter through the matrix elements of the local operators taken between H_Q. For example the total semileptonic $b \to u$ width through order $1/m_b^2$ is given by

$$\Gamma(H_b \to lvX) \propto m_b^5 \frac{1}{2M_{H_b}} \left\langle H_b \left| \bar{b}b - \frac{1}{m_b^2}\bar{b}i\sigma Gb \right| H_b \right\rangle,$$
(1)

with $i\sigma G = i\gamma_\mu \gamma_\nu G_{\mu\nu}$, $G_{\mu\nu}$ being the gluonic field strength tensor.

In this Letter we will expand the general method to treat the lepton *spectra* in the semileptonic decays of heavy flavor hadrons. The issue of how to treat inclusive lepton spectra in heavy flavor decays within QCD was first analyzed in Ref. [3]: A general relationship to an operator product expansion was stated and the structure of nonperturbative corrections was discussed there. A new feature is encountered with spectra: One is dealing with an expansion in powers of $1/(p_Q - p_l)^2$ rather than of $1/m_b^2$, with p_Q and p_l denoting the momenta of Q and the lepton l, respectively. This series is singular at the end point of the lepton spectrum; thus some care has to be applied in interpreting the results. Two dimension-five operators generate the leading nonperturbative corrections of order $1/m_Q^2$: the color magnetic operator $\bar{Q}i\sigma GQ$ and the operator $\bar{Q}[D^2 - (vD)^2]Q$ describing the kinetic energy of Q in the gluon background field; D_μ denotes here the covariant derivative and v_μ is the four-velocity vector of the hadron. Corrections actually arise already on the $1/m_Q$ level; it is crucial that in QCD those can be expressed completely in terms of the difference between the quark and the hadron mass. We will phrase our discussion in terms of b decays with a few added comments on charm decays.

Ignoring gluon bremsstrahlung one obtains a lengthy expression for the lepton spectra in the semileptonic decays of a b-flavored hadron H_b:

$$\frac{d\Gamma}{dy}(H_b \to lvX_q) = \Gamma_0 \theta(1-y-\rho)2y^2 \left\{ (1-f)^2(1+2f)(2-y) + (1-f)^3(1-y) \right.$$

$$+ (1-f)\left[(1-f)\left(2+\frac{5}{3}y - 2f + \frac{10}{3}fy\right) - \frac{f^2}{\rho}[2y + f(12-12y+5y^2)]\right]G_b$$

$$\left. - \left[\frac{5}{3}(1-f)^2(1+2f)y + \frac{f^3}{\rho}(1-f)(10y - 8y^2) + \frac{f^4}{\rho^2}(3-4f)(2y^2 - y^3)\right]K_b \right\},$$
(2)

0031-9007/93/71(4)/496(4)$06.00

VOLUME 71, NUMBER 4   PHYSICAL REVIEW LETTERS   26 JULY 1993

$$\Gamma_0 = \frac{G_F^2 m_b^5}{192\pi^3} |V_{qb}|^2, \quad f = \frac{\rho}{1-y}, \quad \rho = \frac{m_q^2}{m_b^2}, \quad y = \frac{2E_l}{m_b},$$

$$K_b = (1/2M_{H_b})\langle H_b|\bar{b}(i\mathbf{D})^2 b|H_b\rangle/m_b^2, \quad G_b = (1/2M_{H_b})\langle H_b|\bar{b}i\sigma G b|H_b\rangle/2m_b^2,$$

with m_q denoting the mass of the quark q in the final state. Equation (2) represents the master formula containing both relevant cases, namely, $q = u, c$.

For $b \to u l \nu$ transitions with $m_u = 0$ this expression simplifies considerably,

$$\frac{d\Gamma}{dy} = \Gamma_0 2y^2 \left\{ 3 - 2y - \left[\frac{5}{3}y + \frac{1}{3}\delta(1-y) - \frac{1}{6}(2y^2 - y^3)\delta'(1-y) \right]K_b + \left[2 + \frac{5}{3}y - \frac{11}{6}\delta(1-y) \right]G_b \right\}. \tag{3}$$

The δ functions and their derivative reflect the previously mentioned singular nature of the expansion at the end point; their emergence has a transparent meaning (see [4] for details). The spectrum is finite at the end point for $m_u = 0$ and thus contains a step function. The chromomagnetic interaction effectively "shifts" the spectrum by changing the energy in either the initial or final state; the shift in the argument of the step function thus yields a δ function. The singular structure in the K_b term on the other hand reflects the motion of the b quark inside the H_b hadron which Doppler shifts the spectrum; in the second order it generates $\delta'(1-y)$ for the steplike spectrum.

Because of these singular terms the expression given above can be identified with the observable spectrum only *outside a finite neighborhood of the end-point region.* (This distance remains constant in absolute units in the limit $m_b \to \infty$.) Yet even this neighborhood does not represent true "terra incognita:" for integrating our expression over this kinematical region yields a finite and trustworthy result that can be confronted with the data. This can be expressed through the function

$$\Gamma(E_l) = -\int_{E_l}^{E_{max}} dE_l \frac{d\Gamma}{dE_l}, \quad E_l \le E_{QCD} < E_{max}. \tag{4}$$

E_{max} denotes the maximal kinematically allowed energy and E_{QCD} the maximal energy for which one can still trust the QCD expansion in Eqs. (2) and (3); its value depends on the size of K_b and G_b. Clearly $\Gamma(0) = \Gamma_{SL}$ has to hold. This is not a trivial relation, for Γ_{SL} is deduced from a completely regular expansion in $1/m_b$ [see Eq. (1)], whereas $\Gamma(0)$ comes from integrating the expression in Eq. (3) containing singularities; thus the singular terms are essential for recovering the correct decay width.

Similar considerations apply to $b \to c l \nu$ transitions. With m_c as an infrared cutoff there arise no singular terms at the end point $y = 1 - \rho$; yet the expansion parameters $G_b/(1-y)$, $K_b/(1-y)^2$ though finite become large there. The need for "smearing" the spectrum over the end-point region, Eq. (4), thus still exists.

The contributions from higher dimensional operators that we are ignoring have terms of the schematical form $\sim [\mu_{hadr}/(m_b(1-y))]^n$; summing them all up would yield a well-behaved function. As long as these quantities are smaller than unity, we can trust the expressions given above for the unintegrated lepton spectrum.

The size of G_b is easily determined from experiment:

$$G_b = \langle B|\bar{Q}i\sigma GQ|B\rangle/4m_b^2$$
$$\simeq 3[M^2(B^*) - M^2(B)]/4m_b^2 \simeq 0.017;$$

for Λ_b it vanishes. A recent QCD sum rules analysis yields [5] $\langle B|\bar{b}(i\mathbf{D})^2 b|B\rangle \sim (0.4 \text{ GeV}^2)2M_B$ in a qualitative agreement with rather general expectations based on simple quark models of hadrons. For our subsequent discussion we will set $K_b = 0.02$; it should be noted that the dependence on the exact value of K appears to be numerically suppressed as compared to the impact of G unless K_b exceeds the above estimate by a few times. We use $m_b \simeq 4.8$ GeV as deduced from a QCD analysis of the Y system, $m_b - m_c = 3.35$ GeV as inferred from the bottom and charm meson masses which for $m_b = 4.8$ GeV yields $m_c = 1.45$ GeV, and put $m_u = 0$. For $b \to u$ decays we estimate that $E_{QCD} \sim 0.9 m_b/2 \simeq 2.15$ GeV; for $b \to c$ a somewhat smaller "smearing" range near the end point seems to be required, namely, $\Delta E_l \approx 0.15$ GeV and thus $E_{QCD} \simeq 2.0$ GeV. Numerically it implies that in the real world the c quark is relatively heavy in b decays: $m_c^2 > \mu_{hadr} m_b$; therefore the falling edge of the spectrum starts in the calculable region.

For a proper perspective we show the partially integrated spectrum $\Gamma(E_l)$ from three different prescriptions, namely, (a) our QCD expansion; (b) the simple free quark picture with $G_b = K_b = 0$, $m_u = 0$; and (c) the phenomenological treatment of Altarelli et al. [6] (hereafter referred to ACM), where one attempts to incorporate some bound state effects. We have set here $m_u = m_{spect} = 0.15$ GeV, $m_c = 1.67$ GeV, and $p_F = 0.3$ GeV as suggested by a fit to CLEO data; p_F denotes the "Fermi momentum." To be consistent we have ignored gluon bremsstrahlung both in ACM and in our QCD expressions [for fitting the data it should be added to Eqs. (2)-(4)]. We cut the plot for the QCD expansion where the correction to the parton model reaches 30%. In comparing the QCD formula with the ACM prediction one has to note a subtle distinction in the definition of the kinematical variables: in ACM energy is expressed in terms of the mass of the b hadron; yet in Eqs. (2) and (3) y measures the energy in units of m_b, which by itself introduces a shift of order $1/m_b$.

Comparing the results on $\Gamma(E_l)$ with the three ap-

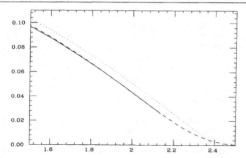

FIG. 1. The function $\Gamma(E_l)$ for $b \rightarrow u$ transitions calculated in the QCD expansion (solid line), the free spectator quark model (dotted line), and the ACM ansatz (dashed line); the ACM curve has been multiplied by a factor 1.07.

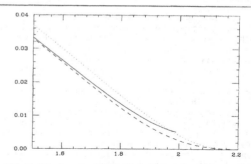

FIG. 2. The function $\Gamma(E_l)$ for $b \rightarrow c$ transitions calculated in the QCD expansion (solid line), the free spectator quark model (dotted line), and the ACM ansatz (dashed line); the ACM curve has been multiplied by a factor of 1.13.

proaches we conclude the following: (i) The *shape* of the QCD, of the free quark model, and of the ACM curves are very close over most of the range of $E_l \lesssim E_{QCD}$ for the $b \rightarrow u$ as well as the $b \rightarrow c$ case. The main difference lies in the overall *normalization*. (ii) The QCD result for $b \rightarrow u$ can be largely simulated by setting $m_u \sim 0.3$ GeV in the free quark spectrum. The nonperturbative corrections thus effectively transform a current quark mass into a "constituent" one of reasonable size. (iii) Once the ACM result is renormalized according to

$$\Gamma(0)_{QCD}(b \rightarrow u) = 1.07 \times \Gamma(0)_{ACM}(b \rightarrow u)$$

and

$$\Gamma(0)_{QCD}(b \rightarrow c) = 1.13 \times \Gamma(0)_{ACM}(b \rightarrow c)$$

the two curves are hardly distinguishable as functions of E_l. (iv) While the QCD results for $\Gamma(E_l)$ are largely independent of the value of K_b for $K_b \sim 0.02$ in $b \rightarrow u$, they are sensitive to it for $b \rightarrow c$; that change can be easily understood [4]. (v) The QCD curves for semileptonic Λ_b decays are somewhat harder than for B meson decays. (vi) The largest differences between the models are found in the end-point region, namely, for $E_l > 1.8$ GeV in $b \rightarrow c$.

Some of these points are illustrated in the figures. In Fig. 1 we show $\Gamma(E_l)$ for the end-point region of $b \rightarrow u$; in Fig. 2 we have plotted these partially integrated spectra for $b \rightarrow c$. The *unintegrated* spectrum $d\Gamma(E_L)/dE_l$ which can be calculated in the QCD treatment for $E_l \lesssim 2$ GeV is shown in Fig. 3. The similarity in the shape of the two spectra is quite striking.

A few comments are in order about charm decays. The lepton spectra are quite different in b and in charm decays, especially in the end-point region, for the charged lepton in b decays is an electron or muon whereas it is an antifermion l^+ in charm decays. Since the lepton spectrum vanishes at the end point in $c \rightarrow q l \nu$ even for $m_q = 0$, the chromomagnetic operator can yield finite terms only while the kinetic energy operator produces a $\delta(1-y)$ but not a $\delta'(1-y)$ term. Yet since the nonperturbative corrections are much larger in charm decays than in b decays—$G_c, K_c \sim 0.2$—the smearing required in y is much larger for charm decays, namely, $\Delta y \sim 0.5$. While this allows us to make meaningful statements about fully integrated quantities such as Γ_{SL}, it appears beyond the scope of our present analysis to treat lepton *spectra* in charm decays.

To summarize, we have shown here how an expansion in $1/(p_b - p_l)$ allows us to incorporate nonperturbative corrections to the lepton spectra in inclusive semileptonic $b \rightarrow c$ and $b \rightarrow u$ decays. Because of the singular nature of this expansion at the end point the spectrum cannot be

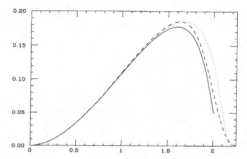

FIG. 3. The lepton spectrum $d\Gamma/dE_l$ for $b \rightarrow c$ transitions calculated in the QCD expansion (solid line), the free spectator quark model (dotted line), and the ACM ansatz (dashed line); the ACM curve has been multiplied by a factor of 1.13.

VOLUME 71, NUMBER 4   PHYSICAL REVIEW LETTERS   26 JULY 1993

computed in detail in a close neighborhood of the end point, yet smeared or partially integrated spectra can. Through order $1/m_b^2$ these expressions are given in terms of the quantities m_b, m_c, G_b, and K_b. These are *not* free parameters — their size can be extracted from the relationship to other observables. Improvements in the precision of their determination, in particular concerning K_b and to a lesser degree m_b, can be anticipated from future progress in theory.

We have compared our QCD results with the ACM description that so far has allowed a decent fit to the data. Despite the obvious differences in the underlying dynamics we have found the shape of the resulting lepton spectra to be remarkably similar, and even more so when one keeps the following in mind:

(a) Initially the scale for the kinematics is different in the two descriptions; for the QCD expansion it is set by the quark mass m_b whereas for ACM by the hadron mass M_B or M_{Λ_b}. The kinematical differences due to $m_b \neq M_{H_b}$ are actually of order $1/m_b$ and formally represent the leading corrections. This suggests an interesting observation: An accurate measurement of the shape of the spectrum allows a determination of the b quark mass m_b free of theoretical uncertainities. Unfortunately, in practice one would have to analyze $b \rightarrow u$; the shape of the $b \rightarrow c$ spectrum is basically determined by $m_b - m_c$ with little sensitivity to m_b.

(b) The ACM ansatz contains three free parameters — m_c, m_{sp}, and p_F — that are to be fitted from the data. There are four quantities that set the scale in the QCD expansion through order $1/m_b^2$, namely, m_b, m_c, G_b, and K_b, but none of them is a free parameter. That our QCD description containing therefore very little "wiggle room" can nevertheless yield a description so similar to that of the ACM ansatz — after the latter has been fitted to the data — has to be seen as quite remarkable.

(c) The ACM prescription can thus be viewed as a simple though smart approximation to a more complete and complex QCD treatment. At the same time it would be incorrect to interpret the fit parameters in ACM literally as real physical quantities; it is thus not surprising that the numerical values for the former differ significantly from what is now known about the corresponding quantities in QCD.

A more detailed analysis shows [4] that QCD provides a natural "home" for the Fermi motion originally introduced phenomenologically in ACM, and it is asymptotically a dominant effect for the end-point shape in b decays. However, it enters in a somewhat different form. Its impact on the total widths is only quadratic in $1/m_b$ (in ACM it produces a linear shift). Nevertheless, the *shape* of the lepton spectrum gets $1/m_b$ corrections (at this point we disagree with the conclusion of Ref. [3]); it will be discussed in detail elsewhere [4].

Despite all similarities there arise relevant numerical differences as well, not only in the overall normalization of the spectra, but also in the end-point region: The QCD spectra place a *higher* fraction of $b \rightarrow c$ events in the end-point region. Such a difference has important consequences for the extraction of $|V_{ub}/V_{cb}|$ from the data. Furthermore, our analysis shows that the lepton spectra in semileptonic Λ_b decays are distinct from those in B decays. These issues will be discussed in a future paper [4].

Another model [7] used for describing semileptonic spectra relies heavily on model-dependent calculations of the form factors for the exclusive final states. In the limit of a heavy c quark, $m_b - m_c \ll m_c$, EHFT yields the necessary form factors to some accuracy; therefore in that limit the prediction of this model coincides with the model-independent QCD spectrum. On the other hand, that is apparently not the case for $b \rightarrow u$ decays where the difference is significant.

On the $1/m_b^3$ level that was not treated here novel effects arise due to "weak annihilation" in the $b \rightarrow u$ channel; for details see [8].

This work was supported in part by the NSF under Grant No. NSF-PHY 92-13313 and in part by DOE under Grant No. DOE-AC02-83ER40105.

*Electronic address: BIGI@UNDHEP,UNDHEP::BIGI
†Electronic address: SHIFMAN@UMNACVX
‡Electronic address:
    URALTSEV@UNDHEP;URALTSEV@LNPI.SPB.SU
§Electronic address: VAINSHTE@UMNACVX
‖ Permanent address.

[1] M. A. Shifman, M. B. Voloshin, Yad. Fiz. **41**, 187 (1985) [Sov. J. Nucl. Phys. **41**, 120 (1985)]; H. Georgi, Phys. Lett. B **240**, 447 (1990).

[2] I. I. Bigi, N. G. Uraltsev, and A. Vainshtein, Phys. Lett. B **293**, 430 (1992); B. Blok and M. Shifman, Nucl. Phys. **B399**, 441 (1993); **B399**, 459 (1993); I. I. Bigi, B. Blok, M. Shifman, N. G. Uraltsev, and A. Vainshtein, Reports No. UND-HEP-92-BIG07 and No. TPI-MINN-92/67-T (to be published).

[3] J. Chay, H. Georgi, and B. Grinstein, Phys. Lett. B **247**, 399 (1990).

[4] I. I. Bigi, M. Shifman, N. G. Uraltsev, and A. Vainshtein (to be published).

[5] V. Braun (private communication); see also the earlier analysis in M. Neubert, Phys. Rev. D **46**, 1076 (1992), that yielded a larger value.

[6] G. Altarelli, G. Corbo, N. Cabibbo, L. Maiani, and G. Martinelli, Nucl. Phys. **B208**, 365 (1982).

[7] B. Grinstein, M. Isgur, and M. B. Wise, Phys. Rev. Lett. **56**, 298 (1986); N. Isgur *et al.*, Phys. Rev. D **39**, 799 (1989).

[8] I. I. Bigi and N. G. Uraltsev, Report No. UNDEHEP-BIG02, 1993 (to be published).

PHYSICAL REVIEW D   VOLUME 56, NUMBER 7   1 OCTOBER 1997

High power n of m_b in b-flavored widths and $n=5\rightarrow\infty$ limit

I. Bigi,[1] M. Shifman,[2] N. Uraltsev,[1,2,3,4] and A. Vainshtein[2,5]

[1]*Department of Physics, University of Notre Dame du Lac, Notre Dame, Indiana 46556*
[2]*Theoretical Physics Institute, University of Minnesota, Minneapolis, Minnesota 55455*
[3]*TH Division, CERN, CH 1211 Geneva 23, Switzerland*
[4]*Petersburg Nuclear Physics Institute, Gatchina, St. Petersburg 188350, Russia**
[5]*Budker Institute of Nuclear Physics, Novosibirsk 630090, Russia*
(Received 21 April 1997)

The leading term in the semileptonic width of heavy flavor hadrons depends on the fifth power of the heavy quark mass. We present an analysis where this power can be self-consistently treated as a free parameter n and the width can be studied in the limit $n\rightarrow\infty$. The resulting expansion elucidates why the small velocity (SV) treatment is relevant for the inclusive semileptonic $b\rightarrow c$ transition. The extended SV limit (ESV limit) is introduced. The leading terms in the perturbative α_s expansion enhanced by powers of n are automatically resummed by using the low-scale Euclidean mass. The large-n treatment explains why the scales of order m_b/n are appropriate. On the other hand, the scale cannot be too small since the factorially divergent perturbative corrections associated with running of α_s show up. Both requirements are met if we use the short-distance mass normalized at a scale around $m_b/n\sim 1$ GeV. A convenient definition of such low-scale operator-product-expansion-compatible masses is briefly discussed. [S0556-2821(97)02619-2]

PACS number(s): 12.38.Bx, 13.20.He, 13.30.Ce

I. INTRODUCTION

A true measure of our understanding of heavy flavor decays is provided by our ability to extract accurate values for fundamental quantities, like the Cabibbo-Kobayashi-Maskawa (CKM) parameters, from the data. The inclusive semileptonic widths of B mesons depend directly on $|V_{cb}|$ and $|V_{ub}|$, with the added bonus that they are amenable to the operator-product-expansion (OPE) treatment [1–3], and the nonperturbative corrections to $\Gamma_{sl}(B)$ are of order $1/m_b^2$ [4–7], i.e., small. Moreover, they can be expressed, in a model-independent way, in terms of fundamental parameters of the heavy quark theory, μ_π^2 and μ_G^2, which are known with relatively small uncertainties.

Therefore, the main problem in a program of precise determination of, say, $|V_{cb}|$ from $\Gamma_{sl}(B)$, is the theoretical understanding of the *perturbative* QCD corrections and the heavy quark masses. The theoretical expression for $\Gamma_{sl}(B)$ depends on a high power of the quark masses m_b (and m_c): $\Gamma_{sl}\propto m_b^n$ with $n=5$. The quark masses are not observables, and the uncertainties in their values are magnified by the power n. For this reason it is often thought that the perturbative corrections in the inclusive widths may go beyond theoretical control.

The quark masses in field theory are not constant, but depend on scale and, in this respect, are similar to other couplings like α_s which define the theory. Of course, m_Q and α_s enter the expansion differently. Moreover, in contrast to α_s in QCD, the heavy quark mass m_Q has a finite infrared limit to any order in the perturbation theory, m_Q^{pole}, which is routinely used in the calculations.

If $\Gamma_{sl}(B)$ is expressed in terms of the pole mass of the b

quark treated as a given number, the expression for $\Gamma_{sl}(B)$ contains a factorially divergent series in powers of α_s,

$$\Gamma_{sl}\sim\sum_k k!\left(\frac{\beta_0}{2}\frac{\alpha_s}{\pi}\right)^k,$$

due to the $1/m_Q$ infrared (IR) renormalon [8–10]. This series gives rise to an unavoidable uncertainty which is linear rather than quadratic in $1/m_b$. The related perturbative corrections are significant already in low orders. They are directly associated with running of α_s and their growth reflects the fact that the strong coupling evolves into the nonperturbative domain at a low enough scale. A strategy allowing one to circumvent this potentially large uncertainty was indicated in Refs. [8, 9]—instead of the pole quark mass, one should pass to a Euclidean mass, according to Wilson's OPE. This mass is peeled off from the IR part; the IR domain is also absent from the width up to effects $1/m_Q^2$, and the (IR-related) factorial divergence disappears. Following this observation, it became routine to express $\Gamma_{sl}(B)$ in terms of the Euclidean quark masses in the modified minimal subtraction ($\overline{\text{MS}}$) scheme, say, $\overline{m}_b(m_b)$.

Eliminating IR renormalons on this route, however, one does not necessarily have a fast convergent perturbative series: in general, it contains corrections of the type $(n\alpha_s)^k$ which are not related to running of α_s, and are present even for the vanishing β function. We will see that such large corrections inevitably appear if one works with the $\overline{\text{MS}}$ masses like $\overline{m}_b(m_b)$ and $\overline{m}_c(m_c)$. Due to the high value of the power n they constitute quite an obvious menace to the precision of the theoretical predictions. The standard alternative procedure of treating inclusive heavy quark decays, which gradually evolved in the quest for higher accuracy, is thus plagued by its own problems.

---

*Permanent address.

0556-2821/97/56(7)/4017(14)/$10.00   <u>56</u>   4017   © 1997 The American Physical Society

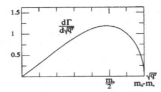 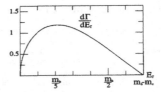

FIG. 1. The decay distribution over the invariant mass of the lepton pair and over energy release $E_r = m_b - m_c - (q^2)^{1/2}$ in $b \to c \ell \nu$ at $m_c / m_b = 0.3$.

The aim of the present work is to turn vices into virtues, by treating n as a free parameter and developing a $1/n$ expansion. In this respect, our approach is conceptually similar to the $1/N_c$ expansion or the expansion in the dimensions of the space-time, etc., which are quite common in various applications of field theory (for a discussion of the virtues of expansion in "artificial" parameters, see [11]). The main advantage is the emergence of a qualitative picture which guides the theoretical estimates in the absence of much more sophisticated explicit higher-order calculations.

A certain unnaturalness of the $\overline{\text{MS}}$ masses normalized, say, at $\mu = m_b$, for inclusive widths is rather obvious before a dedicated analysis, particularly in $b \to c \ell \nu$ inclusive decays. The maximal energy fed into the final hadronic system and available for exciting the final hadronic states, which determines the "hardness" of the process, is limited by $m_b - m_c \approx 3.5$ GeV. Moreover, in a typical decay event, leptons carry away a significant energy $E_{\ell \nu} > \sqrt{q^2}$, since the lepton phase space emphasizes the larger q^2. This is illustrated in Fig. 1 where the distribution over the invariant mass of the lepton pair $\sqrt{q^2}$ and energy release $E_r = m_b - m_c - \sqrt{q^2}$ is shown. The situation is less obvious *a priori* for $b \to u$, but again the typical energy of the final state hadrons is manifestly smaller than m_b. We use the parameter n to quantify these intuitive observations.

Since the power $n = 5$ is of a purely kinematic origin, technically it is quite easy to make n a free parameter. To this end it is sufficient, for instance, to modify the lepton current by introducing fictitious additional leptons emitted by the W boson, along with the standard $\ell \nu$ pair. At the very end, the number of these fictitious leptons is to be put to zero.

The emergence of the large perturbative terms containing powers of n in a general calculation is rather obvious and can be illustrated in the following way. Consider, for example, the $b \to u$ decay rate, and use the pole mass m_b^{pole}. The series will have a factorial divergence in high orders due to the $1/m$ IR renormalon, but we are not concerned about this fact now since we reside in the purely perturbative domain and we do not study high orders $k \sim 1/\alpha_s$ of the perturbation theory. We can consider, for example, the case of vanishing β function, in which nothing prevents one from defining the pole mass with arbitrary precision.

Let us assume then that using the pole mass, m_b^{pole}, the perturbative expansion of the width

$$\Gamma \simeq d_n m_b^n A^{\text{pt}}(\alpha_s) = d_n m_b^n \left[1 + a_1 \frac{\alpha_s}{\pi} + a_2 \left(\frac{\alpha_s}{\pi} \right)^2 + \cdots \right] \quad (1)$$

has all coefficients a_k completely n independent (we will show below that it is the case in the leading-n approximation). Then, being expressed in terms of a different mass $\tilde{m}_b$,

$$\tilde{m}_b = m_b \left(1 - c \frac{\alpha_s}{\pi} \right), \quad (2)$$

one has

$$\Gamma \simeq d_n \tilde{m}_b^n \tilde{A}^{\text{pt}}(\alpha_s) \quad (3)$$

with

$$\tilde{A}^{\text{pt}}(\alpha_s) = \frac{A^{\text{pt}}(\alpha_s)}{[1 - c(\alpha_s / \pi)]^n} = 1 + (nc + a_1) \frac{\alpha_s}{\pi}$$
$$+ \left(\frac{n(n+1)}{2} c^2 + nca_1 + a_2 \right) \left(\frac{\alpha_s}{\pi} \right)^2 + \cdots . \quad (4)$$

Just using a different mass generates n-enhanced terms $\sim (n\alpha_s)^k$. In the case of the $\overline{\text{MS}}$ mass one has $c = 4/3$. In order to have good control over the perturbative corrections in the actual width, one needs to resum these terms (at least, partially). This can be readily done.

These n-enhanced perturbative corrections are only one, purely perturbative aspect of the general large-n picture. The total width is determined by integrating n-independent hadronic structure functions with the n-dependent kinematic factors; the latter being saturated at large $\sqrt{q^2}$ close to the energy release $m_b - m_c$. As a result, the energy scale defining the effective width of integration is essentially smaller than the energy release, and a new, lower momentum scale automatically emerges in the problem at large n.

The question of the characteristic scale of inclusive decays was discussed in the literature. The discussion was focused, however, almost exclusively on the problem of choosing the normalization point for the running coupling α_s. We emphasize that for the inclusive decays it is a secondary question. The principal one is the normalization point for quark masses which run as well. Although their relative variation is smaller, this effect is enhanced, in particular, by the fifth power occurring in the width.

With $n = 5$ treated as a free parameter, we arrive at the following results concerning semileptonic B decays driven by $b \to l \nu q$, $q = c$, or u.

(1) The leading n-enhanced perturbative corrections to the total width are readily resummed by using the low-scale quark masses. No significant uncertainty in the perturbative corrections is left in the widths.

(2) The surprising proximity of the $b \rightarrow c\ell\nu$ transition to the small velocity (SV) limit [12] which seems to hold in spite of the fact that $m_c^2/m_b^2 \ll 1$ becomes understood. The typical kinematics of the transition are governed by the parameter $(m_b - m_c)/nm_c$ rather than by $(m_b - m_c)/m_c$. This "extended SV" parameter shows why the SV expansion is relevant for actual $b \rightarrow c$ decays.

(3) The "large-n" remarks are relevant even if α_s does not run. In actual QCD, the better control of the perturbative effects meets a conflict of interests: the normalization point for masses cannot be taken either very low or too high, thus neither $\overline{MS}$ nor pole masses are suitable. We discuss a proper way to define a short-distance heavy quark mass $m_Q(\mu)$ with $\mu \ll m_Q$ which is a must for nonrelativistic expansion in QCD. A similar definition of the effective higher-dimension operators is also given.

(4) The limits $m_Q \rightarrow \infty$ and $n \rightarrow \infty$ cannot freely be interchanged. Quite different physical situations arise in the two cases, as indicated later. The perturbative regime takes place if $(m_b - m_c)/n \gg \Lambda_{QCD}$, and the process is not truly short distance if the opposite holds. Nevertheless, at $m_c \gg \Lambda_{QCD}$ even in the latter limit the width is determined perturbatively up to effects $\sim \Lambda_{QCD}^2/m_c^2$.

We hasten to add that analyzing only terms leading in n may not be fully adequate for sufficiently accurate predictions in the real world where $n = 5$, which turns out to be too small. Still, it is advantageous to start from the large-n limit and subsequently include (some of the) subleading terms.

II. THE THEORETICAL FRAMEWORK: LARGE-n EXPANSION IN INCLUSIVE SEMILEPTONIC DECAYS

The new tool we bring to bear here is as follows: the actual value $n = 5$ of the power n in the width arises largely through the integration over the phase space for the lepton pair. One can draw up simple physical scenarios where n appears as a free parameter and the limit $n \rightarrow \infty$ can be analyzed. The simplest example is provided by l (massless) scalar "leptons" ϕ, emitted in the weak vertex of the lepton-hadron interaction.

$$\mathcal{L}_{weak} = \frac{G_l}{\sqrt{2}} V_{Qq} \phi^l \bar{\ell} \gamma_\alpha (1-\gamma_5) \nu_\ell \bar{q} \gamma^\alpha (1-\gamma_5) Q, \quad (5)$$

where G_l generalizes the Fermi coupling constant to $l \neq 0$. At $l = 0$, the coupling $G_0 = G_F$. Then by dimensional counting one concludes that $\Gamma(Q) \sim |G_l|^2 \cdot m_Q^n$, where $n = 2l + 5$. It should be kept in mind that the details of the physical scenario are not essential for our conclusions.

We will discuss $b \rightarrow q$ transitions with an arbitrary mass ratio, m_q/m_b, thus incorporating both $b \rightarrow u$ and $b \rightarrow c$ decays. Following Refs. [2, 13], we introduce a hadronic tensor $h_{\mu\nu}(q_0, q^2)$, its absorptive part $W_{\mu\nu}(q_0, q^2) = (1/i)\text{disc}\, h_{\mu\nu}(q_0, q^2)$, and its decomposition into five covariants with structure functions $w_i(q_0, q^2)$, $i = 1, \ldots, 5$. Variables q_0 and q^2 are the energy and the effective mass of the lepton pair. Only w_1 and w_2 contribute when lepton masses are neglected and the semileptonic decay width is then given by

$$\Gamma_{sl} \equiv |V_{cb}|^2 \frac{G_F^2}{8\pi^3}\gamma,$$

$$\gamma = \frac{1}{2\pi} \int_0^{q_{max}^2} dq^2 \int_{\sqrt{q^2}}^{q_{0\,max}} dq_0 \sqrt{q_0^2 - q^2} \left\{ q^2 w_1(q_0, q^2) + \frac{1}{3}(q_0^2 - q^2) w_2(q_0, q^2) \right\}, \quad (6)$$

where

$$q_{max}^2 = (M_B - M_D)^2, \quad q_{0\,max} = \frac{M_B^2 + q^2 - M_D^2}{2M_B}.$$

Note that the upper limits of integration over q^2 and q_0 are determined by vanishing of the structure functions for the invariant mass of the hadronic system less than M_D^2. The lower limit of integration over q_0 and q^2 is due to the constraints on the momentum of the lepton pair. It has nothing to do with the properties of the structure functions. In particular, the structure functions exist also in the scattering channel where the constraint on the lepton momentum would be different. In what follows, we will see that the distinction between the leptonic and hadronic kinematical constraints is important.

By adding l extra scalar "leptons" emitted in the weak vertex, we make $n = 5 + 2l$ a free parameter. The quantity γ then becomes n dependent, $\gamma \rightarrow \gamma(n)$,

$$\gamma(n) = \frac{1}{2\pi} \int_0^{q_{max}^2} dq^2 (q^2)^l \int_{\sqrt{q^2}}^{q_{0\,max}} dq_0 \sqrt{q_0^2 - q^2} \left\{ q^2 w_1(q_0, q^2) + \frac{1}{3}(q_0^2 - q^2) w_2(q_0, q^2) \right\}. \quad (7)$$

In the real world $l = 0$, $n = 5$, but as far as the QCD part is concerned, we are free to consider any value of l. The extra factor $(q^2)^l$ appeared due to the phase space of the scalar "leptons." [1]
A different choice of variables is more convenient for our purposes. Instead of q_0 and q^2 we will use ϵ and T,

$$\epsilon = M_B - q_0 - \sqrt{M_D^2 + q_0^2 - q^2}, \quad T = \sqrt{M_D^2 + q_0^2 - q^2} - M_D. \quad (8)$$

---

[1] Strictly speaking, the extra "leptons" produce a change in the leptonic tensor, in particular, making it nontransversal. This variation leads, in turn, to a different relative weight of the structure function w_2, and the emergence of the structure functions w_4 and w_5 in the total probability. These complications are irrelevant, and we omit them. We also omit an overall l dependent numerical factor, which merely redefines G_l.

4020   I. BIGI, M. SHIFMAN, N. URALTSEV, AND A. VAINSHTEIN   <u>56</u>

The variable T is simply related to the spatial momentum $\vec{q}$. More exactly, T is the minimal kinetic energy of the hadronic system for the given value of $\vec{q}$. This minimum is achieved when the D meson is produced. The variable ϵ is the excitation energy, $\epsilon = (M_X - M_D) + (T_X - T)$, where $T_X = \sqrt{M_X^2 + q_0^2 - q^2} - M_X$ is the kinetic energy of the excited state with mass M_X. The following notations are consistently used below:

$$\Delta = M_B - M_D, \quad \vec{q}^2 = q_0^2 - q^2, \quad |\vec{q}| = \sqrt{\vec{q}^2}.$$

In terms of the new variables one obtains

$$\gamma(n) = \frac{1}{\pi} \int_0^{T_{max}} dT(T + M_D)\sqrt{T^2 + 2M_D T} \int_0^{\epsilon_{max}} d\epsilon(\Delta^2 - 2M_B T - 2\Delta\epsilon + 2T\epsilon + \epsilon^2)^l \left\{ (\Delta^2 - 2M_B T - 2\Delta\epsilon + 2T\epsilon + \epsilon^2)w_1 \right.$$

$$\left. + \frac{1}{3}(T^2 + 2M_D T)w_2 \right\}, \tag{9}$$

where

$$T_{max} = \frac{\Delta^2}{2M_B}, \quad \epsilon_{max} = \Delta - T - \sqrt{T^2 + 2M_D T},$$

it is implied that the structure functions $w_{1,2}$ depend on ϵ and T.

A. Cancellation of the infrared contribution

Before submerging in the large-n limit, we will address a question which naturally comes to mind immediately upon inspection of Eq. (6) or Eq. (7). Indeed, these expressions give the total decay probabilities in terms of an integral over the physical spectral densities, which is in turn over the physical phase space. Both factors depend on the meson masses, and know nothing about the quark mass. This is especially clear in the case of the phase space, which seems to carry the main dependence for large n. And yet, in the total probability, the dependence on the meson masses must

disappear, and the heavy quark mass must emerge as a relevant parameter. In other words, the large-distance contributions responsible for making the meson mass out of the quark mass, must cancel each other.

It is instructive to trace how this cancellation occurs. We will study the issue for arbitrary value of l, assuming for simplicity that the final quark mass vanishes (i.e., $b \to u$ transition). This assumption is not crucial; one can consider an arbitrary ratio m_c/m_b, with similar conclusions. Since the main purpose of this section is methodical, we will limit our analysis to terms linear in $\overline{\Lambda}$. The basic theoretical tool allowing one to trace how M_B^n in the decay probability is substituted by m_b^n, as a result of cancellation of the infrared contributions residing in M_B and in the integrals over the spectral densities, is the heavy flavor sum rules presented in great detail in Ref. [14], see Eqs. (130)–(135).

We start by expanding the integrand in Eq. (9) in ϵ, keeping the terms of the zeroth and first order in ϵ,

$$\gamma(n) = \frac{1}{\pi} \int_0^{M_B/2} dT \, T^2 M_B^l (M_B - 2T)^l \left[M_B(M_B - 2T) \int_0^{M_B - 2T} d\epsilon \, w_1 - 2(l+1)(M_B - T) \int_0^{M_B - 2T} d\epsilon \, \epsilon w_1 \right]$$

$$+ \frac{1}{3\pi} \int_0^{M_B/2} dT \, T^4 M_B^{l-1}(M_B - 2T)^{l-1} \left[M_B(M_B - 2T) \int_0^{M_B - 2T} d\epsilon \, w_2 - 2l(M_B - T) \int_0^{M_B - 2T} d\epsilon \, \epsilon w_2 \right]. \tag{10}$$

The integrals over ϵ are given by the sum rules mentioned above,

$$\frac{1}{2\pi} \int d\epsilon \, w_1 = 1, \quad \frac{1}{2\pi} \int d\epsilon \, \epsilon w_1 = M_B - m_b = \overline{\Lambda}, \tag{11}$$

and

$$\frac{1}{2\pi} \int d\epsilon \, w_2 = \frac{2m_b}{T}, \quad \frac{1}{2\pi} \int d\epsilon \, \epsilon w_2 = \frac{2m_b}{T}\overline{\Lambda}. \tag{12}$$

Expressions (11) and (12) imply that the integration over ϵ is saturated at small ϵ, of order $\overline{\Lambda}$. After substituting the sum rules in Eq. (10) and integrating over T we arrive at

$$\gamma(n) = \frac{M_B^{2l+5}}{2(l+4)(l+3)(l+2)} \left[1 - (2l+5)\frac{\overline{\Lambda}}{M_B} \right] + \frac{m_b M_B^{2l+4}}{2(l+4)(l+3)(l+2)(l+1)} \left[1 - (2l+4)\frac{\overline{\Lambda}}{M_B} \right]. \tag{13}$$

The first term here comes from w_1 and the second from w_2. This expression explicitly demonstrates that the meson mass is substituted by the quark mass at the level of $1/m$ terms we consider: $M_B(1 - \Lambda/M_B) = m_b$.

Thus, the cancellation of the infrared contribution is evident. It is worth emphasizing that the infrared contributions we speak of need not necessarily be of a nonperturbative nature. The cancellation of infrared $1/m_b$ effects takes place for perturbative contributions as well, as long as they correspond to sufficiently small momenta, $\lesssim m_b/n$.

Let us parenthetically note that Eq. (13) is valid for any l; in particular, we can put l equal to zero. One may wonder how we got, in this case, a nonvanishing correction associated with $\int d\epsilon \, \epsilon w_2$, while the weight factor for w_2 in the original integrand (9) at $l=0$ seemingly has no dependence on ϵ at all. This case is actually singular: the domain of $T \rightarrow M_B/2$ contributes to the integral at the level Λ_{QCD}/M_B. For such T, the interval of integration over ϵ shrinks to zero, but nevertheless, the integral of w_2 stays finite (unity) according to Eq. (12); this signals a singularity. It is directly seen from Eq. (10) where at $l=0$ one observes logarithmic divergence at $T \rightarrow M_B/2$. The simplest way to deal with w_2 at $l=0$, is to insert a step function $\theta(M_B - 2T - \epsilon)$ in the integrand instead of the upper limits of integration. This step function does provide an ϵ dependence, and the term linear in ϵ is $\epsilon \delta(M_B - 2T)$. At nonvanishing values of l the δ function above produces no effect. As usual, analytic continuation from the nonsingular case $l > 0$ to $l=0$ leads to the same result.

In principle, it is instructive to trace the l dependence of other nonperturbative terms, e.g., those proportional to μ_π^2 and μ_G^2 which are known for $b \rightarrow u$ at arbitrary l from explicit calculations of Refs. [4–6], and have a very simple form. We will not dwell on this issue here.

Our prime concern in this work is the interplay of the infrared effects in the perturbative corrections. The main new element appearing in the analysis of the perturbative corrections is the fact that the integrals over ϵ are not saturated in the domain $\epsilon \sim \Lambda$. That is why we need to introduce the normalization point μ which will divide the entire range of the ϵ integration into two domains, $\epsilon > \mu$ and $\epsilon < \mu$.

The domain $\epsilon < \mu$ provides us with a clear-cut physical definition of Λ, which becomes μ dependent,

$$\overline{\Lambda}(\mu) + \frac{\mu_G^2}{2|\vec{q}|} + \frac{\mu_\pi^2 - \mu_G^2}{3|\vec{q}|}\left(1 - \frac{|\vec{q}|}{m_b}\right) + O\left(\frac{\Lambda_{\text{QCD}}^3}{\vec{q}^2}\right)$$

$$= \frac{\int_0^\mu d\epsilon \, \epsilon w_1^{b \rightarrow u}(\epsilon, |\vec{q}|)}{\int_0^\mu d\epsilon \, w_1^{b \rightarrow u}(\epsilon, |\vec{q}|)}, \qquad (14)$$

where the variable ϵ is defined by Eq. (8) with M_D set equal to zero, the second variable $T = |\vec{q}|$ in the $b \rightarrow u$ transition. Equation (14) is a consequence of the sum rules derived in Ref. [14]. There, the integrals over the spectral densities are given in the form of a "condensate" expansion. The currents inducing given spectral densities (and the spectral densities themselves) acquire certain normalization factors due to short-distance renormalization of the weak vertex, which are irrelevant for our present purposes. To get rid of these normalization factors we consider the ratio of the sum rules. The power corrections to the denominator in the right-hand side of Eq. (14) shows up only in $O(\Lambda_{\text{QCD}}^3/\vec{q}^2)$ terms.

The quark mass, $m_b(\mu) = M_B - \Lambda(\mu)$, becomes in this way a well-defined and experimentally measurable quantity. A very similar definition of $\Lambda(\mu)$ also exists in the $b \rightarrow c$ transition, through the integral over $w_1^{b \rightarrow c}$, see Ref. [14]. This definition generalizes the Voloshin sum rule [15] to the arbitrary values of $\vec{q}$, and includes corrections $O(\Lambda_{\text{QCD}}^2/m)$. See Sec. IV for further discussion of the $\Lambda(\mu)$ definition.

B. The large-n limit

Now that we understand how the infrared parts cancel, we are ready to address the practical issue of the particular normalization point μ which is convenient to use in the analysis of the total widths. Certainly, the theoretical predictions can be given for any value of $\mu \gg \Lambda_{\text{QCD}}$. Our goal is to choose μ in such a way that the domain of momenta above μ does not give the enhanced perturbative corrections. Then all perturbative contributions enhanced by large n will be automatically included in the definition of $m_b(\mu)$.

To this end we invoke the large-n limit, as was explained in the Introduction. For $l \gg 1$ one has

$$(\Delta^2 - 2M_B T - 2\Delta \epsilon + 2T\epsilon + \epsilon^2)^l$$

$$\approx \Delta^{2l} \exp\left[-2l\left(\frac{TM_B}{\Delta^2} + \frac{\epsilon}{\Delta} - \frac{T\epsilon}{\Delta^2} - \frac{\epsilon^2}{2\Delta^2}\right)\right],$$

and the dominant domains in the integration variables are given by

$$T \lesssim \frac{1}{l}\frac{\Delta^2}{M_B}, \qquad \epsilon \lesssim \frac{1}{l}\Delta.$$

All expressions then simplify, and the two integrations decouple

$$\gamma(n) = \frac{1}{8\pi}\frac{\Delta^{8+2l}}{M_B^3}\left\{\int_0^\infty d\tau \, \tau^{1/2}(\tau + 2\tau_0)^{1/2}(\tau + \tau_0)e^{-(l+1)\tau}\int_0^\infty d\epsilon \, e^{-2(l+1)(\epsilon/\Delta)}w_1\left(\epsilon, \frac{\tau\Delta^2}{2M_B}\right)\right.$$

$$\left. + \frac{\Delta^2}{12M_B^2}\int_0^\infty d\tau \, \tau^{3/2}(\tau + 2\tau_0)^{3/2}(\tau + \tau_0)e^{-l\tau}\int_0^\infty d\epsilon \, e^{-2l(\epsilon/\Delta)}w_2\left(\epsilon, \frac{\tau\Delta^2}{2M_B}\right)\right\}, \qquad (15)$$

where

$$\tau \equiv \frac{2 M_B T}{\Delta^2}, \qquad \tau_0 \equiv \frac{2 M_D M_B}{\Delta^2}.$$

The integrals over the excitation energy ϵ

$$J_i(T;\sigma) \equiv \frac{1}{2\pi} \int_0^\infty d\epsilon \, e^{-\epsilon/\sigma} w_i(\epsilon, T) \tag{16}$$

are a combination of the sum rule considered in Refs. [16,14]. They are related to the Borel transform of the forward scattering amplitude of the weak current off the B meson [17]. The quantity σ is the Borel parameter, and expanding $J_i(T;\sigma)$ in $1/\sigma$ gives us the first, second, and so on sum rules of Ref. [14]. The Borelized version has an advantage, however, of providing an upper cutoff of ϵ in a natural way. This is important in the analysis of the perturbative corrections; the second argument of J defines the normalization point.

It is apparent from Eq. (15) that only a fraction $\sim 1/n$ of the total energy release is fed into the final hadronic system. The three-momentum carried by the final state hadrons is likewise small, namely of order $\sqrt{m_c \Delta/n}$ for $b \to c$ and Δ/n for $b \to u$, respectively. The latter case is obtained from the former by setting $M_D = 0$. This implies that by evaluating the various quantities at a scale $\sim \Delta/n$ rather than the Δ one includes the potentially large higher-order corrections. These conclusions can be clarified by considering two limiting cases.

(a) If m_c stays fixed, the limit $n \to \infty$ leads to the SV regime, as is expressed by $v \sim (m_b - m_c)/n m_c \ll 1$. Neglecting τ compared to τ_0, one gets

$$\gamma(n) = \Delta^{2l+5} \left(\frac{M_D}{M_B} \right)^{3/2} \int_0^\infty d\tau \, \tau^{1/2} e^{-(l+1)\tau} J_1\left(\frac{\tau \Delta^2}{2 M_B}; \frac{\Delta}{2(l+1)} \right) + \frac{1}{3} \Delta^{2l+5} \left(\frac{M_D}{M_B} \right)^{5/2} \int_0^\infty d\tau \, \tau^{3/2} e^{-l\tau} J_2\left(\frac{\tau \Delta^2}{2 M_B}; \frac{\Delta}{2l} \right). \tag{17}$$

The behavior of Γ_{sl} at $n \to \infty$ then is determined by the integrals $J_{1,2}$ near the zero argument. The second term is clearly subleading because the weight function contains an extra power of τ. It is not difficult to see that the axial current contribution is dominant, and one obtains for the reduced width

$$\gamma(n)|_{n \to \infty} \simeq \Delta^n \sqrt{2\pi} \left(\frac{M_D}{M_B} \right)^{3/2} \frac{1}{n^{3/2}} \xi_A\left(\frac{\Delta}{n} \right), \tag{18}$$

where $\xi_A(\mu) = 1 + O[\alpha_s(\mu)]$ is the coefficient function of the leading operator in the first sum rule for the axial current at zero recoil (for further details see Ref. [14]). For the degree of accuracy pursued here (power corrections are switched off so far), one has $\xi_A \simeq \eta_A^2$ where the factor η_A incorporates the perturbative corrections to the axial current at zero recoil.[2] Equation (18) provides a decent approximation of the true width even for $n = 5$. The vector current contribution is subleading; to take it into account at $n = 5$ we need a refined expansion to be described in the Appendix.

We note that the leading-n expansion of the width yields $(M_B - M_D)^n$ which *differs* from $(\overline{m}_b - \overline{m}_c)^n$ by the targeted terms $(n\alpha_s)^k$, but coincides with this accuracy with $[m_b(\mu) - m_c(\mu)]^n$ if $\mu \lesssim m_b/n$,

$$m_b(\mu) - m_c(\mu)$$

$$\simeq (M_B - M_D) + (m_b - m_c) \frac{\mu_\pi^2(\mu) - \mu_G^2(\mu)}{2 m_b m_c} + \cdots$$

$$\simeq (M_B - M_D)\left(1 - \frac{4\alpha_s}{3\pi} \frac{\mu^2}{2 m_b m_c} + \cdots \right). \tag{19}$$

In the last line we used the perturbative expression for $\mu_\pi^2(\mu) - \mu_G^2(\mu)$.

(b) Another case of interest is $m_q \lesssim m_b/n$, when the final state quark is ultrarelativistic. In this case, the axial and vector currents contribute equally. The reduced width now takes the form

$$\gamma(n) = \frac{1}{4} M_B^n \int_0^\infty d\tau \, \tau^2 e^{-(l+1)\tau} J_1\left(\frac{\tau M_B}{2}; \frac{M_B}{2(l+1)} \right). \tag{20}$$

The quantity $J_1(\tau M_B/2)$ stays finite (unity at the tree level) at $\tau \to 0$. The second structure function, again, formally yields only subleading in $1/n$ contributions. Thus

$$\gamma_n(b \to u)|_{n \to \infty} \simeq M_B^n \frac{4}{n^3} \xi_u(M_B/n), \tag{21}$$

where

$$\xi_u(n) = J_1\left(\frac{M_B}{n}; \frac{M_B}{n} \right), \qquad \xi_u^{\text{tree}} = 1. \tag{22}$$

This width decreases faster with n than for massive quarks due to the higher power of $|\vec{q}| \sim m_b/n$. This underlies the fact that the numerical factor in Γ_{sl} in front of Δ^5 is much smaller for $b \to u$ than in the SV limit, namely, $1/192$ vs

---

[2] The quantity $\eta_A^2 \equiv \lim_{\mu \to 0} \xi_A^{\text{pert}}(\mu)$ is ill defined once power corrections are addressed; at the same time $\xi_A(\mu)$ is even then a well-defined quantity provided that $\mu \gg \Lambda_{\text{QCD}}$. Yet, to any finite order in perturbation theory, one can work with η_A.

1/15. For the same reason, the simple expansion described above provides a poor approximation for $b \rightarrow u$ when evaluated at $l = 0$, i.e., $n = 5$.

(c) One can consider the third large-n regime when $m_c/m_b \sim 1/n$. Although it appears to be the closest to the actual situation, we would not dwell on it here: the analysis goes in the very same way, but expressions are more cumbersome since one has to use the full relativistic expression for the kinetic energy of the D meson. No specific new elements appear in this case.

(d) Let us discuss now the normalization point for quark masses at large n. Equations (18) and (22) show the kinematically generated dependence on the masses. The nontrivial hadronic dynamics are encoded in the factors ξ_A and ξ_u. Our focus is the heavy quark masses. In the SV case (a), the dependence on the masses is trivial as long as low-scale masses are employed. In the case of the $b \rightarrow u$ transitions, we showed in Sec. II A, that the factor $\xi_u(M_B/n)$ actually converted M_B^n into $m_b(\mu)^n$, and, thus, effectively established the normalization point for the latter, $\mu \lesssim m_b/n$. Using the effective running mass $m_b(\mu \sim m_b/n)$ does not incorporate, however, the domain of the gluon momenta above $\sim m_b/n$. This contribution has to be explicitly included in the perturbative corrections. It is not enhanced by powers of n. Physically, it is nothing but a statement that for $\Gamma(b \rightarrow u)$ the proper normalization scale *for masses* is $\mu \sim m_b/n$.

We hasten to emphasize that the statement above is *not* a question of normalization of α_s used in the perturbative calculations. One can use α_s at any scale to evaluate $m_b(\mu)$ as long as this computation has enough accuracy. The fact that using inappropriate α_s can lead to an apparent instability in $m_b(\mu)$, is foreign to the evaluation of the width.

Let us summarize the main features which are inferred from analysis of the straightforward large-n expansion of inclusive widths introduced via Eq. (15). The integral over the lepton phase space carries the main dependence on n. The QCD corrections—the real theoretical challenge—are only indirectly sensitive to n, for kinematics determines the energy and momentum scales at which the hadronic part has to be evaluated.

The characteristic momentum scale μ for the inclusive decay width is smaller than the naive guess $\mu \sim m_b$. In the large-n limit, it scales like m_b/n. This momentum defines the relevant domain of integration of the structure function in q_0, i.e., the scale at which the forward transition amplitude appears in the total decay rate. In particular, by evaluating the quark masses that determine the phase space at this scale $\sim m_b/n$, one eliminates the strongest dependence of the radiative corrections on n.

The expansion in n derived from Eq. (15) allows a transparent discussion of the underlying physics. Unfortunately it yields, as already stated, a decent numerical approximation only for very large values of n; i.e., for $n = 5$, the nonleading terms are still significant. Not all those terms are dominated by kinematical effects, and their treatment poses nontrivial problems. A more refined expansion can be developed that effectively includes the kinematics-related subleading contributions and leads to a good approximation already for $n = 5$. This treatment, however, is rather cumbersome and less transparent. It will be briefly described in the Appendix. The purpose is only to demonstrate that the essential features of our expansion leading to the qualitative picture we rely on— that the characteristic momentum scale is essentially lower than m_b—hold already for the actual case $n = 5$. The reader who is ready to accept this assertion, can skip the technicalities of the refined 1/n expansion in the Appendix.

III. APPLICATIONS

We now briefly consider the consequences of the 1/n expansion for a few problems of interest.

A. The extended SV limit

In many respects, one observes that the inclusive $b \rightarrow c$ decays seem to lie relatively close to the SV limit. Most generally, the various characteristics of the decay depend on the ratio m_c^2/m_b^2; although it is rather small, the actual characteristics often do not differ much from the case where it approaches 1, when the average velocity of the final state c quark was small. The proximity of the inclusive $b \rightarrow c$ decays to the SV limit has two aspects. First, it is obvious that the nonperturbative effects work to suppress the effective velocity of the final state hadrons; the most obvious changes are kinematical replacement $m_b \rightarrow M_B$ and $m_c \rightarrow M_{D,D^*,D^{**}} \ldots$ when passing from quarks to actual hadrons. The impact of increasing the mass is much more effective for the final state charm than for the initial state beauty. This decreases the velocities of the final state hadrons significantly at $\Delta = m_b - m_c \approx 3.5$ GeV. Yet this simple effect would not be numerically large enough were it not considerably enhanced by the properties of the lepton phase space, which can be easily seen comparing it, for example, with the one for semileptonic decays at fixed $q^2 = 0$.

On the other hand, without any nonperturbative effects, theoretical expressions at the purely parton level are known to favor the SV kinematics even for actual quark masses. The simplest illustration is provided by the tree-level phase space, z_0:

$$z_0(m_b, m_c) = m_b^5 \left(1 - 8 \frac{m_c^2}{m_b^2} - 12 \frac{m_c^4}{m_b^4} \ln \frac{m_c^2}{m_b^2} + 8 \frac{m_c^6}{m_b^6} - \frac{m_c^8}{m_b^8} \right). \tag{23}$$

It is most instructive to analyze the sensitivity of this expression to m_b and $\Delta = m_b - m_c$ rather than m_b and m_c, as expressed through

$$\kappa_\Delta \equiv \frac{\Delta}{z_0} \frac{\partial z_0(m_b, \Delta)}{\partial \Delta}, \quad \kappa_b \equiv \frac{m_b}{z_0} \frac{\partial z_0(m_b, \Delta)}{\partial m_b}, \tag{24}$$

where

$$\kappa_\Delta + \kappa_b = 5.$$

In the light quark limit—$m_c^2/m_b^2 \rightarrow 0$—one has $z_0|_{\text{light}} = m_b^5 [1 - O(m_c^2/m_b^2)]$ and, thus,

$$\kappa_\Delta|_{\text{light}} = 0, \quad \kappa_b|_{\text{light}} = 5. \tag{25}$$

In the SV limit, on the other hand, one finds

4024   I. BIGI, M. SHIFMAN, N. URALTSEV, AND A. VAINSHTEIN   <u>56</u>

$$z_0|_{SV} \approx \frac{64}{5}(m_b - m_c)^5. \qquad (26)$$

Therefore, the SV limit is characterized by

$$\kappa_\Delta|_{SV} = 5, \quad \kappa_b|_{SV} = 0. \qquad (27)$$

For actual quark mass values, $m_c/m_b \approx 0.28$, one finds

$$\kappa_\Delta \approx 3, \quad \kappa_b \approx 2, \qquad (28)$$

i.e., even for $m_c^2/m_b^2 \approx 0.08 \ll 1$ one is still closer to the SV than the light-quark limit. The "half-way" point, $\kappa_\Delta = \kappa_b = 2.5$, lies at $m_c^2/m_b^2 \approx 0.05$. This is a consequence of the large parameter $n = 5$. A similar pattern persists also for the low-order perturbative corrections [18–19], and on the level of the $1/m^2$ power corrections [4,6,7].

The relevance of the SV approximation in the actual inclusive decays thus finds a rational explanation in the $1/n$ expansion.

B. Resummation of large perturbative corrections

In computing the perturbative corrections for large-n one can potentially encounter large n-dependent corrections of the type $(n\alpha_s/\pi)^k$, which makes them sizable for the actual case $n = 5$. The related practical concern of certain numerical ambiguity of the one-loop calculations of the widths was raised in Ref. [20]. Fortunately, the leading subseries of these terms can readily be summed up. The summation essentially amounts to using the quark masses normalized at a low scale, $\mu \sim (m_b, \Delta)/n$. As long as one does not go beyond a relatively low order in α_s one can ignore renormalon divergences and simply use the "kth-order pole mass." Quite obviously, the terms $\sim (n\alpha_s/\pi)^k$ appear only if one uses deep Euclidean masses. If the result is expressed in terms of low-scale Euclidean masses (i.e., those normalized at $\mu \sim \Delta/n$ or $\mu =$ several units$\times \Lambda_{QCD}$) the terms $\sim (n\alpha_s/\pi)^k$ enter with coefficients proportional to μ/Δ and, therefore, contribute only on the subleading level.

The above assertion is most easily inferred by applying Eq. (15) to w_i computed through order k (with $k < n$); they are self-manifest in Eqs. (18) and (22). Upon using the thresholds in the perturbative w_i to define m_b and m_c, the moments of w_i are smooth and independent of n in the scaling limit $m, \Delta \sim n$, which ensures the absence of the leading power of n. The perturbative thresholds, on the other hand, are given just by the pole masses as they appear in the perturbation theory to the considered order.

It is worth clarifying why these terms emerge if one uses different masses. The reason is that the moments of the structure functions (or more general weighted averages in the "refined" expansion) intrinsically contain the reference point for the energy q_0 and q_{max}^2. At $q^2 > q_{max}^2$ the integral over q_0 merely vanishes whereas for $q^2 < q_{max}^2$ it is unity plus corrections. A similar qualification applies to the evaluation of the integral over the energy q_0 itself. These properties which single out the proper 'on-shell' masses, were tacitly assumed in carrying out the expansion in $1/n$.

On the formal side, if one uses a mass other than the perturbative pole mass, the perturbative w_i contain singular terms of the type

$$\delta(\epsilon - \delta m) = \delta(\epsilon) + \sum_{k=1}^{\infty} (-\delta m)^k \frac{\delta^{(k)}(\epsilon)}{k!}, \qquad (29)$$

where $\delta m \sim \alpha_s \cdot \Delta$ is the residual shift in the energy release. (A similar shift in the argument of the step functions emerges for the perturbative continuum contributions.) Each derivative of the δ function generates, for example, a power of n in the integral over ϵ in the representation (15). For $b \to u$ widths considered in detail in Sec. II, using $m_b(m_b)$, would yield $-\delta m \approx (4\alpha_s/3\pi) m_b$ leading to the series $(4n\alpha_s/3\pi)^k/k!$ which sums into $e^{n4\alpha_s/3\pi}$ and converts $m_b(m_b)^n$ into $m_b^n(\mu)$ with $\mu \ll m_b$.

A more detailed analysis can be based on the refined $1/n$ expansion described in the Appendix, Eqs. (A2)–(A4); it will be presented elsewhere. For our purposes here it is enough to note that using masses normalized at a scale $\sim \Delta/n$ does not generate terms $\sim (n\alpha_s/\pi)^k$.[3]

A qualifying comment is in order. Including higher-order terms in the expansions in $1/n$ and in α_s, one cannot find a universal scale that is appropriate for *all* perturbative corrections simultaneously, including the normalization of the strong coupling. Even upon resumming all n-dependent effects, one would end up with the ordinary corrections to the structure functions, e.g., those describing the short-distance renormalization of the vector and axial currents. The relevant scale for such effects is obviously $\sim m_b$; we are not concerned about them here since they are not enhanced by the factor n and, thus, not large.

C. Power corrections and duality in inclusive widths

Using n as an expansion parameter one can estimate the importance of the higher-order nonperturbative corrections by summing up the leading terms in n. The simple estimates show, however, that these effects do not exceed a percent level, and thus are not of practical interest.[4] It is still instructive to mention a qualitative feature drawn from this analysis.

The characteristic momentum scale decreases for large n; at $\Delta/n \sim \Lambda_{QCD}$ one is not in the short-distance regime anymore. Correspondingly, the overall theoretical accuracy of the calculations naturally seems to deteriorate with increasing n (and/or increasing m_c up to m_b). However, using the sum rules derived in [14] it is not difficult to show that even in the limit $n \to \infty$ the width is defined by short-distance dynamics unambiguously up to $1/m_Q^2$ terms. Moreover, in this

---

[3]It is important that here we consider large n for a given order in the perturbative expansion.

[4]The $b \to c\tau\nu$ decays represent a special case: due to the sizable τ mass the energy release is smaller than in $b \to c\mu\nu$ and $b \to ce\nu$ and a SV scenario is realized in a rather manifest way. The n-enhanced nonperturbative corrections can then reach the 10% level, but are readily resummed using the exact meson masses in the phase space factors.

limit a straightforward resummation of the leading nonperturbative effects yields the width in terms of the zero-recoil form factor $|F_{D^*}|^2$ ($|F_D|^2$) and the physical masses M_B and $M_{D^{(*)}}$, without additional uncertainties. Thus, in the *worst* limit for inclusive calculations, the theoretical computation of the width merely reduces to the calculation of the exclusive $B \to D^*$ transition, the best place for application of heavy quark symmetry [12,21].

The large-n arguments may seem to suggest that the width can depend on the meson masses rather than on the quark masses, in contradiction to the OPE. In fact, since in the $b \to c$ transitions one encounters the SV limit when $n \to \infty$, the leading term is given by $(M_B - M_D)^n$ which differs from $(m_b - m_c)^n$ only by $O(\Lambda^2_{\rm QCD}/m_Q^2)$, in accordance with the OPE. The problem can emerge in the next-to-leading order in $1/n$. These corrections can be readily written up in terms of the transition form factors at small velocity transfer, and take the form of the sums of the transition probabilities which one encounters in the small velocity sum rules [14]. The latter ensure that the terms $\sim \Lambda_{\rm QCD}/m_Q$ are absent. By the same token, using the third sum rule, one observes that the leading-n term, $n \mu_\pi^2/m_Q^2$, is absent from the width, although it is obviously present in $(M_B - M_D)^n$ (we assume that the quark masses are fixed as the input parameters). Similar properties hold to all orders in $1/n$, however, they are not very obvious in this expansion.

If the infrared part of the quark masses cancels in the widths, why is it not the short-distance masses like $\overline{\rm MS}$ that emerge? The answer is clear: the excited final states in the width enter with the weight $\sim \exp(-n\epsilon/\Delta)$, and the Coulomb part of the mass cancels in the width only up to momenta $\sim \Delta/n$.

The above consideration only interprets the known OPE results. The general question at which scale duality is violated is more instructive. We immediately see that, generically, the characteristic momenta are given by the scale m_Q/n rather than by m_Q. For example, considering the calculation of the relevant forward transition amplitude for $b \to u$ in coordinate space [4,22], we have

$$T \sim 1/x^{n+4}. \tag{30}$$

Its Fourier transform $\sim p^n \ln p^2$ (+ terms analytic in p^2) is saturated at $|x| \sim r_0 \sim p/n$, corresponding to a time separation $t_0 \sim n/m_Q$. In the case of the massive final state quarks, one has $t_0 \sim n/\Delta$.

In the semileptonic $b \to c$ decays, one may naively conclude that the scale is too low. However, here the heavy quark symmetry ensures vanishing of all leading corrections even at $\Delta \to 0$, i.e., in the SV limit. This applies as well to the ESV limit. The nontrivial corrections appear only at the $1/\Delta^2$ or $n/(m\Delta)$ level and they are still small.

The situation is, in principle, different in the nonleptonic decays, where the color flow can be twisted (instead of b to c, it can flow from b to d). In the nonleptonic $b \to c u \bar{d}$ decay, with $\Delta \simeq 3.5$ GeV, one is only marginally in the asymptotic freedom domain and *a priori* significant nonperturbative corrections as well as noticeable violations of duality could be expected. In this respect it is more surprising that so far no prominent effects have been identified on the

theoretical side [22].[5] We note that a particular model for the violation of duality in the inclusive decays suggested in Ref. [22] explicitly obeys the large-n scaling above. The importance of the duality-violating terms is governed by the ratio Δ/n,

$$\frac{\Gamma_{\rm dual.\ viol.}}{\Gamma_{\rm tree}} \sim \exp\left[-\Delta \cdot \rho\left(1 - \frac{n}{\Delta \cdot \rho} \ln \frac{n}{\Delta \cdot \rho}\right)\right], \tag{31}$$

with ρ being a hadronic scale parameter. The corrections blow up when $\Delta/n \sim \mu_{\rm had}$. We hasten to remind the reader that this consideration is only qualitative, since we do not distinguish between, say $n = 5$ and $n - 2 = 3$. The arguments based on the refined expansion in the ESV case suggest that the momentum scale is governed by an effectively larger energy than is literally given by Δ/n with $n = 5$.

IV. LOW-SCALE HEAVY QUARK MASS; RENORMALIZED EFFECTIVE OPERATORS

Addressing nonperturbative power corrections in b decays via Wilson's OPE one must be able to define an effective nonrelativistic theory of a heavy quark normalized at a scale $\mu \ll m_Q$; the heavy quark masses in this theory (which enter, for example, equations of motion of the Q fields) are low-scale (Euclidean) masses $m_Q(\mu)$. Moreover, even purely perturbative calculations benefit from using them, as was shown above: it allows one to solve two problems simultaneously—resum the leading n-enhanced terms and avoid large but irrelevant IR-renormalon-related corrections. For the perturbative calculations *per se*, therefore, the question arises of what is the proper and convenient definition of a low-scale Euclidean mass? Of course, one can always define it through an off-shell quark propagator. Generically this will lead to a mass parameter which is not gauge invariant. This feature, though not being a stumbling block, is often viewed as an inconvenience.

Usually, one works with the mass defined in the $\overline{\rm MS}$ scheme, which is computationally convenient. However, due to the well-known unphysical features of this scheme it is totally meaningless at the low normalization point μ = several units $\times \Lambda_{\rm QCD}$. For example, if we formally write the $\overline{\rm MS}$ mass at this point

$$m_Q^{\overline{\rm MS}}(\mu) \simeq m_Q^{\overline{\rm MS}}(m_Q)\left(1 + \frac{2\alpha_s}{\pi} \ln \frac{m_Q}{\mu}\right) \tag{32}$$

we get large logarithms $\ln m_b/\mu$ which do not reflect any physics and do not exist under any reasonable definition of the heavy quark mass.

Thus, one must find another definition. We suggest the most physical one motivated by the spirit of the Wilson's OPE, i.e., the mass which would be seen as a "bare" quark mass in the completely defined effective low-energy theory.

---

[5]*A priori* the effect of the chromomagnetic interaction in order $1/m_Q^2$ could have been significant, but it gets suppressed due to the specific chiral and color structure of weak decay vertices [4], properties that are external to QCD.

The basic idea is to follow the method suggested in Ref. [14], and is based on the consideration of the heavy flavor transitions in the SV kinematics, when only the leading term in the recoil velocity, $\sim \vec{v}^2$, is kept. Note that in Sec. II A we have used for a similar purpose a different kinematical limit when the final particle is ultrarelativistic.

Let us consider the heavy-quark transition $Q \to Q'$ induced by some current $J_{QQ'} = Q'\Gamma Q$ in the limit when $m_{Q,Q'} \to \infty$ and $\vec{v} = \vec{q}/m_{Q'} \ll 1$ is fixed. The Lorentz structure Γ of the current can be arbitrary with the only condition that $J_{QQ'}$ has a nonvanishing tree-level nonrelativistic limit at $\vec{v}=0$. The simplest choice is also to consider $Q'=Q$ (which, for brevity of notation, is assumed in what follows). Then one has [14]

$$\overline{\Lambda}(\mu) \equiv \lim_{m_Q \to \infty} [M_{H_Q} - m_Q(\mu)]$$

$$= \lim_{\vec{v}\to 0} \lim_{m_Q \to \infty} \frac{2}{\vec{v}^2} \frac{\int_0^\mu d\epsilon\, \epsilon\, w(\epsilon, m_Q v)}{\int_0^\mu d\epsilon\, w(\epsilon, m_Q v)}, \qquad (33)$$

where $w(\epsilon, |\vec{q}|)$ is the structure function of the hadron (the probability to excite the state with the mass $M_{H_Q} + \epsilon$ for a given spatial momentum transfer $\vec{q}$); M_{H_Q} is the mass of the initial hadron.

The OPE and the factorization ensures the following.

(a) For the given initial hadron the above $\overline{\Lambda}(\mu)$ does not depend on the chosen weak current.

(b) The value $m_Q(\mu) = M_{H_Q} - \overline{\Lambda}(\mu)$ does not depend on the choice of the hadron H_Q. This holds if μ is taken above the onset of duality.

To define perturbatively the running heavy quark mass one merely considers Eq. (33) in the perturbation theory, to a given order:

$$m_Q(\mu) = [M_{H_Q}]_{\text{pert}} - [\overline{\Lambda}(\mu)]_{\text{pert}}. \qquad (34)$$

In the perturbation theory H_Q is the quasifree heavy quark state and M_{H_Q} is the pole mass m_Q^{pole} to this order; $w(\epsilon)$ at $\epsilon > 0$ is a perturbative probability to produce Q and a number of gluons in the final state which belongs to the perturbative continuum. It is a continuous function [plus $\delta(\epsilon)$ at $\epsilon = 0$]. Both numerator and denominator in the definition of $\overline{\Lambda}$ are finite. All quantities are observables and, therefore, are manifestly gauge invariant. For example, to the first order

$$[\overline{\Lambda}(\mu)]_{\text{pert}} = \frac{16}{9}\frac{\alpha_s}{\pi}\mu, \quad m_Q(\mu) = (m_Q^{\text{pole}})_{\text{one-loop}} - \frac{16}{9}\frac{\alpha_s}{\pi}\mu,$$

$$m_Q(\mu) = m_Q^{\overline{\text{MS}}}(m_Q)\left(1 + \frac{4}{3}\frac{\alpha_s}{\pi} - \frac{16}{9}\frac{\alpha_s}{\pi}\frac{\mu}{m_Q}\right). \qquad (35)$$

The IR renormalon-related problems encountered in calculation of m_Q^{pole} cancel in Eq. (34) against the same contribution entering $\overline{\Lambda}(\mu)$ in Eq. (33).

Similarly, one can define matrix elements of heavy quark operators renormalized at the point μ [14,23,24]. The zeroth moment of the structure function is proportional to $\langle H_Q | \overline{Q}Q | H_Q \rangle / (2 M_{H_Q})$ and to the short-distance renormal-

ization factor. In the limit $m_Q \to \infty$ this matrix element equals unity up to terms μ^2/m_Q^2. Then it is convenient to define matrix elements of higher dimension operators in terms of the ratios (for which all normalization factors go away):

$$\mu_\pi^2(\mu) = \frac{\langle H_Q | \overline{Q}(i\vec{D})^2 Q | H_Q \rangle_\mu}{\langle H_Q | \overline{Q}Q | H_Q \rangle_\mu}$$

$$= \lim_{\vec{v}\to 0} \lim_{m_Q \to \infty} \frac{3}{\vec{v}^2} \frac{\int_0^\mu d\epsilon\, \epsilon^2 w(\epsilon, m_Q v)}{\int_0^\mu d\epsilon\, w(\epsilon, m_Q v)}, \qquad (36)$$

$$\rho_D^3(\mu) = -\frac{1}{2} \frac{\langle H_Q | \overline{Q}(\vec{D}\vec{E})Q | H_Q \rangle_\mu}{\langle H_Q | \overline{Q}Q | H_Q \rangle_\mu}$$

$$= \lim_{\vec{v}\to 0} \lim_{m_Q \to \infty} \frac{3}{\vec{v}^2} \frac{\int_0^\mu d\epsilon\, \epsilon^3 w(\epsilon, m_Q v)}{\int_0^\mu d\epsilon\, w(\epsilon, m_Q v)}, \qquad (37)$$

and so on. All these objects are well defined, gauge invariant, and do not depend on the weak-current probe.

Returning to the heavy quark mass, the literal definition given above based on Eq. (34) is most suitable for $\mu \ll m_Q$ when the terms $\sim (\alpha_s/\pi)(\mu^2/m_Q)$ are inessential. One can improve it including higher-order terms in μ/m_Q, if necessary, in the following general way.

Consider the $1/m_Q$ expansion of the heavy hadron mass

$$(M_{H_Q})_{\text{spin av}} = m_Q(\mu) + \overline{\Lambda}(\mu) + \frac{\mu_\pi^2(\mu)}{2m_Q(\mu)} + \frac{\rho_D^3(\mu) - \rho^3(\mu)}{4m_Q^2(\mu)}$$

$$+ \cdots . \qquad (38)$$

We took the average hadronic mass over the heavy quark spin multiplets here. Besides the μ_π^2 and ρ_D^3 defined above, the new quantity ρ^3 enters the $1/m_Q^2$ term. It is a nonlocal correlator of the two operators $\overline{Q}(\vec{\sigma}i\vec{D})^2 Q$ defined in Eq. (24) of [14]. We apply the relation (38) to the perturbative states and get

$$m_Q(\mu) = m_Q^{\text{pole}} - [\overline{\Lambda}(\mu)]_{\text{pert}} - \frac{[\mu_\pi^2(\mu)]_{\text{pert}}}{2m_Q(\mu)}$$

$$- \frac{[\rho_D^3(\mu)]_{\text{pert}} - [\rho^3(\mu)]_{\text{pert}}}{4m_Q^2(\mu)} - \cdots . \qquad (39)$$

For example, through order $1/m_Q$ one has

$$[\mu_\pi^2(\mu)]_{\text{pert}} \simeq \frac{4\alpha_s}{3\pi}\mu^2. \qquad (40)$$

To renormalize the nonlocal correlators entering the definition of ρ^3, one represents them as a QM sum over the intermediate states and cuts it off at $E_n \geq M_{H_Q} + \mu$. For example, $[\rho^3(\mu)]_{\text{pert}} \simeq (8\alpha_s/9\pi)\mu^3$. One subtlety is worth mentioning here: including higher orders in $1/m_Q$ and going beyond one-loop perturbative calculations one must be careful calculating the coefficient functions in the expansion Eq. (38).

The general definition above is very transparent physically and most close to the Wilson's procedure. It corresponds to considering the QM nonrelativistic system of slow heavy flavor hadrons and literally integrating out the modes with $\omega = E - m_Q > \mu$. Considering the SV limit and keeping only those terms which are linear in $\vec{v}^2$ is important. In this case, no problem arises with choosing the reference energy in ϵ or in imposing the cutoff (energy versus invariant mass, etc.) since they are all equivalent in this approximation, and the differences appear starting terms $\sim \vec{v}^4$. The conceptual details will be further illuminated in subsequent publications.

V. DETERMINATION OF $|V_{cb}|$ $(|V_{ub}|)$ FROM THE INCLUSIVE WIDTHS

We now address the central phenomenological problem— how accurately one can extract $|V_{cb}|$ and $|V_{ub}|$ from inclusive semileptonic widths. $O(\alpha_s)$ terms are known, and higher-order terms are expected to be small for b decays. However radiative corrections to the mass get magnified due to the high power with which the latter enters the width. This concern has been raised in [20]. The effect of large n indeed shows up as the series of terms $\sim (n\alpha_s/\pi)^k$. Therefore one needs to sum up such terms. The resummation of the running α_s effects in the form $(\alpha_s/\pi)[(\beta_0/2)(\alpha_s/\pi)]^k$ has been carried out in [10] (the term with $k=1$ in this series was earlier calculated in [25]); with $\beta_0/2 = 4.5$, accounting for the large-n series seems to be at least as important. We emphasize that the BLM-type [26] improvements leave out these large-n terms.

In this paper we discussed a prescription automatically resumming the $(n\alpha_s/\pi)^k$ terms. This eliminates the source of the perturbative uncertainty often quoted in the literature as jeopardizing the calculations of the widths.[6] Let us demonstrate this assertion in a simplified setting.

Consider two limiting cases, $m_q \ll m_b$ and $m_q \approx m_b$. Inessential overall factors $G_F^2 |V_{qb}|^2/(192\pi^3)$ $(G_F^2 |V_{qb}|^2/(15\pi^3))$ can be dropped from the widths to get

$$\Gamma \approx \Delta^n \left[1 + a_1 \frac{\alpha_s}{\pi} + a_2 \left(\frac{\alpha_s}{\pi} \right)^2 + \cdots \right], \qquad (41)$$

with $a_1 = -(2/3)(\pi^2 - \frac{25}{4})$ (or $a_1 = -1$, respectively) at $n = 5$. As was shown here, these coefficients do not contain factors that scale like n, nor a_2 contains n^2, etc., and we can neglect them altogether. However, when one uses, say, the $\overline{\text{MS}}$ masses, one has

$$\overline{m}_Q \approx m_Q \left(1 - \frac{4}{3} \frac{\alpha_s}{\pi} \right), \qquad (42)$$

and to the same order in α_s one arrives, instead, at a different *numerical* estimate:

$$\Gamma = \Delta^n \left(1 - \frac{4}{3} \frac{\alpha_s}{\pi} \right)^n \left(1 + n \frac{4}{3} \frac{\alpha_s}{\pi} \right)$$

$$\to \overline{\Delta}^n \left[1 - \frac{(n+1)n}{2} \left(\frac{4}{3} \frac{\alpha_s}{\pi} \right)^2 \right.$$

$$\left. + \frac{(n+1)n(n-1)}{3} \left(\frac{4}{3} \frac{\alpha_s}{\pi} \right)^3 - \cdots \right]. \qquad (43)$$

$\overline{\Delta}$ is written in terms of the $\overline{\text{MS}}$ scheme masses. The expressions in Eq. (41) and Eq. (42) are equivalent to order α_s, yet start to differ at order α_s^2 due to n-enhanced terms. The size of the difference has been taken as a numerical estimate of the impact of the higher-order (non-BLM) corrections which then seemed to be large indeed: the coefficient in front of $(\alpha_s/\pi)^2$ is more than 25; only such a huge enhancement could lead to an uncertainty in the width $\sim 15\%$ from the higher-order corrections.

Since we are able to resum these terms, the width Eq. (41) has small second- (and higher-) order corrections. Alternatively, using the $\overline{\text{MS}}$ masses one is bound to have large higher-order perturbative coefficients, and their resummation returns one to using the low-scale masses.

We thus conclude that the uncertainty associated with the higher-order perturbative corrections discussed in Ref. [20] is actually absent and is an artifact of the approach relying on the $\overline{\text{MS}}$ masses without a resummation of the large-n effects. A similar numerical uncertainty, in a somewhat softer form, appeared as a difference between the "OS" and the "$\overline{\text{MS}}$" all-order BLM calculations of Ref. [10]; it is likewise absent if one uses the proper low-scale masses.

In our asymptotic treatment of the large-n limit we cannot specify the *exact* value of the normalization scale μ to be used for masses. It can equally be 0.7 GeV or 1.5 GeV; the exact scale would be a meaningless notion. We also saw that the position of the saddle point varies depending on the structure function considered. All these nuances are rather unimportant in practice. The impact of changing μ has been studied in Ref. [19], and it was found that the values of $|V_{qb}|$ extracted from the widths vary by less than $\pm 1\%$ for any reasonable value of μ.

The large factor of 5 also enhances the sensitivity of $|V_{cb}|$ to the expectation value of the kinetic energy operator μ_π^2 in the currently used approach where $m_b - m_c$ is related to M_B, M_{B^*}, M_D, M_{D^*}, and μ_π^2. This is certainly a disadvantage, but the emerging dependence of $|V_{cb}|$ on μ_π^2 is practically identical (although it differs in sign) to the one in the determination of $|V_{cb}|$ from the zero recoil rate $B \to D^* \ell \nu$ (for details see the recent review [24]). We do not see, therefore, a way to eliminate the uncertainty due to μ_π^2 other than to extract its numerical value from the data. Having at hand its proper field-theoretic definition, one does not expect significant theoretical uncertainties in it. In the exclusive case, unfortunately, even this would not remove the sizable element of model dependence.

The $1/n$ expansion can also be applied when the final state quark mass is practically zero as in $b \to u$. The large corrections of order α_s^2 that appear when m_b is taken at the high scale m_b, turn out to be quite small when m_b is evaluated at $\Delta/n \sim 1$ GeV with Δ denoting the energy release. Using the analysis of Ref. [19],[7] one can calculate the inclusive total semileptonic width $\Gamma(B \to l\nu X_u)$ quite reliably in terms of $|V(ub)|$ and $m_b(1 \text{ GeV})$ as inferred from Υ spectroscopy. Numerically one has [24]

$$|V_{ub}| \simeq 0.004\ 15 \left(\frac{B(B \to X_u \ell \nu)}{0.0016} \right)^{1/2} \left(\frac{1.55 \text{ ps}}{\tau_B} \right)^{1/2}. \quad (44)$$

The theoretical uncertainty here is smaller than the experimental error bars which can be expected in the near future.

VI. CONCLUSIONS

There are three-dimensional parameters relevant for semileptonic decays of beauty hadrons: m_b, the energy release $\Delta_{bq} = m_b - m_q$, and Λ_{QCD}. Since m_b, $\Delta_{bq} \gg \Lambda_{QCD}$, the heavy quark expansion, in terms of powers of $\Lambda_{QCD}/\Delta_{bq}$, yields a meaningful treatment, with higher-order terms quickly fading in importance. It is natural then that uncertainties in the perturbative treatment limit the numerical reliability of the theoretical predictions. It was even suggested that the latter uncertainties are essentially beyond control.

Our analysis shows that such fears are exaggerated. First, we have reminded the reader that the dependence of the total width on the fifth power of m_b largely reflects the kinematics of the lepton pair phase space. We then used an expansion in n to resum higher-order contributions of kinematical origin by identifying unequivocally the relevant dynamical scales at which the quark masses have to be evaluated. We found that the relevant scale is not set by the energy release, but is lower, parametrically of order Δ_{bq}/n.

We demonstrated that the expansion in $1/n$ works reasonably well in simple examples. This is not the main virtue, in our opinion. More important is the fact that this expansion allows one to have a qualitative picture of different types of corrections based on the scaling behavior in n. In particular, it shows the emergence of new, essentially lower scales relevant to the semileptonic decays.

Even without a dedicated analysis, it is obvious that the typical scale of the energy release in semileptonic b decays lies below m_b. Without a free parameter at hand it is unconvincing to defend say, the scale $m_b/2$ as opposed to $2m_b$. Needless to say, this scale variation has a noticeable impact on the final numbers. Using n as an expansion parameter the ambiguity is resolved—the appropriate normalization scale for masses is even below $m_b/2$—and already this, quite weak, statement is extremely helpful [19] in numerical estimates.

Based on the large-n expansion, we arrived at a few concrete conclusions.

One can resum the dominant n^k terms in the perturbative expansion of the inclusive widths, merely by using the Eu-

clidean low-scale quark masses (e.g., normalized at the scale $\sim \Delta/n$). Therefore, the previous calculations of the width [16,19] are free of the large uncertainties noted in [20], and which were later claimed to be inherent to the inclusive widths.

We introduced the so-called "extended" SV limit. We showed that m_c need not be close to m_b for the SV regime to emerge in the $b \to c$ inclusive decays. Large n helps ensure this regime, which gives a rationale for the relevance of the SV consideration in the actual inclusive $b \to c$ decays, both at the quark gluon and at the hadronic levels.

Recently, the complete second-order corrections to the zero-recoil form factors have been computed [27]; the nontrivial non-BLM parts proved to be small. The arguments based on the large-n expansion suggest then that the non-BLM perturbative corrections not computed so far for $\Gamma_{sl}(b \to c)$ are not large either.

The approach suggested here is applicable to other problems as well. For example, we anticipate that the second-order QED corrections to the muon lifetime which have not been calculated so far, will not show large coefficients if expressed in terms of the muon mass normalized at the scale $\sim m_\mu/3 - m_\mu/2$. Moreover, using α_{em} at a similar scale (in the V scheme) is also advantageous, though, clearly, it does not matter in practice in QED.

Note added. When this paper was prepared for publication the full two-loop calculation of the perturbative correction in $b \to c\ell\nu$ at another extreme kinematic point $q^2 = 0$ was completed [28]. The result suggests that the non-BLM perturbative corrections to the width are indeed small if one relies on the low-scale quark masses, in accordance with our expectations.

ACKNOWLEDGMENTS

N.U. is grateful to V. Petrov for invaluable discussions and participation in this analysis at its early stages. We are grateful to the hospitality of the CERN Theory Division, the Theory Division of MPI Werner-Heisenberg-Institut, and to Isaac Newton Institute for Mathematical Sciences, Cambridge, England, where various parts of this project have been worked out. M.S. and A.V. benefited from participation in the program *Non-Perturbative Aspects of Quantum Field Theory* organized by the Isaac Newton Institute for Mathematical Sciences. This work was supported in part by NSF under Grant No. PHY 92-13313 and by the U.S. DOE under Grant No. DE-FG02-94ER40823.

APPENDIX: THE REFINED $1/n$ EXPANSION

Let us start from Eq. (7) expressed in terms of the quark masses m_b and m_c, with $\Delta = m_b - m_c$, and introduce variables η and ω defined by

$$\eta = 1 - \frac{\sqrt{q^2}}{\Delta}, \quad \omega = \frac{1}{\Delta}[\sqrt{m_b^2 - 2m_b q_0 + q^2} - m_c].$$

$$(A1)$$

Thus, ω measures the effective mass M_X in the final hadron state, $\omega = (M_X - m_c)/\Delta$, (in particular, $\omega = 0$ represents the free quark decay) whereas η determines the difference between q^2 and Δ^2. Then we have

---

[7]A more complete BLM calculation has been carried out in [10]; the numerical results were quoted, however, only for the OS and $\overline{\text{MS}}$ schemes; the latter is expected to suffer from large n-related corrections, whereas the OS scheme is affected by the IR renormalons when higher-order BLM corrections are included; the BLM resummation does not cure it completely, again due to the n^2-enhanced terms in the spurious $1/m^2$ corrections.

TABLE I. Comparing widths in the refined n expansion with the exact results at $n=5$; the cases $m_c=0$ and $m_c \geq 0.2 m_b$ are treated separately.

m_c/m_b	0	0.2	0.3	0.4	0.6	0.8	1
$B_{1,n=5}^{(A)}$	0.223	0.524	0.462	0.462	0.476	0.488	0.494
$B_{1,\text{exact}}^{(A)}$	0.25	0.396	0.434	0.459	0.486	0.497	0.5
$B_{1,n=5}^{(V)}$	0.229	0.093	0.060	0.0374	1.26×10^{-2}	2.48×10^{-3}	0
$B_{1,\text{exact}}^{(V)}$	0.25	0.104	0.066	0.041	1.36×10^{-2}	2.65×10^{-3}	0
$B_{2,n=5}^{(A,V)}$	0.171	0.248	0.24	0.233	0.225	0.219	0.215
$B_{2,\text{exact}}^{(A,V)}$	0.25	0.25	0.25	0.25	0.25	0.25	0.25
$\gamma_{\text{sl},n=5}/\gamma_{\text{sl,exact}}$	0.80	1.11	1.00	0.97	0.94	0.93	0.924

$$\gamma(n)=2\Delta^n \sqrt{\frac{\Delta}{m_b}} \left(\int_0^1 d(1-\eta)^{n-2} \left\{ \eta \left(\eta + \frac{2m_c}{\Delta} \right) \left[1 - \left(1 - \frac{m_c}{2m_b} \right) \eta + \frac{\Delta}{4m_b} \eta^2 \right] \right\}^{1/2} v_1(\eta) \right.$$

$$\left. + \frac{\Delta^2}{3m_b^2} \int_0^1 d\eta \, \eta^{3/2} (1-\eta)^{n-4} \left(\eta + \frac{2m_c}{\Delta} \right)^{3/2} \left[1 - \left(1 - \frac{m_c}{2m_b} \right) \eta + \frac{\Delta}{4m_b} \eta^2 \right]^{3/2} v_2(\eta) \right) \qquad \text{(A2)}$$

where functions $v_i(\eta)$ are defined through

$$v_1(\eta)(2m_c\eta + \eta^2)^{1/2} \left[1 - \left(1 - \frac{m_c}{2m_b} \right) \eta + \frac{\eta^2 \Delta}{4m_b} \right]^{1/2}$$

$$= \frac{\Delta}{m_b} \int_0^\eta d\omega (m_c + \omega\Delta)(\eta-\omega)^{1/2}(2m_c + \eta\Delta + \omega\Delta)^{1/2} \left[1 - \left(1 - \frac{m_c}{2m_b} \right) \eta - \frac{m_c}{2m_b} \omega + \frac{\Delta(\eta^2-\omega^2)}{4m_b} \right]^{1/2} \frac{w_1}{2\pi}, \qquad \text{(A3)}$$

$$v_2(\eta)(2m_c\eta + \eta^2\Delta)^{3/2} \left[1 - \left(1 - \frac{m_c}{2m_b} \right) \eta + \frac{\eta^2\Delta}{4m_b} \right]^{3/2}$$

$$= \frac{\Delta}{m_b} \int_0^\eta d\omega (m_c + \omega\Delta) \left\{ (\eta-\omega)(2m_c + \eta\Delta + \omega\Delta) \left[1 - \left(1 - \frac{m_c}{2m_b} \right) \eta - \frac{m_c}{2m_b} \omega + \frac{\Delta(\eta^2-\omega^2)}{4m_b} \right] \right\}^{3/2} \frac{w_2}{2\pi}. \qquad \text{(A4)}$$

Equation (A2) represents an identity with two gratifying features. First, the width is expressed through the quantities v_i. Being weighted integrals of the structure functions w_i, they are smoother analytically than w_i themselves. In a certain respect, v_i are generalizations of the moments I_i, that are relevant for calculating inclusive width.

Second, the form of Eq. (A2) is particularly well suited for deriving the large-n expansion. One typically encounters integrals of the form $\int_0^1 d\eta \, \eta^a (1-\eta)^k$. At $k \to \infty$ the integral is saturated at $\eta_0 \approx 1 - a/(k+a) \approx 1 - a/k$, i.e., $1 - \eta_0 \ll 1$. (We loosely refer to this saturation as a saddle point evaluation, although it is not really a saddle point calculation in its standard definition.) This was actually the approximation used in the simple large-n expansion. However, for $k=0,2$ and $a \sim 1/2-3/2$ the ansatz $\eta_0 \ll 1$ is quite poor numerically. On the other hand, $\int_0^1 d\eta \, \eta^a (1-\eta)^k$ still has a reliable "saddle point" even for large a and $k=0$, since the width of the distribution is governed by $ak/(a+k)^3$ and does not become large. The point only shifts somewhat upward as compared to the "naive" approach. This is the idea lying behind the improved expansion. Its purpose is merely to determine the essential kinematics in the process at hand, and to use the hadronic averages expanded around it. Of course, the phase space integrals can always be taken literally for any n, if necessary.

As it was with the simple n expansion, one finds that the explicit dependence on n is contained in the kinematical factors that can be treated separately from the QCD dynamics contained in the hadronic averages $v_{1,2}(\eta)$. There is some residual dependence on n entering the value of the scale η_0 at which $v_{1,2}(\eta)$ are to be evaluated (i.e., the exact shape of the weight functions). Again, one has to treat the vector and axial current contributions separately for $b \to c$ and $b \to u$.

In Table I we compare the exact results with those obtained from the refined $1/n$ expansion for $n=5$ as a function of m_c/m_b in the simplest setting of the tree-level decay where the cumbersome factors in Eq. (A2) are replaced by their values at the "saddle" point. We have used the following notation there. The quantities $\gamma_{1,2}^{(V,A)}$ denote the width factors for the vector and axial contributions obtained by integrating w_1 and w_2, respectively. We then have $\gamma_{\text{sl}} = \gamma_1^{(A)} + \gamma_1^{(V)} + \gamma_2^{(A)} + \gamma_2^{(V)}$. One defines branching ratios normalized to the exact ($n=5$) tree level expression:

$$B_{1,2}^{(A,V)}(m_c/m_b) = \frac{\gamma_{1,2}^{(A,V)}(m_c/m_b)}{\gamma_{\text{sl, exact}}(m_c/m_b)} \qquad \text{(A5)}$$

and calculates them using, on the one hand, the expansion in n evaluated at $n=5$, and on the other hand the exact $n=5$ results. Since the axial and vector parts of w_2 coincide in the tree approximation, only one of them is shown in Table I. One sees that the n expansion works with a typical accuracy of about 10% for the inclusive width, as expected. The weight of the nonleading terms decreases as $m_c \to m_b$, i.e., in the SV limit. It is remarkable that the expansion based on the SV kinematics works so well down to a rather small mass ratio of $m_q/m_b \simeq 0.2$. This illustrates the observation made above that it is the parameter $(m_b - m_c)/nm_c$ that describes the proximity to the SV limit. For further analysis we note that at $m_c/m_b = 0.3$ the "saddle points" for $n=5$ occur at $\eta_* \equiv 1 - \sqrt{q^2}/(m_b - m_c)$ equal to 0.16, 0.28, and 0.4 for $\gamma_1^{(A)}$, $\gamma_1^{(V)}$ and γ_2, respectively; for the $b \to u$ case η_* is 0.33 and 0.55 for $\gamma_{1,2}$, respectively.

Another crosscheck of the numerical reliability of the refined $1/n$ expansion at $n=5$ is a comparison with the known perturbative expression at the one-loop level. In the SV regime one finds, through order α_s/π,

$$\gamma_{\text{sl,SV}} \simeq \frac{8}{15}\Delta^5 \left(0.924 - 0.945 \cdot \frac{\alpha_s}{\pi} \right)$$

$$= \frac{8}{15}\Delta^5 \cdot 0.924 \left(1 - 1.023 \cdot \frac{\alpha_s}{\pi} \right) \qquad \text{(A6)}$$

to be compared to the exact result to that order

$$\gamma_{\text{sl,exact}} \simeq \frac{8}{15}\Delta^5 \left(1 - \frac{\alpha_s}{\pi} \right), \qquad \text{(A7)}$$

i.e., a difference of only 2%.

From these comparisons we conclude that the refined $1/n$ expansion yields good numerical results already for the physical value $n=5$.

[1] M. Voloshin and M. Shifman, Yad. Fiz. **41**, 187 (1985) [Sov. J. Nucl. Phys. **41**, 120 (1985)].

[2] J. Chay, H. Georgi, and B. Grinstein, Phys. Lett. B **247**, 399 (1990).

[3] I. Bigi and N. Uraltsev, Phys. Lett. B **280**, 271 (1992).

[4] I. Bigi, N. G. Uraltsev, and A. Vainshtein, Phys. Lett. B **293**, 430 (1992); **297**, 477E (1993).

[5] B. Blok and M. Shifman, Nucl. Phys. **B399**, 441 (1993); **B399**, 459 (1993).

[6] I. Bigi, B. Blok, M. Shifman, N. Uraltsev, and A. Vainshtein, in *The Fermilab Meeting*, Proceedings of the 1992 DPF Meeting of APS, C. H. Albright *et al.* (World Scientific, Singapore, 1993), Vol. 1, p. 610.

[7] I. Bigi, M. Shifman, N. Uraltsev, and A. Vainshtein, Phys. Rev. Lett. **71**, 496 (1993).

[8] I. Bigi, M. Shifman, N. Uraltsev, and A. Vainshtein, Phys. Rev. D **50**, 2234 (1994).

[9] M. Beneke, V. Braun, and V. I. Zakharov, Phys. Rev. Lett. **73**, 3058 (1994).

[10] P. Ball, M. Beneke, and V. M. Braun, Phys. Rev. D **52**, 3929 (1995).

[11] E. Witten, in *The 1/N Expansion in Atomic and Particle Physics*, Proceedings of the Cargese Summer Institute Recent Developments in Gauge Theories, edited by G. 't Hooft *et al.* (Plenum, New York, 1980).

[12] M. Voloshin and M. Shifman, Yad. Fiz. **47**, 801 (1988) [Sov. J. Nucl. Phys. **47**, 511 (1988)].

[13] B. Blok, L. Koyrakh, M. Shifman, and A. Vainshtein, Phys. Rev. D **49**, 3356 (1994).

[14] I. Bigi, M. Shifman, N. Uraltsev, and A. Vainshtein, Phys. Rev. D **52**, 196 (1995).

[15] M. Voloshin, Phys. Rev. D **46**, 3062 (1992).

[16] M. Shifman, N. Uraltsev, and A. Vainshtein, Phys. Rev. D **51**, 2217 (1995).

[17] A. Grozin and G. Korchemsky, Phys. Rev. D **53**, 1378 (1996).

[18] M. Shifman and N. G. Uraltsev, Int. J. Mod. Phys. A **10**, 4705 (1995).

[19] N. G. Uraltsev, Int. J. Mod. Phys. A **11**, 515 (1996).

[20] P. Ball and U. Nierste, Phys. Rev. D **50**, 5841 (1994).

[21] N. Isgur and M. Wise, Phys. Lett. B **232**, 113 (1989); **237**, 527 (1990).

[22] B. Chibisov, R. Dikeman, M. Shifman, and N. G. Uraltsev, Int. J. Mod. Phys. A **12**, 2075 (1997).

[23] N. G. Uraltsev, Nucl. Phys. **B491**, 303 (1997).

[24] I. Bigi, M. Shifman, and N. Uraltsev, Report No. TPI-MINN-97/02-T, hep-ph/9703290.

[25] M. Luke, M. Savage, and M. Wise, Phys. Lett. B **343**, 329 (1995); **345**, 301 (1995).

[26] S. J. Brodsky, G. P. Lepage, and P. B. Mackenzie, Phys. Rev. D **28**, 228 (1983); for more references to the original approach and the extended summation see, e.g., [10].

[27] A. Czarnecki, Phys. Rev. Lett. **76**, 4124 (1996); A. Czarnecki and K. Melnikov, hep-ph/9703277.

[28] A. Czarnecki and K. Melnikov, Phys. Rev. Lett. **78**, 3630 (1997).

Printed in the United States
By Bookmasters